*The Invention of the Modern Dog*

T0146192

ANIMALS, HISTORY, CULTURE

Harriet Ritvo, *Series Editor*

# The
# INVENTION

## *of the*

# MODERN DOG

*Breed and Blood in Victorian Britain*

MICHAEL WORBOYS
JULIE-MARIE STRANGE
& NEIL PEMBERTON

Johns Hopkins University Press
*Baltimore*

Johns Hopkins Paperback edition, 2022
2  4  6  8  9  7  5  3  1

Johns Hopkins University Press
2715 North Charles Street
Baltimore, Maryland 21218-4363
www.press.jhu.edu

*The Library of Congress has cataloged the hardcover edition
of this book as follows:*

Names: Worboys, Michael, 1948–, author.
Title: The invention of the modern dog : breed and blood in Victorian Britain /
Michael Worboys, Julie-Marie Strange, and Neil Pemberton.
Description: Baltimore : Johns Hopkins University Press, 2018. |
Series: Animals, history, culture | Includes bibliographical
references and index.
Identifiers: LCCN 2018002310 | ISBN 9781421426587 (hardcover : alk. paper) |
ISBN 1421426587 (hardcover : alk. paper) | ISBN 9781421426594 (electronic) |
ISBN 1421426595 (electronic)
Subjects: LCSH: Dogs—Breeding—Great Britain—History—19th century. |
Great Britain—History—Victoria, 1837–1901.
Classification: LCC SF427.2 .W67 2018 | DDC 636.7/082—dc23
LC record available at https://lccn.loc.gov/2018002310

A catalog record for this book is available from the British Library.

ISBN 9781421443294 (paperback)

*Special discounts are available for bulk purchases of this book.
For more information, please contact Special Sales at specialsales@jh.edu.*

# CONTENTS

*Acknowledgments   vii*

INTRODUCTION   1

**PART I   1800–1873**

1
Before Breed, 1800–1860   23

2
Adopting Breed, 1860–1867   54

3
Showing Breed, 1867–1874   90

**PART II   1873–1901**

4
Governing Breed   115

5
Improving Breed I: Experience   139

6
Improving Breed II: Science   160

7
Whither Breed   184

CONCLUSION
The Past in the Present   221

*Notes   229*
*Index   275*

*Plates appear following page 86*

We would like to thank the Arts and Humanities Research Council for their funding of the project (AH/H030808/1), which included support of a major outreach event, entitled "Breed: The British and Their Dogs," at the Manchester Museum in October 2012. The event's success was due to the imagination and hard work of the museum's staff, especially Stephen Welsh, Anna Bunney, and Nick Merriman, its partners Redman Design, our project artist Jo Longhurst, and Leonarda Pogodzinski (Association of Bloodhound Breeders).

Our research involved extensive library and archive work, and we are particularly grateful for the support we enjoyed at the Kennel Club Library. Ciara Farrell could not possibly have done any more to help, being ably assisted by her colleagues, especially Colin Sealy. We would also like to thank the staff of the following libraries: the Natural History Museum, the British Library, the John Rylands Library, the University of Manchester Library, and the Stockport Public Library, which has an extensive collection of nineteenth-century dog books. Our archival research was mainly at the Kennel Club, the Imperial College London (Huxley Papers), and the University College Library (Francis Galton Papers). A valuable source of material on individual dog breeds was made available by breed club archivists and website managers. Of particular assistance with specific breeds were Linda Skerritt (Basset Hounds) and Glem Blumke and Elizabeth Murphy (Irish Wolfhounds). Our thanks to Kath Ward for researching the Robinson family, Alice Cornwell (Mrs. Stennard-Robinson), and Phil Robinson. The Kennel Club (Heidi Hudson), Tate Modern, the British Library, the National Archives, Mrs Elizabeth Murphy, and Paul Dodd supplied or gave permission for the images.

We have benefited enormously throughout the project from the enthusiastic support and wide knowledge of Valerie Foss, who corrected our mistakes and gave us new leads on numerous occasions. Colleagues at Manchester and other institutions have read and commented on drafts of the book, and we thank Roger Wood, Duncan Wilson, Elizabeth Toon, Tom Quick, and Vladimir Jankovic. We have also been helped by conversations with Philip Howell, Greger Larson, and above all Harriet Ritvo, our series editor and inspiration for histories of human-animal relations. At Johns Hopkins University Press we have enjoyed the support and expertise of Matt McAdam, William Krause, Debby Bors, and Juliana McCarthy, and special thanks to our superb copyeditor Julia Ridley Smith, who sharpened our text and saved us from many errors.

We have given a number of seminar and conference papers on different parts of the book and would like to thank everyone who contributed to discussions at seminars at the Universities of Kent, Leicester, Manchester, Sheffield, and York, and to conference papers in Rennes (Representing Animals in Britain), London (Cosmopolitan Animals), Royal Holloway (Nineteenth Century Numbers), and Manchester (Manchester Histories Festival).

*The Invention of the Modern Dog*

# Introduction

In 2013, the novelist and columnist Jilly Cooper asked, "Why are mongrels a dying breed?" She was highlighting a new trend in dog ownership in Britain, the emergence of new crossbreeds, such as the Labradoodle, Cockerpoo, and Puggle. These dogs are both material and cultural inventions. Material in the sense that breeders have produced dogs that combine desirable features from different breeds; for example, the Labradoodle has the look and temperament of the Labrador, with the nonshedding coat of a poodle. Cultural in the sense that breeders have rebranded mongrels as crossbreeds or "designer dogs," such that owners celebrate the breed components of their dog, down to fine fractions—for example, their pet might be half one breed, three-eighths another, and one-eighth another. The influence of this change can be seen today. A recent survey of dog ownership in Britain found that 64 percent of British owners said they had a pedigree breed; 31 percent, a crossbreed; 2 percent, a designer crossbreed; and 3 percent, "not sure." Seemingly, "no breed" was not an option. Many companies now meet a growing demand for genetic tests to determine the breed components of dogs. For Cooper, this trend signals the end of the "proper mongrel," a unique individual, produced by promiscuous mating and with an unspecifiable lineage.[1]

It is significant that the headline of Cooper's article considers mongrels to be a *breed*.[2] The change to dogs being seen principally in terms of breed and as a number of distinct breeds began in the mid-Victorian period and was profound. It saw not only the predominance, even fetishization, of the notion of breed but the material remodeling by selective breeding of dog varieties into standardized *conformations,* the term used to describe dogs' external physical form or morphology.[3] In turn, this development led to the production of sharper distinctions between the new breeds and a more uniform look across each. There was also an increasing degree of "internal" genetic uniformity, due to breeding within limited pedigree stock and a concern with pure blood.

In this book, we ask when, where, why, and how breed became the principal way of thinking about and producing dogs that were modern. Beginning in 1860s Britain, competitions at dog shows and changing fashions in dog ownership led to selective, cross- and inbreeding to conformation standards by, to use an epithet favored by Victorians, "doggy people."[4] These breeders, owners, sportsmen, exhibitors, dealers, and fanciers materialized breed in individual dogs and then worked to make whole breed populations more uniform, a goal captured in the word *conformation*. Over time this way of seeing and producing dogs spread from show dogs to all dogs, be they sporting, working, pet, assistance or, indeed, mongrel.

For centuries there had been different dog types, which had emerged in different places from breeding for work, sport, or companionship. There was variation in "look" within the types, and they merged into each other at the margins, with gradations in size, shape, color, coat, and so on. In the fifth volume of his *Histoire Naturelle generale et particulière* in 1755, the French natural philosopher Georges-Louis Leclerc, Comte de Buffon, wrote, "In the same country, one dog differs greatly from another; and, in different climates, the very species seems to be changed. From these causes, the number and mixture of races are so great, that it is almost impossible to recognize or enumerate them."[5] The gradation of forms disappeared with the adoption of *breed* as the principal way of thinking about and categorizing dogs from the mid-Victorian period onward. Breeders reimagined and remodeled dogs, reducing the variation within types and at the same time producing and proliferating discrete, differentiated, standardized breeds. The difference between pre- and postbreed dogs can be compared to how colors appear in a rainbow versus on a modern paint sample card. The former has distinct colors, but these vary in hue and shade, bleeding into one another where they meet. The latter consists of distinct, separate, and uniform blocks of color. If the small number of dog varieties in the early nineteenth century can be seen as akin to the seven colors of the rainbow, then the 204 breeds now recognized by the British Kennel Club (or the 332 listed by the Fédération Cynologique Internationale) are like the numerous colors presented on a paint sample card.

An example of the effect of the adoption of breed over the Victorian period can be seen in the ways in which certain dogs from Newfoundland became *the* Newfoundland. In the 1830s, Newfoundlands were a well-known variety from two popular sources: Byron's epitaph to his dog Boatswain, who had died in 1808, and Edwin's Landseer's painting, *Distinguished Member of the Humane Society,* first shown in 1835.[6] Byron's verse, which was engraved on the dog's monument on the grounds of Newstead Abbey and has been included in many editions of his poems, begins

Near this Spot
are deposited the Remains of one
who possessed Beauty without Vanity,
Strength without Insolence,
Courage without Ferocity,
And all the virtues of Man without his Vices.[7]

Landseer's painting was a portrait of Paul Pry, a dog that represented Bob, a well-known character on the Thames in London, who had allegedly saved twenty-three people from drowning in the previous fourteen years (see plate 1).[8] The painting marked Bob's receipt of a Royal Humane Society medal, which recognized his ability in water and his benevolent service to Londoners. Both works celebrated the character rather than the form of Newfoundlands; Byron never mentioned Boatswain's color, size, or shape, while Landseer's painting was a portrait of an individual. Bob's coat was principally white, one variant of the combinations of white and black normal for Newfoundlands at the time. Dogs like Bob and Paul Pry had been improved in Britain, not in form but in becoming more civilized, gaining nobility and kindness once away from the rough and cruel treatment they suffered in Newfoundland.

William Youatt's *The Dog,* first published in 1845, was the most influential book on canines in the early Victorian period. Youatt celebrated Newfoundlands' lifesaving abilities, courage, noble disposition, and amiability. He quoted one owner as saying of these dogs, "They seem only to want the faculty of speech, in order to make their good wishes and feelings understood and they are capable of being trained for all the purposes for which every other variety of the canine species is used." There were wide differences in form and character among Newfoundlands. Some were "comparatively small, but muscular, strong, and generally black" and had been used as retrieving gundogs. There was also a larger variety, "admired on account of his stature and beauty, and the different colours with which he is often marked," but not so good-natured and manageable. Youatt closed his description of the Newfoundland with lines from Byron's epitaph to Boatswain.[9]

By the 1870s, a preference for black Newfoundlands had developed in the dog show fancy and was adopted in the breed standard. John Henry Walsh, who published under the pseudonym "Stonehenge," had assumed Youatt's place as the leading authority on dogs. He agreed that black Newfoundland dogs had a superior conformation and that other colors were inferior. Walsh wrote in 1877 that white-and-black Newfoundlands were "open in their frames, weaker in their middles, and generally displaying a more shambling and ungraceful gait in walking." They were said to have a bad coat that "often resembles the wool on the back of a Shropshire sheep," and "they generally fall off in their hindquarters, being tucked up in the loin and leggy, and lack quality and type."[10]

Toward the end of the century, the dominance of black Newfoundlands was established, but white-and-black dogs had been given a secondary place and called Landseer Newfoundlands. In the 1890s, Rawdon B. Lee became the most influential author on dogs, and in his book *A History and Description of Modern Dogs,* the Landseer Newfoundland was shown standing behind and largely obscured by the ideal black one (fig. I.1).[11] In the next edition, just five years later, the ideal black dog had changed, drawn as heavier and with a larger head (fig. I.2). The Landseer variant had been demoted further, shown sitting in a pose similar to that in the "Great Humanitarian," perhaps indicating that although its conformation was poor, it still had indomitable character. The name of these dogs had always been capitalized because it referred to the place of origin, but by this time it became the norm to capitalize the names of all breeds. This made breed names proper nouns, acknowledging that breeds were seen as unique and individual entities; hence, contemporaries wrote about the Bulldog, not of bulldogs. The Newfoundland had not only been remade physically in Britain, but its *blood* had allegedly been preserved and improved. Lee claimed that "in the island which gave him his name, [the dog] is a sad mongrel creature," and only in Britain was "purity" to be found.[12] By "purity" he meant some continuity of heredity with the best dogs from Newfoundland, enhanced by crossbreeding with English dogs and then "fixed" by inbreeding.

By the end of the Victorian era, aficionados of the breed had decided that bodily form, size, and color mattered most, but there were many other conformation features defined in the breed standard. For example, the head "should be broad and massive, flat on the skull, the occipital bone well

**Figure I.1.** The Newfoundland. By Arthur Wardle, in R. B. Lee, *A History and Description of the Modern Dogs of Great Britain and Ireland: Non-sporting Division* (London: Horace Cox, 1894), opp. p. 79

**Figure I.2.** The Newfoundland. By Arthur Wardle, in R. B. Lee, *A History and Description of the Modern Dogs of Great Britain and Ireland: Non-sporting Division*, 3rd ed. (London: Horace Cox, 1899), 78

***Figure I.3.*** The Newfoundland, 2017. "Ch. Merrybear D'Artagnan." Winner of the Working Group Class at Crufts Dog Show, 2017. © Paul Dodd Photography

developed." The high degree of specification extended to the tail, which, Lee explained, "should be of moderate length, reaching down a little below the hocks; it should be of fair thickness, and well covered with long hair, but not to form a flag. When the dog is standing still and not excited, it should hang downwards with a slight curve at the end; but when the dog is in motion it should be carried a trifle up, and when he is excited straight out with a slight curve at end. Tails with a kink in them, or curled over the back, are very objectionable."[13] The breed has continued to be changed, with the current standard quite different from either of the illustrations in Lee's *Modern Dogs* (see fig. I.3).[14]

## The Modern Dog

Our claim that the adoption of *breed* in the Victorian era defines the "modern" dog is not novel. Most Victorian authorities, from Youatt in the 1840s to Lee in the 1890s, claimed that the breeds of the era were modern, in the sense of being recently created by being improved externally in form and internally with better blood.[15] Historians have tended to place the emergence of the modern dog earlier, either with domestication or the coming of the pet dog. However, the domestication of canines as companion or working animals living with humans occurred over twenty thousand years ago and is only modern on an evolutionary timescale. And many historians, most notably Keith Thomas, have shown that dogs as companions were common in Britain for centuries.[16]

In this book we argue that breed makes dogs modern because it was a new way of thinking about, defining, and increasing the variety of forms within the species *Canis lupus familiaris*.[17] Dog breeds were thoroughly Victorian inventions, influenced by industrialization, commercialization, class and gender attitudes, the rise of leisure, and evolutionary thinking. The principal arena for the invention of dog breeds was the dog show, where competitions selected the best dogs, whose owners gained prestige and also some income, not from the token prizes awarded but from sales and stud fees. With dogs, competition drove *specialization,* by emphasizing the differentiation of varieties; *standardization,* by adapting physical form to new designs; *objectification,* by viewing dogs' bodies as being made up of quantifiable parts (so-called points); *commodification,* by promoting dogs as tradable goods; and *differentiation,* by proliferating breeds. The form of each breed was "improved," which meant anything from returning to its imagined historic form to changing its form to suit specific needs or fashionable tastes. The templates for breed standards drew upon history, art, natural history, physiology and anatomy, and aesthetics. However, there was a tension in breeding between earned and inherited worth, that is, between "best in breed" winners in meritocratic conformation competitions, as opposed to dogs with good pedigrees showing inheritance of pure blood. This tension points to the importance in our story of other Victorian values. Breed and breeds were examples of order and hierarchy, and of tradition, even though much, if not all, were invented.

The world of pedigree dog breeding in Victorian Britain looked to other breeding cultures, with participants fashioning themselves as improving entrepreneurs in a landscape of competition and markets. The terms *invention* and *manufacture* were used by contemporary critics of dog shows to capture the fact that not only were new, "unnatural" dog types being created but that older types were being standardized and commodified, with their value determined in competitive shows and the market for stud fees and sales. This situation pointed to other typically Victorian divisions: gentlemen and traders—amateurs and professionals. The former, predominantly from the landed and upper classes, defined themselves as primarily interested in the long-term improvement of the nation's dogs, for which pedigree and "good blood" were all important. They saw themselves in a struggle against men of trade, who they styled as "dog dealers" whose only interests were in short-term profit and social success.

## Breed

The essence of breed was and is the division of a domesticated species by form rather than function. With livestock and poultry, form is typically a proxy for function, indicating ability to produce meat, milk, eggs, and so on. The term *breed* was used first for purebred livestock, commercial poultry, and Thoroughbred horses, then with sporting and fancy animals, and finally with companion animals.[18] The application of the term was more than a change of name for varieties of domesticated and companion animals, however; breed was associated with five changes.

1. Breeds were defined by, and bred to, a conformation, physical standard, with this delineated by subdividing their body into points (fig. I.4).
2. Within a breed population, there was a drive to achieve greater uniformity of conformation, and the previous normal distribution of size, color, and so on diminished or all but disappeared.
3. Breeds were made more distinct from each other, with a tendency to develop embellished points to demarcate the differences between breeds. Previously the physical forms of varieties of domesticated animals existed on a continuum; as breeds these varieties became divided or segmented with no intermediate forms, though over time, and especially very recently with dogs, crossbreeds have occupied some of the spaces between breeds.
4. Breed characteristics had to be inheritable, though breeders accepted that inheritability was often unpredictable.
5. The goal of having a standardized, uniform population coexisted and often conflicted with that of improving and hence altering conformation to meet the changing ideals of fancy breeders.

Breed and breeds were, and still are, coproduced socially and materially.

Historians of livestock have often struggled with the origins of types of domesticated cattle and sheep. In part this uncertainty is because of the dearth of written sources and the variations in pictorial representations, which reflect contemporary artistic conventions and technologies of reproduction. But primarily the difficulty lies in the fact that historians have been

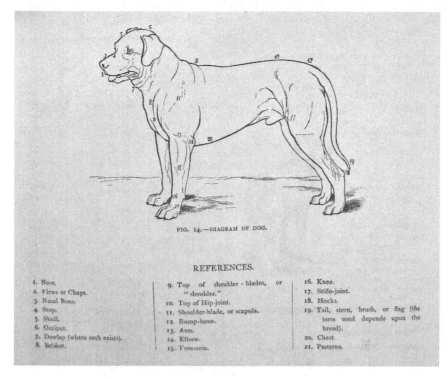

FIG. 14.—DIAGRAM OF DOG.

REFERENCES.

| | | |
|---|---|---|
| 1. Nose. | 9. Top of shoulder-blades, or "shoulder." | 16. Knee. |
| 2. Flews or Chaps. | | 17. Stifle-joint. |
| 3. Nasal Bone. | 10. Top of Hip-joint. | 18. Hocks. |
| 4. Stop. | 11. Shoulder-blade, or scapula. | 19. Tail, stern, brush, or flag (the term used depends upon the breed). |
| 5. Skull. | 12. Rump-bone. | |
| 6. Occiput. | 13. Arm. | |
| 7. Dewlap (where such exists). | 14. Elbow. | 20. Chest. |
| 8. Brisket. | 15. Fore-arm. | 21. Pasterns. |

**Figure I.4.** Diagram of Dog. V. Shaw, *The Illustrated Book of the Dog* (London: Cassell, Petter, Galpin & Co., 1881), 37

using *breed* ahistorically, seeking conformation types before they had been invented conceptually and materially. In his study of animal breeding in early modern England, Nicholas Russell recognizes the modernity of breed. He shows how before the nineteenth century, within most named types of live-stock, there was variation in color, markings, size, and shape, with little uniformity even within herds, let alone between them. In describing sheep types in the eighteenth century, he muses, "It may be that the concept of 'breed' in the twentieth-century sense of a group of domestic animals sharing a large number of common morphological features by virtue of genetic homogeneity, is wholly inapplicable to the regional forms outlined here." He goes on to observe, citing a survey of European primitive breeds, that "even when modern relict breeds under primitive management remain isolated, the management and selection pressures working on them seem to favour the survival of diverse morphologies rather than tending towards similarity of appearance."[19] For the late seventeenth century, Russell offers a classification of sheep in seven groups, defined loosely by size, face color, horns, and fleece type that he links to regions and topography, but he concludes that from the mid-eighteenth century this classification was "dramatically altered" by the spread of the breeding methods of Robert Bakewell, a farmer from Dishley in Leicestershire.[20]

Bakewell is widely celebrated as the founder of modern livestock breeding for his use of inbreeding, strict selection of the best progeny, and "progeny testing"—determining the parents that were best at passing on to their offspring desired traits of conformation, ability, or both. His ideas for improving sheep, horses, and cattle were widely publicized in his lifetime and after, and were taken up across Europe and North America.[21] Bakewell is also credited with having invented and popularized the modern notion of breed, which the economic historian John R. Walton has characterized as "an ingenious marketing and publicity mechanism." This mechanism worked for breeders on many levels.

> Certain identifiable physical characteristics were imprinted in animals of a particular strain, and prospective purchasers were then encouraged to associate those markers with some attribute or attributes of productivity which, it was claimed, such animals also possessed: rapid weight gain, larger size, high food conversion rates, better distribution of meat, heavier milk yields and so on. The success of a breed depended to some extent on the visual impact of the chosen marker or trademark, and the ease of its transmission from one generation to the next, to some extent on the degree to which the claims made for the breed's performance were thought to be valid.[22]

In other words, *breeds* were "brands." Bakewell's "New Leicester" was claimed to be a better-value product, with its name and design differentiating it from its competitors. But typically for a brand it had more than economic value. Those who acquired or bred from the New Leicester were buying into good, if not pure, blood and gaining an association with an improving ideology. Harriet Ritvo makes a similar point with regard to the aristocracy's fascination with, and investment in, prize bulls in improving agriculture in the nineteenth century, which she sees as a metaphor for landowners' elite position based on genealogy and visible power. With the dog fancy, however, she shows that there was a contradiction: "The prize-winning pedigreed dogs of the late nineteenth century seemed to symbolize simply the power to manipulate and the power to purchase—they were ultimately destabilising emblems of status and rank as pure commodities."[23]

This direct linkage between Bakewell's practices and the emergence of "recognisable [dog] breeds" has also been made by Martin Wallen in his account of the development of foxhounds and curs. Like Russell and Ritvo, he discusses the adoption of breed for varieties of dog, linking it also to Enlightenment projects of classification and discourses on race. However, his approach is through an animal studies lens and considers how "modern human–canine relations cannot help but be framed through breed."[24] He stresses how breed naturalizes differences in canines, concealing the labor that has gone into their creation and maintenance, and offers models of natural variation that can be mapped on human racial, ethnic, and class differences.

## Blood

By the early nineteenth century, farmers and other breeders were adopting Bakewell's practices. The results appear in stud books, which recorded pedigrees. These were often set out like those for royal or aristocratic families, which defined lineages for the inheritance of titles and property, and perhaps character. For breeders, pedigrees were important in recording "bloodlines." While "good blood" was valued, "pure blood" was even better because it was more likely to breed "true," that is, to reproduce type in form and function. The breeders' term for this was *prepotency*. Blood then was akin to the popular notion of genes today. The use of the word came from the humoral model of the body, which was applied to all animals and was only given up in the nineteenth century, first in science and medicine, and then in popular understandings. The four humors were blood, black bile, yellow bile, and phlegm. When the four were balanced, people and animals were healthy; imbalances due to excesses or deficiencies produced ill health and disease. Blood was both the fluid that contained "the vital principle" that animated the body, giving and sustaining life, and supplying the material continuity of life between generations. In breeding, conception was seen to result from the mingling of female blood and male semen, itself derived from blood, in the womb. The dual character of blood allowed for the inheritance of the acquired characteristics of parents, a view of heredity that was common until the beginning of the twentieth century and the development of genetics. The discovery of the fusion of human eggs and spermatozoa changed understandings of the mechanism of conception, demonstrating the continuity of life from generation to generation.[25]

Breeders of livestock and other animals were largely ignorant of the changing understanding of the biology of conception, focusing instead on the practical matters of achieving successful matings and producing good stock. Their guiding principles were that like begat like and that pure blood bred true. Both tenets were problematized at all levels by experience, anecdote, and tacit knowledge, as well as advice in printed texts. Like Bakewell, the breeders and owners of racehorses had sought improvement of their stock. The key innovation was importing Arabian horses in order to combine their lightness and speed with the strength and stamina of English mares to produce a horse that was fast and durable.[26] The "thorough" in Thoroughbred meant, and still means, that a horse's inheritance is confined to bloodlines from the foundation stallions and to a lesser extent selected mares. Sales of Thoroughbred horses are still referred to as "bloodstock sales."[27]

## Victorian Britain

This book has a dual focus on social and animal history in the coproduction of dog breeds materially and culturally. It offers a history of the Victorian period from the perspective of particular human-dog interactions, and while many themes will be familiar, we give a novel take on these. The preoccupation of some Victorians with dog breeds and breeding offers a unique perspective through which to view a rapidly changing society. If the doggy world at the beginning of this period was driven by landed rural and sporting interests on the one hand and an urban proletariat on the other, its transformation into a mass, consumer-oriented spectacle at the end of the Victorian era highlights manifold changes in the social, cultural, and economic makeup of British society. In following dogs through the Victorian period, we map much broader changes, such as the overall rise in plebeian standards of living and, its corollary, a burgeoning consumer culture in which commercial interests provided organized leisure pursuits. The dog shows that began as a novelty at agricultural fairs or as pub room meetings had by the late nineteenth century become major ticketed events held in cutting-edge venues. Akin to the great trade exhibitions of the age, dog shows were promoted as education and entertainment for the masses, and as shop windows for breeders and producers of doggy consumer goods.[28]

The coterie of those the late Victorians termed "doggy people" expanded too. While the landed classes kept their large kennels and bloodlines, the plebeian (under) world of dog fighting and dog dealing never entirely went away. Among breeders and exhibitors at the end of the century there was a significant presence of the middle and, to a lesser extent, the upper-working classes. The promotion of shows as popular entertainment should not eclipse the strong rhetoric around competition, fairness, and meritocracy; in theory at least, the dog fancy represented a microcosm of an industrial society in which hard work and improvement paid dividends. The expanding doggy press, itself a mirror of technological and commercial change, liked to feature breeders at home with their kennels against a backdrop of country houses and estates, though increasingly there were features on those who had made it in trade or the professions.

If breeding and owning dogs might have been seen as becoming more democratic, particular "brands" of breed in certain historical moments carried specific associations that mirrored and reinforced other kinds of cultural and social capital. The Russian Borzoi, barely known in Britain in the mid-nineteenth century had, by the turn of the twentieth, become a symbol of eastern exoticism, as romantic and avant-garde as Sergei Diaghilev and the Ballets Russes.[29] Imports of the Borzoi to Britain through aristocratic networks added to the exclusivity of the breed. By the outbreak of World War I, the Borzoi had become popular at shows and freighted with feminine, oriental, and patrician qualities, much like the noble ladies who showed them. As the Borzoi example indicates, human appropriation of dog breeds to express social identity extended beyond class to include gender and nationality. The doggy world throws light on other dimensions of social and political change, as well. The formation of the Ladies Kennel Association in 1894 crystalized doggy women's increasing agitation to be recognized as breeders and exhibitors in their own right and to have their say in canine affairs. That this was concurrent with a growing women's rights movement should not surprise us. Similarly, women were typically the early champions of hygiene and welfare in the show world, and the Ladies Kennel Association was a strong advocate of banning the cruel practices associated with particular breeds, notably, cropping ears and docking tails.

For all the rhetoric of improvement that was attached to contemporary and historical commentaries on dog breeding, our study also provides a lens through which to appreciate the slipperiness of social relations throughout the period. The earliest debates on conformation standards, and about who should judge them and how, highlighted tensions between different social groups. Inherited landed estate interests promoted a language of improving the nation's dogs and were intensely suspicious of urban, "new money" breeders. The large-scale show itself, with organizers' stated desire to turn a profit, was sufficient to call into question the veracity of the proceedings. The shifting geography of breed similarly indicates the contested redistribution of wealth and power over the period, with rural sporting privilege losing ground to centers of trade and industry and London competing with cities such as Birmingham and Manchester. The world of breeding accommodated, often uneasily, relationships between these interests. In his analysis of how the First World War created an anti-dog climate and exposed divisions and tensions within the dog fancy, Philip Howell reminds us that the status of the dog fancy and pedigree dogs is political and open to challenge.[30] More generally, the dog fancy's ideas about breed were framed by the context of British anxieties about empire, savagery, and civilization. As this book shows, the languages and conceptions of breeds and breeding, and domestic and foreign dogs, overlapped with understandings of human race and the imperial project. The Victorian dog fancy has often been aligned with eugenics, with some writers lambasting fanciers for their overuse of inbreeding and selection of "superior" features, leading to the detriment of dog health.[31] But our account shows that the application of science to dog breeding in this period was uneven, tending to rely upon individual popularizers of ideas, though their views remained contested to the end of the Victorian period—and beyond.

### Dogs and Their Historians

Our approaches and methods are those of social history and the history of biology. Rather than writing about changes in scientific knowledge and institutions, however, we instead consider the material transformation of the morphology of dogs and, as a consequence, their anatomy. We chart the social and biological coproduction of breed as a way of thinking about dogs, acting upon dogs, and interacting with dogs, along with the changing meanings given to this enterprise. Invention is a key term for us because it points to the social processes of producing, negotiating, and agreeing on "designs" for breeds as new types of canines. Dogs are not passive objects in our story; in the Victorian period they were active agents through the possibilities and limits of their biology, which in form at least proved to be remarkably mutable, as seen in the size difference between a Great Dane and a Pomeranian. Dogs were also active in the type of companionship relations that developed with humans at shows, at work, in the home, in the field, and in sports, where potential and limitations were negotiated.

The argument developed in this book is novel, and we draw on sources little used previously, though of course we build on the work of our fellow historians. There is a huge literature on the history of individual dog breeds in the books and other publications produced by and for dog fanciers, breeders, and owners.[32] These authors are mostly amateur historians, but their work is typically based on primary sources, and we have benefited greatly from their thorough research and valuable insights. However, from the perspective developed in this book, their view of breed is often presentist and essentialist, as they seek the historic origins of breeds. Making this point in her seminal book *The Animal Estate,* Ritvo writes that there is "little evidence of relation between ancient animals and modern breeds" and that "the very notion of *breed* as it was understood by Victorian dog fanciers . . . may have been of relatively recent origin."[33]

Our contention is that there is no *may* about it: the modern dog was a Victorian invention. Like all historians of animals in Victorian Britain, we are in Ritvo's debt, and in many ways this book is an expansion of the chapter "Prize Pets" in *The Animal Estate.*[34] A key argument in that book and Ritvo's other writings is that discourses on animals in nineteenth-century England, and by implication in other times and places, were also about human social structures and relations. With regard to the invention of dog breeds, she argues that "this elaborate system of categories metaphorically expressed the hopes and fears of the fanciers about issues like social status and the need for distinctions between classes" and "offered a vision of a stable, hierarchical society where rank was secure and individual merit, rather than inherited position, appreciated."[35] We agree with Ritvo that social class was reflected in dog fanciers' ideas and practices; in the culture of exhibitions, trade, publishing and practices they developed; and in the meanings they gave to their canine creations. Our study develops this insight further, exploring in greater depth and breadth differences within classes, across regions, between town and country, and by gender.[36] We are also able to discuss the effect of individuals, many of whom were great characters, plus the many canine groupings that formed—and not just the Kennel Club.

Other historians have seized on Ritvo's insights about animals in Victorian Britain and examined in greater detail livestock breeding, zoos, animal welfare, animal diseases, and hunting.[37] There are also resonances with her approach in histories of the dog in the eighteenth century and earlier, no doubt also encouraged by Keith Thomas's influential *Man and the Natural World: Changing Attitudes in England, 1500–1800.*[38] However, historians following Thomas have moved away from studies of human ideas about and attitudes toward animals, to explore the coproduction of human-animal relations.[39] In this context, it is impossible not to acknowledge Donna Haraway's work and the debate on companionship animals, which dominates scholarship in animal studies and emphasizes cooperation and interdependencies in human and nonhuman animal lives.[40] We give a new historical dimension to this topic by considering the shifting histories and entwined nature of human and dog lives in the Victorian period.

There is a growing body of work by environmental historians and geographers on the coevolution of humans and animals.[41] The American environmental historian Edmund Russell has recently published a long history of the greyhound, which shows how the form and character of these dogs was repeatedly changed from the twelfth to the late nineteenth century in response to the forces of modernization.[42] His argument is very similar to that advanced in this book. He shows that the conformation-based and essentialist characterization of "the" greyhound as a breed, what he terms the "statue view," was a creation of the Victorian era. Philip Howell's *At Home and Astray: The Domestic Dog in Victorian Britain* covers the same period as our book and should be read as a companion volume.[43] He too writes of the "invention of the modern dog" but refers to new places of domestication and says little on breed, except by implying its absence in street dogs. Surprising to some—though not to Howell, who shows that there is little historical evidence that Britain was a nation of dog lovers in the Victorian era—there is perhaps more scholarly work on breeds in other countries.[44] Margaret Derry, who is sensitive to the changing meanings and materialities of breed, has written extensively and impressively on animal breeding, principally in the United States, with studies that are interdisciplinary and span the eighteenth to the twenty-first centuries.[45] She combines histories of science, livestock, and professional and amateur breeding, and the economic history of commercial and fancy breeds, including a detailed study of Collies.[46]

A claim close to ours about breed and the invention of the modern dog is made briefly in John Homans's *What Are Dogs For?* (2012) and echoed in Michael Brandow's polemical *A Matter of Breeding*, published in 2015.[47] Homans makes the claim within a larger narrative about the Labrador, while Brandow is a canine activist, whose agenda is indicated by his subtitle: "A Biting History of Pedigree Dogs and How the Quest for Status Has Harmed Man's Best Friend." In his foreword to Brandow's book, the well-known ethologist and campaigner Marc Bekoff writes that "While the modern cult of pedigree and the very concept of officially recognized breeds are inventions of Victorian England, these ideas have taken root in our culture with profound consequences for dogs and people alike."[48] Our book, through its detailed historical narrative and critical-analytical discussion, stresses that the invention of the modern dog was not a singular event; rather, it was a process, as we showed with the Newfoundland, wherein the modern dog's meaning and materialities in individual breeds were reinvented throughout the Victorian period and up to the present day. We demonstrate too that the triumph of form over function was patchy at best, touching different breeds to different degrees, and that conflicts over "improving" dogs meant that, for some fanciers at least, function and character remained important corollaries to form. Our wider canvas also allows us to offer new insights in the history of the social, cultural, and (un)natural history of the Victorian era.

Our narrative and analysis draws not only on new sources but also on reading familiar sources in new ways, sensitive to material remodeling and the changing cultural meanings of canines. While most of our sources are textual, we have also critically explored pictorial representations of dogs and reconstructed the practices of breeding, showing, judging, trading in, and living with dogs. The invention of the modern dog was a very public affair, acted out in shows; the social and economic networks of doggy people; in newspapers, journals, and books; and, of course, on the bodies of dogs. In the early Victorian period, dogs were discussed regularly in the sporting press, especially *Bell's Life of London and Sporting Chronicle* and the *Field,* and publications serving breeders of livestock and fancy animals in general. From the 1880s, a growing number of dog-only publications reported on shows and published editorials, letters, and advertisements for sales, stud services, food, and medicines. The discussion of breeds and their conformation was extensive and dense, with correspondence on a particular breed point—say, the shape of an ear—often continuing for months. Typically, letters were signed using aliases (often witty), though it is likely that editors knew the correspondents' names. Later dog-only publications also had gossip columns in which the micropolitics and personal rivalries of the fancy were aired.

Among our key sources have been books on dogs and discussions of dogs in the wider sporting literature and that on domesticated or fancy animals more generally. Much of this literature began in natural histories of animals, but by the end of the century it took the form of accounts of breed histories and specifications of ideal conformations. The extensive detail in these accounts allows us to follow, typically through successive editions of the same volume, how conformations and the materiality of dogs were altered. Authors were exacting in their descriptions and demanded that breeders met, and show judges enforced, standards. We have also drawn upon the many illustrations in these volumes, being sensitive to the changing conventions of representation and print technologies. One very noticeable feature of Victorian dogs books was that authors made no use of photography, preferring to rely upon paintings and the precise, sharp lines of drawings. By the end of the Victorian period, there were many books specifically on dogs and especially on breeds; earlier, such discussions appeared in manuals on hunting, shooting, and coursing, and in natural history encyclopedias. The range of interests producing these texts and the diverse intended audiences have allowed us, through critical reading, to triangulate to give a richer understanding of the meanings associated with emergent breeds and of how dogs with relations with humans were altered by changing ideas and practices.

Our main archival sources were those held at the Kennel Club, which include records of its affairs and the extensive collection of unpublished material and miscellanea built up by the club's library. We have also benefited from the records held by national and local breed clubs, from meeting minutes to photographic collections. However, our use of archival sources has been limited because our primary concern is with the materiality and meaning of breed in the public sphere. The Kennel Club's records are principally administrative, reflecting the club's assumed role in canine affairs, and do not bear directly on the development of breed and breeds. We show in later chapters how the club delegated breed standards to specialist clubs and to competition at shows and the market in sales and stud fees. Second, there was little of interest in the archive on breed that was not in the public realm. The club published the minutes of its meeting; indeed, the archive records were often cut and pasted from published works. Also, the extensive correspondence and gossip in the canine and sporting press meant key issues were well aired; at the same time Kennel Club affairs, including the actions of individuals, were closely and publicly followed. Thus, we have mainly used published sources because we are principally interested in the circulation and translation of the ideas and practices.

## The Book

Our book is in two parts. Chapters 1 to 3 are broadly chronological, covering the period from 1800 to 1873, while chapters 4 to 7 are thematic, dealing with events from the mid-1870s to 1900. In chapter 1, "Before Breed, 1800–1860," we discuss how types of dog were classified by natural philosophers in the eighteenth and early nineteenth centuries, showing the varied systems used. Many terms were used for the types, with the term *breed* having no special connotation and being used interchangeably with *kind, race, sort, strain, type,* and *variety.* We then consider the practices and ways of thinking of social groups that bred the different types for sport, work, or companionship. In chapter 2, "Adopting Breed, 1860–1867," we consider the development of dog shows as institutions for exhibition, competition, and sociality from the late 1850s to 1867. In the mid-1860s, John Henry Walsh began to promote defined physical standards, or points, for the varieties of dog, importing the notion of breed and the practices for inventing and sustaining breed standards from livestock, poultry, and pigeon breeding and exhibiting. Dog shows grew rapidly in number and scale, becoming great exhibitions of canine diversity and human inventiveness, but with variable economic success. There were conflicts between the gentlemanly values of sporting owners, breeders, and exhibitors, who claimed only to be interested in the improvement of the nation's dogs, and those of so-called dog dealers motivated by short-term profit. At one level, chapter 3, "Showing Breed, 1867–1874," sets out a revisionist history of the founding of the Kennel Club, showing that its emergence was more contested and hard fought than has been previously recognized. This political history was the context for the invention of breed standards, which saw the institutionalization of the cultural practices that supported the remodeling of dogs into newly segmented, uniform conformation types.

Chapter 4, "Governing Breed," discusses the issues and opposition that the Kennel Club faced in securing its leadership position in the nation's canine affairs. Old struggles continued, and two new rivals emerged: a Fourth Estate with the growth of the canine and sporting press, and specialist clubs that represented the interests of those breeding and trading particular breeds. In chapter 5, "Improving Breed I: Experience," we consider the practices of breeders: How did they manufacture dogs to the physical specifications of breed conformation standard? One aspect of improvement was the differentiation and proliferation of breeds, mostly by dividing existing types but at other times by reviving "extinct" breeds, manufacturing entirely new ones, or importing (and then improving) foreign breeds. In these endeavors, breeders and exhibitors drew upon the ideas and practices of livestock and poultry breeders, adapting them to the biology of canines and drawing also upon wider assumptions and understandings of inheritance. Chapter 6,

"Improving Breed II: Science," explores how breeders drew upon the ideas and methods of the biological and medical sciences in their thinking and practices. Needless to say, the debates that followed the publication of Charles Darwin's developing ideas on evolution by natural selection were an important context. Like all animal breeders, dog breeders had an ambivalent attitude to science and mostly preferred to see their enterprise as an "art." An exception was Everett Millais, the eldest son of Effie Grey and the Pre-Raphaelite painter John Everett Millais. We discuss his effort to promote "rational breeding," based on the ideas of Francis Galton, and his experimental studies of dog distemper. Better health and a better diet were increasingly encouraged as necessary additional ways to improve the nation's dogs.

Chapter 7, "Whither Breed," discusses the growing body of criticism of the new dog breeds and shows. The attacks included campaigns against specific practices, such as cropping ears and docking tails; questions about the viability of individual breeds; and fundamental critiques of an enterprise that had produced allegedly unnatural, exaggerated, and unhealthy dogs. The criticisms made at the end of the nineteenth century were similar to those made of pedigree dogs today. In the conclusion, "The Past in the Present," we reflect on the meaning of our history of the invention of the modern dog for current debates on dog breeds. Indeed, we see our book as a contribution to the contemporary debate, showing how the very notion and materiality of modern breeds was the invention of a particular time and place—Victorian Britain. With dogs in the Victorian period, the notion of breed, the creation of individual breeds, and the definition of breed standards were all contested at every level; hence, the establishment of individual breed conformations and the whole panoply of breeds was contingent on time and place. History shows that breed could have been differently conceived and dog breeds differently made.

# PART ONE

# 1800–1873

# CHAPTER ONE

# Before Breed, 1800–1860

Before the mid-nineteenth century, writers on the dog worked in the fields of natural history, rural sports, or veterinary medicine, with their thinking focused, respectively, on questions of classification, working abilities, and health. Early Victorian writers assumed a framework of flexible Creation. By then most theological scholars had long rejected the literal biblical story, which made the Earth just over four thousand years old; rather, they believed in continuous creation on a longer timescale, with plants and animals changing in different environmental conditions, such as domestication by humans.[1] Thus, the dog, or a closely related animal, had been created many, many thousands of years ago and had adapted to occupy roles in different societies across the world. However, there was no agreement on what the original canine was like in form or character, or whether it had been domesticated once and then spread across the world, or had been domesticated in many places.

Before the livestock notion of breed was applied to dogs in the 1860s, early Victorians used many terms for the variety of dogs and did so permissively, indicating that they were not that significant. The terms were *varieties, types, tribes, strains, sorts, races, kinds,* and *breeds.* Within these groupings, variation in form was normal and expected, with ranges of size, shape, color, and other features. Also, function was related to form, with ability, character, and temperament being key features of each variety. For example, the height of foxhounds was matched to particular topographies and styles of hunting, while the coat of retrievers, smooth or wavy, was chosen to aid their work in different types of vegetation. Thus, there was no expectation of uniformity in varieties across the country.

Particular social groups bred and kept different types of dog for work, sport, and companionship. With the exception of the ladies who kept lap-dogs, these social groups were skewed toward the masculine cultures of the field and the tavern. The smallest group of breeders was made up of landown-ers and their masters of hounds, who maintained packs for hunting foxes, a pursuit that was about both pest control and sport. A second, larger group were those who bred greyhounds for elite coursing events and gun dogs to find and retrieve game and wildfowl on private estates. A third group, with a wider class and geographical profile, bred greyhounds for public coursing—where catching the hare was incidental to the competition between the pairs of hounds following its twists and turns—and for betting on the result. Fourth, there was a working-class community, associated with the tavern, pugilism, and betting, which was styled "The Fancy." These men had bred and kept dogs for fighting, but when this activity was banned by the Cruelty to Animals Act in 1835, they turned to ratting competitions and canine beauty contests.[2] A fifth group, which overlapped with the working-class groups, were the breeders and suppliers of toy dogs, who produced ladies' pets, such as King Charles and Blenheim spaniels, small terriers, and Italian grey-hounds. Finally, there were dogs for whom no group of people was respon-sible, whose existence was due to social indifference and neglect—street dogs and strays that, if they had a type, were curs.

## Origins and Classification

Early Victorians thought about the origins of varieties of dog as they did about human races; that is, they thought that the first creation had been altered by acclimatization to different environments.[3] The names given to varieties derived from where they came from geographically, as with the Newfoundland, the St. Bernard, the spaniel (Spain), and the Skye terrier; what they did, as with the foxhound, pointer, retriever, or sheep dog; or what they looked like, as with the pug and the mastiff. While function was more important than form, breeders and owners assumed that dogs' morphology, anatomy, and physiology had developed from their role. Thus, the long snout of a hound was a proxy for good scenting, and a deep chest meant good lungs and stamina for the chase. Greyhounds were light and streamlined for speed, broad chested for wind, and had eagle eyes on the front of their skull for spot-ting and following their prey. Retrievers had to have a good nose, but more essential was a soft mouth so that game was brought back undamaged. Rat-ting dogs had to be small and have the speed and aggression that enabled the very best to achieve incredible feats of catching and killing. Toy dogs also had to be small, so as to be able to be nestle on a lady's lap and even in her handbag, while tractable enough to tolerate a cosseted and confined indoor life.

In the early nineteenth century, there were two main authorities on the origin and varieties of dog: John Caius's *Of Englishe Dogges, the diuersities, the names, the natures and the properties,* published in 1576, and volume 5 of Georges-Louis Leclerc, Comte de Buffon's *Histoire Naturelle generale et particulière,* published in 1755.[4] Both were republished many times, with new editions. Caius was a physician and naturalist who wrote his book on English dogs as a contribution to a planned larger work, by the Swiss naturalist Conrad Gesner, on the natural history of the animal kingdom. Caius set out five dog groups based on their work or social function: (1) hunting beasts, (2) those good at finding game, (3) gentle comforting companions, (4) farmers' assistants with livestock, and (5) the "mongrels and rascal sort," largely used as guard dogs. Caius's text named seventeen "sorts" and elaborated upon their character. The following is his description of those that were "a gentle comforting companion":

> These dogges are litle, pretty, proper, and fyne, and sought for to satisfie the
> delicatenesse of daintie dames, and wanton womens wills, instrumentes of
> folly for them to play and dally withall, to tryfle away the treasure of time, to
> withdraw their mindes from more commendable exercises, and to content
> their corrupted concupiscences with vaine disport. . . . These puppies the
> smaller they be, the more pleasure they prouoke, as more meete play fellowes
> for minsing mistrisses to beare in their bosoms, to keepe company withal in
> their chambers, to succour with sleepe in bed, and nourishe with meate at
> bourde, to lay in their lappes, and licke their lippes as they ryde in their
> waggons, and good reason it should be so, for coursnesse with fynenesse hath
> no fellowship, but featnesse with neatenesse hath neighbourhood enough.[5]

Gesner did not use Caius's work, but it appeared as a separate publication and was picked up in the nineteenth century, when it was frequently drawn upon by writers on dogs to show the long ancestry of certain types that could be claimed as British.

Early Victorian writers on dogs drew most heavily on the reprinted work of the eighteenth-century French natural philosopher Comte de Buffon. His writings are representative of Enlightenment science, especially its classifying imperative and interest in the role of the environment in shaping animal form.[6] One question that Buffon discussed, which fascinated his contemporaries, was: Were wolves and dogs distinct species (the French term was *races*), or was the dog a domesticated wolf? Species were created and fixed but were alterable to varying degrees by acclimatization to different environments and activities. Interspecies hybrids were "unnatural," typically infertile (as with the mule produced by the horse-donkey cross) or degenerate.[7] Buffon reported success in making viable wolf-dog crosses; however, fertility was eventually lost in subsequent generations. Most of the crosses acted like domesticated dogs, but this behavior was unreliable and again could disappear in later generations. These findings led Buffon to conclude that the dog and the wolf were distinct though closely related species. A key difference was the dog's companion relationship with humans, which Buffon stated was almost unique to the whole animal kingdom. "Of all animals, the Dog is also most susceptible of impressions, most easily modified by moral causes, and most subject to alterations occasioned by physical influence. His temperament, faculties, and habits, vary prodigiously; and even the figure of his body is by no means constant." Here Buffon confirmed that the variation in the form of dogs was continuous, with gradations within and between major types. He goes on: "To the same causes must be attributed those remarkable varieties in size, figure, length of muzzle, form of the head, length and direction of the ears and tail, colour, quantity of hair, &c. In a word, nothing seems to be permanent in these animals but their internal organization, and the faculty of procreating together. As those which differ most from each other are capable of intermixing and of producing fertile individuals, it is evident, that all dogs, however diversified, constitute but one species."[8]

The Bible stated that everything in Nature had a purpose. The dog had been created to serve man; hence, the common assumption that the original type was the shepherd's dog. Buffon supposed that as dogs had spread around the world, they had been changed by environmental factors and the uses to which they had been put. Over time, these changes had produced the thirty "races" known to Europeans. He created a genealogical tree, with the shepherd's dog shown as "nearer to the primitive race than any of the other kinds," and then imagined how it had been transformed. Buffon contended that the original shepherd's dog, the Pomeranian, the Siberian, the Iceland, and the Lapland dogs were all herders; the Irish greyhound, the common greyhound, the mastiff, and the great Danish dog hunted with their eye; the hound, the harrier, the terrier, the spaniel, and the water dog hunted with their nose. The other races were "nothing but mongrels produced by the commixture of the above seventeen races"; thus, the mastiff was the product of a cross between a bulldog and an Irish greyhound, and the pug came from a bulldog cross with a small Dane.[9] At this time *mongrel* was sometimes used in a pejorative sense, but mostly it was a descriptive term for mixed-variety dogs. Indeed, such mixing or crossbreeding was widely practiced to refine physical form and add abilities to produce dogs for specific roles.

The first modern British challenger to Caius and Buffon was Sydenham Edwards in his book *Cynographia Britannica: Consisting of Coloured Engravings of the Various Breeds of Dogs Existing in Great Britain; Drawn from the Life,* published in 1800. His aim was to record the dogs found in England in a twelve-part publication; only six were published. Edwards was a renowned natural history illustrator and broke new ground in his publications with detailed color illustrations, each a portrait of "some distinguished individual Dog of each particular breed." Although he used "breed" in the title, he also referred to *kinds, variety,* and *races,* and accepted variation of form within each grouping. Edwards idealized dogs as moral animals and sought to capture this conception in his illustrations. He wrote that "in the execution of the portrait much study and attention has been paid, to represent as strongly as possible the *peculiar character and manners* of each respective race" (emphasis added). In the text, there was just one paragraph on the form of the bulldog and another on its colors, which were "black, salmon, brindled, and white," but he then devoted many pages to its "properties," above all its "matchless courage and perseverance, even to death" (see plate 2).[10]

Through Britain's history, types of dog had come and gone. For example, the turnspit, a dog with a long body and short legs that ran inside a wheel to turn the spit on which a joint of meat roasted over a fire, was the most recent casualty, having been superseded by mechanization (the roasting jack).[11] Edwards maintained that England's dogs, along with its horses, had long been superior due to the country's climate or "the pains taken in their breeding, education and maintenance" by its gentlemen. However, reflecting contemporary views of the vagaries of acclimatization, he observed that "out of this island they lose their properties in a few years."[12]

Natural histories on dog varieties were supplemented in the first half of the nineteenth century by manuals on dog diseases and their treatment, which included discussions on the nature of canines. These were essentially teach-yourself books for owners, as veterinary surgeons' work was dominated by the horse and, to a lesser extent, farm animals. Veterinarians saw the treatment of dogs as unprofitable and demeaning. One exception was the London veterinary surgeon Delabere Blaine, who in 1817 published *Canine*

*Pathology.* He opened with a chapter on the natural history and moral qualities of the dog, and agreed with Buffon that all types came from one origin, though his was not the shepherd's but either the Asiatic or Indian dog. The varieties were explained "by occasional causes, as change of climate, &c.; and by artificial means, as propagating from duplicates of accidental variety." Setting out contemporary assumptions about heredity, he continued:

> The progeny of all animals have a general tendency to bear the similitude of their species; but they have also a particular tendency to imitate their immediate parents. Should the father and mother, therefore, in any instance, happen by chance, or selection, to possess a variation of form from the common stock, it will usually happen that the same will be continued in the succeeding family branches. By future selections, likewise, of such as possess the deformity, or variation, in the most remarkable degree, and by propagating from these only, the external form of the body may be, eventually, very greatly altered. These varieties have doubtless been greatly increased by the effect of climate, which we know has a powerful operation on the animal frame.[13]

It followed, therefore, that "from these united sources, there is no reason to doubt, have sprung all the different breeds of dogs with which we are acquainted." Moreover, "the mental qualities of the animal have also been altered and cultivated in an equal degree with the personal varieties." The moral qualities of the dog were bravery, fidelity, attachment, gratitude, and intelligence. He acknowledged that dogs had a natural ferocity but commented that "the determined perseverance in battle, the contempt of pain, danger, and death, that characterise the bull-dog, is altogether a cultivated property."

The most widely cited book on the dog in the early Victorian period was by Blaine's former business partner William Youatt. His book *The Dog* was first published in 1845 and was for many decades *the* reference work on canine origins, types, care, and veterinary treatment. Worried about his debts, Youatt committed suicide in 1847, but the book was reprinted and revised many times, the last time in 1904.[14] Youatt was a leading London veterinarian, supporter of the Royal Society for the Prevention of Cruelty to Animals, and an evangelical Christian who wrote books on the horse, cattle, and sheep for the Society for the Diffusion of Useful Knowledge (SDUK). *The Dog* was illustrated with engravings of the different varieties, drawn to show character or ability. The mastiff was drawn prone, "on watch" with his head

THE MASTIFF.

*Figure 1.1.* The mastiff. W. Youatt, *The Dog* (London: Charles Knight and Co., 1845), 100

held high, "aware of all the duties required of him, and [which] he punctually discharges."[15] The pointer was drawn on the move, and the Scotch terrier with a rat that it has just killed (figs. 1.1, 1.2, and 1.3).

Youatt was in no doubt that the dog had a special place in Creation: "The dog, next to the human being, ranks highest in the scale of intelligence, and was evidently designed to be the companion and the friend of man." Among all the animals that man had domesticated there was "one animal only whose service was voluntary, and who was susceptible of disinterested affection and gratitude." Following Buffon, Youatt maintained that climate, food, education, and crossbreeding explained the divergence from the dog's common origin, and he assumed, as did his contemporaries, that the inheritance of acquired characteristics allowed for rapid changes. The possibilities of adaptation were illustrated by, at one extreme, the tiny Blenheim, cultivated by one of the Dukes of Marlborough for the drawing room, and at the other, the large Newfoundland, which had developed to pull carts and rescue drowning

THE POINTER.

***Figure 1.2.*** The pointer. W. Youatt, *The Dog* (London: Charles Knight and Co., 1845), 92

sailors in Canada. Varieties of dog were far from stable. The sturdy and fearless British bulldog did not suit the tropics: "When transported to India he becomes, in a few years, greatly altered in form, loses all his former courage and ferocity, and becomes a perfect coward." Furthermore, Youatt stated, varieties had become more numerous with the progress of civilization, as "the ingenuity of man has devised many inventions to increase his comforts: he has varied and multiplied the characters and kinds of domestic animals, whereas the ruder, or savage tribes possess but one form."[16]

Youatt prefaced his classification with a review of historical reports of different types in ancient civilizations. He reflected the Victorian fascination with ancient Egypt, giving an account of their veneration of greyhound-like dogs. Addressing natural history concerns in his observations on wild and feral dogs, he speculated whether modern domesticated dogs had developed from wild ancestors or descended from types that had always enjoyed human company and tutelage. Youatt believed that the different forms and characters of the domestic dog had both developed from natural adaptation to climate and from human choices in selection and training. For example, the Scottish greyhound or deerhound had long hair to keep warm in the High-

**Figure 1.3.** The Scotch terrier. W. Youatt, *The Dog* (London: Charles Knight and Co., 1845), 103

lands, and the "perfection of its scent" allowed it to follow a wounded deer for days. However, the English greyhound "exhibits so little power of scent . . . because he has never been taught to use it, or has been cruelly corrected when he has attempted to exercise it." The Irish wolf-dog, reputedly the tallest of all varieties, had been bred to chase and bring down a wolf but was now extinct; its role had disappeared with the elimination of the wolf from Ireland in the eighteenth century. Choice was evident in the size, beauty, and good nature of the Italian greyhound, which made it "a deserved favourite in the drawing room." The longest entry in Youatt's description of varieties was on the foxhound. He discussed its form and character and went on to elaborate upon the best sorts for hunting, as well as pack management and kennel design. As an urban veterinarian, Youatt would have had little experience of hunting; hence, this part of his book relied on established authorities, such as Peter Beckford's *Thoughts on Hunting,* first published in 1781, Robert Vyner's *Notitia Venatica: A Treatise on Fox-hunting* (1841), and Nimrod's *The*

*Horse and the Hound* (1842). Nonetheless, the attention given to the foxhound indicated the importance of hunting in British landed culture.[17]

For his classification of dogs, Youatt mostly followed Georges-Frédéric Cuvier, the younger brother of the leading French naturalist at that time, Georges Cuvier. Georges-Frédéric had been head of the menagerie at the Muséum d'Histoire Naturelle in Paris from 1804 to 1838.[18] Following the anatomical fashion of the 1830s and early 1840s, Youatt set out three divisions of the dog, determined by skull shape.[19] Specifically, he argued that the different types were determined "according to the development of the frontal sinuses and the cerebral cavity, or, in other words, the power of scent, and the degree of intelligence."[20] The principle was not *form* as such but had to do with how anatomy both reflected and shaped behavior and character.

Youatt identified sixty-eight "varieties" or "breeds," terms that he used interchangeably. His classification was global in scope; he described wild dogs from all continents and English, Scotch, Irish, Russian, Grecian, Turkish, Persian, and Italian greyhounds. The first division of twenty-five dogs, of which ten were "wild" and fifteen "domesticated," had elongated heads that narrowed toward the top, the greyhound being typical. The second division of thirty-four—typified by spaniels, setters, pointers, and hounds—had moderately elongated heads, with a rising forehead that enlarged the cerebral cavity and the frontal sinus, indicating intelligence and scenting ability. They were "the most pleasing and valuable division." The final division of nine, mostly terriers, had a shortened muzzle, enlarged frontal sinus, and a raised forehead with diminished capacity. It was "an inferior and brutal division," headed by the bulldog. His opposition to bullbaiting and dogfighting no doubt influenced Youatt's categorization; he was, after all, the first veterinarian appointed to the Royal Society for the Prevention of Cruelty to Animals. Also pervading his writing were associations between varieties of dog and different social classes and races.[21]

The latter half of Youatt's *The Dog* was on diseases and their treatment; it also included descriptions of anatomy and advice on the use of medicines. The book soon had a direct rival, Edward Mayhew's *Dogs: Their Management* (1854), which contained Mayhew's own, much admired illustrations. He was the brother of Henry Mayhew, the famous social reformer, journalist, and cofounder of *Punch*. Edward Mayhew had trained at the Royal Veterinary College, after a career on the stage and as fine arts critic for the *Morning Post*. His later books on the horse and stable management were more successful, and his book on the dog was updated by Frank Forester and published in

America in 1870, bound together with *Dink's Sportsman's Vade Mecum* and William Hutchinson's *Dog Breaking*.[22]

Youatt's book had a few other contemporary rivals, such as Charles Hamilton Smith's *The Natural History of Dogs* (1840), Hugh Richardson's *Dogs: Their Origin and Varieties* (1847), and the canine parts of David Low's *On the Domesticated Animals of the British Islands* (1853).[23] Richardson's volume was the most influential and included brief sections on dog management and diseases. A British army captain, Richardson had settled in Dublin, where he had taken up dog breeding and writing on agricultural affairs, producing books on bees, domestic fowl, horses, pigs, and farm pests, as well as on dogs. He maintained that "the dog was created prophetically by the Creator in order that he might be a friend and assistant to man." Richardson divided domestic varieties into classes: greyhounds, hounds (which included terriers and spaniels), and mastiffs, plus the "mongrel races," meaning crossbred dogs, that had become "settled varieties."[24] His accounts of the varieties mixed speculation on origins, with descriptions of dogs' physical form and character; thus, the bloodhound was said to have "sprung directly" from the Talbot hound, a variety that had all but died out because its slowness did not suit modern hunting. On the bloodhound, all Richardson wrote was that

> [it] is a tall, showy hound; but in a state of purity seldom attains, and certainly never exceeds, twenty-eight inches in height at the shoulder. . . . The ears are singularly long and pendulous, and should, in a perfect specimen, be within an inch or two of the animal's height, from tip to tip across the head. . . . The colour of the Bloodhound is tan, or black and tan, like an English terrier; if white be present, the breed is impure. The jowl of the bloodhound is deep, and his air majestic and solemn. The vertex of the head is remarkably protuberant, and this protuberance is characteristic of high breeding.[25]

Most of the description was about the bloodhound's scenting ability, with readers recommended to look at Landseer's many portraits to appreciate its "grave and dignified majesty."

Richardson also gave advice on breeding, acknowledging the variation within different kinds of dog and how features had been and still were being manipulated by breeders. He warned of relying on the appearance of potential parents and recommended obtaining dogs of the correct age, in good health and with strong pedigrees. On training, Richardson recommended "gentleness" over "violence" and "cruelty," and advised that hard-to-train puppies were best gotten rid of. The character and ability of varieties were mostly assumed to be inherited. Typical of dog writers of the time, he used anecdotes to illustrate character. Richardson told of a pointer that had been loaned to a poor shot and lost patience, becoming so frustrated that he "turned boldly round, placed his tail between his legs, gave one howl, long and loud, and set off as fast as he could to his own home."[26] Such stories gave insight into assumptions about human-dog relations and more often than not revealed the ability and character of a well-bred dog to be superior to that of a woeful human.

Dogs were also discussed in books and pamphlets on domestic animals. From the early nineteenth century, many books on keeping domestic animals had been published. Initially these were about food production (cows for milk, poultry for eggs and meat, pigs and rabbits for meat), but around mid-century their range expanded, as in James Rogers's *A Complete Directory for the proper treatment, breeding, feeding and management of all kinds of Domestic Poultry, Pigeons, Rabbits, Dogs, Bees, etc.*, published in 1854. As we discuss in chapter 2, beginning early in the century there was a craze or "mania" for fancy poultry and pigeons. On "the dog tribe," Rogers argued that its "innumerable varieties" were unique and that in no other animal was there seen such a range of size and appearance.[27]

One of the first books on pets alone was Jane Loudon's *Domestic Pets: Their Habits and Management* (1851). She observed that there were "more than a hundred kinds of dog" and, focusing on character, wrote that some approached the wolf, while others were "so mild and gentle in their manners" and seemed "almost like intellectual beings." There were many distinctions between kinds, but Loudon was clear that all "have the genuine characteristic of their race [being] affectionate and submissive to their master." She wrote very little on form, just some remarks on ears, but did mention Youatt's three divisions, though elaborated these in terms of intelligence, ability to learn, and fidelity.[28]

Not all works on dogs were didactic. Arguably the mid-Victorian period's most appreciated book on dogs was Edward Jesse's *Anecdotes of Dogs,* first published in 1846. Jesse enjoyed the patronage of the nobility and was awarded the sinecure post of deputy surveyor of the royal parks and palaces. He wrote extensively on natural history topics for newspapers and journals and published several books, which one biographer described as "colourful . . . rather than systematic." *Anecdotes of Dogs* had chapters full of stories of the "character, sensibilities and intellectual faculties" of dogs. Reviewers commented that Jesse's tales were "apocryphal," if not incredible, but had a consistent theme: celebrating the fidelity and intelligence of all types of dog. Jesse even found good words for the cur, writing, "He is, however, valuable to the cottager; he is a faithful defender of his humble dwelling; no bribe can seduce him from his duty; and he is a useful and an effectual guard over the clothes and scanty provisions of the labourer, who may be working in some distant part of the field. All day long he will lie upon his master's clothes seemingly asleep, but giving immediate warning of the approach of a supposed marauder."[29]

There was a sense throughout Jesse's book that dogs' behavior was an exemplar to their owners, who should have been ashamed if they even dreamt of ill treating their companion.

### Breeding and Keeping

In the remainder of this chapter, we consider the social groups that invested time and money in breeding, the varieties they produced, and how they understood canine form and function.

### *Hunting and Hounds*

Harriet Ritvo has suggested that foxhounds were probably the first dogs for which pedigrees were kept.[30] Owners of landed estates and their masters of hounds followed the same principles of mate selection for dogs as breeders did for livestock and Thoroughbred horses. Writers on hunting and dogs at the beginning of the nineteenth century agreed that English foxhounds had been changed with the development of a new, faster style of hunting. In medieval and early modern England, hunting involved a slow chase, starting at dawn with riders on heavier horses, accompanied by less athletic though more durable hounds that could follow a scent all day. The favored hound had been the Talbot, a white dog allegedly brought to England by the Normans. Mid-nineteenth-century authorities on hunting, such as Robert Vyner, suggested that the new foxhound had been produced through Talbot-greyhound crosses, the latter giving lightness and speed, without loss of "nose."[31] The new style was associated with Hugo Meynell, the so-called father of modern foxhunting, who was the master of hounds of the Quorn Hunt in Leicestershire between 1753 and 1800.[32] The new hunting required agile horses to jump the hedges and fences of enclosed fields, as well as hounds able to cover the ground at speed and negotiate the same obstacles. In his book on dogs, Richardson observed that "fox hunting is no longer hunting—it is nothing but steeple-chasing." Faster hunting took less time, fitting a pursuit that had become a sporting and social event more than a means of pest control.[33]

Hounds were improved by selective breeding within packs but also by outcrossing between packs. There was a breeding network among landed estates, with bitches sent around the country to be "put with" hounds to bring in new blood that would add to the home pack missing abilities, such as athleticism, scenting, or stamina. Landowners and masters of foxhounds published listings of their hounds, advertising the qualities and availability for breeding.[34] Typically, packs consisted of fifteen to twenty couples (hounds were counted in pairs), with a pyramid structure: a wide base of first-year hounds, on through the generations, and up to one or two hounds with eight seasons behind them. Young hounds were trained and also learned by following their seniors. For aesthetic reasons, masters liked to maintain a uniform look of size and color in their pack, but practicalities also demanded the same in equal abilities in the field. Quality was also kept up by regularly "drafting out," that is, putting down unfit and ill-behaved animals. The hounds of a particular estate or hunt were typically referred to as a "strain," signifying a particular bloodline. Those kept by Meynell had "Quorn blood," while at a nearby estate, the hounds were identified with its owner and said to have figuratively embodied "Lord Segrave's blood."[35] Masters wanted to improve the pack as a whole, and if breeding in desired features failed, they would acquire a new strain by purchasing part of or a whole pack.

Meynell's estate was just five miles from that of Robert Bakewell at Dishley. Although there is no direct evidence of their collaboration on breeding practices, it is not improbable that they spoke on such matters. Robert Vyner wrote in 1841 that in the late eighteenth century: "Hound-breeding was *at that period* as scientifically pursued as sheep-breeding, and the successful perseverance of Mr. Meynell and [his friendly rival] the first Lord Yarborough will ever be deserving of the warmest gratitude from true sportsmen, for lighting up what might be justly termed the *dawn of science in the chase*."[36]

In 1839, Frederick Delmé Radcliffe published *The Noble Science,* arguing that fox hunting remained "uninjured by the march of innovation" and flourished "not only in pristine purity, but in maturity of excellence." However, he accepted that hounds had been changed to meet new hunting styles: "The breed of hounds has arrived, if not at absolute perfection, at such a degree as may content its votaries; nor is there any lack of goodly steeds." There was also an association between hound and horse breeding. The favored dogs of the leading packs were referred to as "stallions," and, just as all racehorses were alleged to have been bred from just four Arabian horses, it was claimed that half the foxhounds in the country were descendants of two hounds: Hugo Meynell's Gusman and Lord Yarborough's Ranter.[37] If true, this dominance would have been aided by the rail network, which made stud visits and exchanges easier.[38] Seemingly, the railway brought more uniform hounds, as it did uniform time. A further move toward national standards came with the establishment of the national Foxhunting Committee, a loose federation of masters of hounds that met for the first time at Boodles Club in London in 1856. Foxhound packs were first exhibited at agricultural shows, with the first competition held at Redcar on 9 September 1859, ten weeks after (and probably prompted by) the Newcastle upon Tyne sporting-dog show.[39]

## Coursing and Greyhounds

Coursing enthusiasts believed that dogs' acclimatization to place and role accounted for the many varieties of greyhound. Rough coats suited the cold of Ireland and the Scottish Highlands; smoother coats were fine for temperate southern Britain. The heat of Greece produced delicate animals, diminutive Italian greyhounds did well indoors, and the hairless type lived in the tropical heat of Brazil and Africa.[40] With unsurprising chauvinism, British coursing men presented the English greyhound as "perfection," though liable to degeneration in other latitudes. Across Britain, physical forms and performance varied with geography, and particular strains were associated with Newmarket, Wiltshire, Leicestershire, Yorkshire, and Scotland. Greyhounds hunted by sight and had acute vision. They were considered to be descendants of ancient Gazehounds, *greyhound* being a corruption of this name and not a reference to color. In the 1830s, following the assumptions of craniology, if not phrenology, the streamlined shape of the skull, with its low, sloping brow, meant that the dog was assumed to have limited intellectual faculties, enabling it to be single-minded in sighting and tracking hares. The ideal greyhound would have stoutness, speed, honesty, and good working powers, that is, features of character and ability from which the best physical shape would follow.[41] The aim was for a greyhound "to be symmetrical" and "have a long neck, deep shoulders, thin withers, broad loins and back, flat sides, deep gaskins, thin feet, straight legs, short from the hock, and have a small delicately shaped head."[42] This description, from John Mills's *The Sportman's Library* published in 1846, uses several terms from horse breeding: "withers," the highest part of the back; "sides," the area between ribs and hip; "gaskins," muscular part of the hind leg between the stifle (cf. knee); and "hock," the joint between the knee and the fetlock.

Coursing had been a recreation of landowners.[43] Typically, they set out with their greyhounds into open country to seek out hares; when spotted, the dogs were slipped (released) to chase down and kill their prey. In the early Victorian era, the new pastime of public coursing became part of the developing leisure economy.[44] These events, still held on private land, were open across social classes and promoted as "democratic." In the new sport, catching and killing the hare was incidental to the performance of the greyhound in tracking its twists and turns.[45] The greyhounds were slipped in pairs and competed against each other for style. Judges assessed on a points scale the dogs' acceleration from the slips, their straight-line speed, and the quality of their turns along the course. Most events were knockout competitions, with the winners from each round or heat going through, and the best two dogs of the day contesting a grand final.

Coursing grew as a spectator sport, though betting was perhaps the main attraction. What became the premier national meeting, the Waterloo Cup, was first held on 9 March 1836 at the Altcar estate of Lord Molyneux, Earl of Sefton. The winner was Melanie, owned by the event's organizer William Lynn, proprietor of the Waterloo Tavern in Liverpool. He had organized the city's "first steeple-chace" [sic] nine days previously on Lord Molyneux's Aintree estate, and in subsequent years the coursing cup and a subsidiary steeplechase, now the famous "Grand National," were run in the same week. That a publican ran the events on a landed estate, at which there was betting and drinking, showed the cross-class and commercial character of public coursing, although it no doubt bolstered the evangelical middle classes' dismay that aristocrats and plebeians shared the same "low" pleasure culture. As the number of coursing events grew, they became even more "democratic," given that all that was needed was open ground, droves of hares, and men with sporting dogs. Around midcentury an alternative type of coursing, with small greyhounds or whippets, developed among the working class in the Midlands and North of England. Rather than chasing hares, whippets were raced against each other, being released together to run over two hundred yards to their owners, who waved flags to attract them.[46]

The breeding of greyhounds also drew upon ideas and conventions from horse racing. Charles Butler's 1819 book *The Complete Dog-Fancier's Companion* described each variety in verse. The entry for the English greyhound was typical of the pre-breed era in grouping running form, ability, and character together.

His sinewy limb, symmetric form and grace,
Vie with the hind, and equal him in pace;
Elastic, bounding o'er the flowery mead,
Unmatched his spring, agility and speed.
His color black or white, red, blue or grey;
The latter, the original, they say;
Long tapering nose, small ear and piercing eye,
His destined game unerringly descry.
With common hound, disdains to scent his prey;
By sight alone, he speeds his light'ning way;
The trembling hare, for safety lies in vain,
And fruitless seeks the shelt'ring copse to gain.
On Albion's shore, where hunting is the rage,
His feats and pedigree grace every page;
Grandsire and Grandam, all is there recorded,
And to the fleetest, is the prize awarded.[47]

Butler's claim that the dog's "feats and pedigree grace every page" points to the public recording of competition results and lineage. In both horse racing and coursing, such information allowed comparisons and calculations that enabled breeders, sellers, buyers, and betting men to see beyond physical appearance and competition history to the quality of a dog's blood. John Henry Walsh, who became the most influential figure in canine affairs in the Victorian era, published his book *The Greyhound: On the art of breeding, rearing, and training greyhounds for public running, and their diseases and treatment* under the pseudonym "Stonehenge" in 1853. It was based on articles he had published in *Bell's Life of London, and Sporting Chronicle,* a newspaper that at first mainly covered prize fighting but grew to become "Britain's leading sporting newspaper, without which no gentleman's Sunday was quite complete." Walsh went to great lengths to secure accurate drawings of top competition and stud greyhounds, which showed that "greyhounds can run in all forms." He stressed the importance of pedigree in breeding, observing that "few coursers now-a-days would be content to rear a puppy without knowing more of him than the shape of his tail, or his colour and conformation." It is not without significance that in coursing and in horseracing, the record of the performance in competition was referred to as its "form."[48]

Greyhound pedigrees had been published since the 1840s in *Thacker's Courser's Annual Remembrancer and Stud Book,* modeled on the Jockey Club stud book. The pedigrees that Walsh cited for greyhounds were, however, quite short, going back just two or three generations. Typically, the owner's name was given as well as the dog's, as the value of a pedigree depended on the trust accorded to the person submitting the information. Walsh reproduced, among others, the pedigrees of the most famous greyhounds of the day, the mare Tollwife and sire King Cob. He remarked on their degree of inbreeding with second and third cousins, but it was not something that worried him. He cited the pedigree of the highly successful Mustard, who was inbred to King Cob on three bloodlines, as an example where careful inbreeding had not led to "loss of constitution, size, or bone." The term *strain* was used within pedigrees to refer to the lineage of a particular dog or bitch, as with the King Cob strain of "fast, clever and stout dogs." One practice that Walsh was doubtful about was crossing greyhounds with bulldogs to increase aggression and courage, making resulting dogs "inclined to run cunning but not slack." His experience was that any advantage in their performance only lasted two years; thereafter, the crosses lost their zest and ability, which illustrated again the short-term nature of hybrid vigor.[49]

Walsh drew a distinction between inbreeding in animals, wild or domesticated, and humans. "Breeding 'in and in' cannot be shown to be injurious in wild animals, since it is well known that, in gregarious wild animals, the strongest male retains his daughters, grand-daughters, and often great-grand-daughters, as part of his train. Nor is there reason to believe that it is so prejudicial in the domestic animals as in man, with whom the adoption of the practice seems especially to be avoided, and has been forbidden by most of our lawgivers, human and divine."[50] Doctors regularly pointed to the fact that the children of family intermarriage, or consanguinity ("together" + "blood"), were predisposed to many physical and mental diseases, especially deafness and mutism, idiocy and sterility. However, there was a social-class dimension to such observations. In the early 1860s, the French psychiatrist and educationalist Édouard Séguin, then based in New York, observed that, while defects and weaknesses were multiplied by intermarriage in most families, "in alliances between members of a family endowed with a good constitution, there will be augmentation of the vital forces of the offspring."[51]

There was an effective national network of greyhound breeders who met at coursing events. As well as competing, they could see the abilities and physiques of other dogs and make arrangements for bitches to visit favored "stallions," whose services were also advertised in sporting newspapers. Weekly show reports and results in *Bell's Life* gave the owners' names and often carried details about the dogs' parents. Writing in the late 1870s, Walsh reflected on how the form of the smooth English greyhound had become more uniform across the country: "Up to within the last thirty years each coursing district had its peculiar breed of greyhounds, best suited to the country over which it was used." On the flat downs of Newmarket, fast, stout hounds were needed for long courses, whereas in Wiltshire and Berkshire, smaller, slower animals suited the more undulated ground. However, it was no longer necessary to accommodate the dogs to the landscape because the development of organized public coursing had led to shorter courses on similar terrain. This change, combined with the improvement of hares by better feeding and breeding, meant that the demand was now for dogs bred just for speed and not stamina. Moreover, the railway had facilitated national events and allowed for the interchange of breeds across the country, such that "there is really no locality in which a strain peculiar to itself can be said to exist."[52]

### Shooting: Pointers, Setters, and Spaniels

Pointers, setters, and spaniels had been employed because of their scenting and hearing abilities to find game birds for individual hunters. They worked, usually in pairs, scouting an area and, on finding game, freezing short of it, thus showing its whereabouts and allowing the guns to prepare for the bird's flight. The dogs had to have a soft mouth and not be gun-shy. In the early nineteenth century there was a new demand for dogs that only retrieved game, due to the growth of the new pastime of battue shooting, a style of hunting suited for an industrial age. Battue parties did not shoot wild game. Rather, game was bred on a large scale by the gamekeeper of an estate and then released onto moorland to graze and grow. After the "glorious twelfth"—the start of the season on 12 August—farmworkers and other local men would be hired to walk in lines, beating the ground and driving the released gamebirds toward lines of men standing hidden in butts with technologically advanced, rapid reloading, and accurate weapons. Previously, men had returned from a day on the moors with a few birds; after a battue shoot they returned with hundreds.[53]

Writers on dogs assumed that pointers were descended from the old type of foxhound, selected or crossbred for a shorter muzzle and longer ears. As

Youatt observed, other countries had their varieties, notably Portugal, France, Russia, and Italy, some of which were seen by the British to follow the national stereotypes applied to people. Thus, the Spanish and Portuguese pointers were said to be quarrelsome and soon tired, while the Russian was "a rough, ill-tempered animal." In choosing a dog, form was secondary to ability and that was best secured by recommendation from trusted owners. But care was needed. Incidences of dog theft rose dramatically in the 1830s and 1840s, with snatched animals either being ransomed or sold. In *The Sportsman's Library,* John Mills wrote of the difficulty of securing good dogs and told how he had once been taken in, buying a brace of pointers for thirty guineas from "a man very like a gamekeeper in his costume and person," who told a story about his squire giving up shooting because of ill health but who was likely one of London's notorious "'dog-dealers' and 'dog-stealers.'"[54]

The English pointer was widely believed to be a modification of the Spanish pointer, made lighter when imported dogs were found to be too large and heavy for the local conditions. These pointers were improved by crossing with new and lighter English foxhounds. An effective pointer was both born and made. Good scenting and efficient pointing came from instinct but could be improved by training or, to use the contemporary term, "breaking." The term came from work with horses, where it meant to break a horse's wild spirit, habituating or taming it to be ridden or used for draft work. Francis Butler, in his book on breeding and training dogs for work with game, *The Breeding, Training, Management, Disease &c. of Dogs* (published in New York in 1857), wrote that *"breaking* is certainly a very appropriate term for pounding bad habits out of canine pelts; habits, which never could have been contracted, had ordinary attention been paid to early tuition. When animals commence their training after months of entire freedom from restraint, severity is often necessary, as they become so naturally self-willed, as absolutely to require *Breaking.*"[55]

However, the typical advice in books on breaking pointers was that cruelty was mostly counterproductive, and that severity and punishment needed to be used judiciously with a view to the temperament of the dog. Indeed, Butler explained that a mix of rewards and reprimands was needed: "Encouragement, rather than threats should be the basis of all our attempts to direct the instinctive developments of the sporting-dog; our rebukes and punishments should be reserved for enforcing submission, and of testifying our disapproval of conscious disobedience." This advice echoed that of Col.

William N. Hutchinson in his influential 1848 volume on dog breaking; he stressed knowing the disposition and temper of each dog and the importance of the character of the trainer.[56]

Breeders and owners assumed that by generation-upon-generation selection and training, pointers had acquired specific instincts and become more tractable through inheritance. Such views accorded with prevailing ideas on evolution and particularly those of the French natural historian Jean-Baptiste Lamarck, who saw all domestic dogs as descended from the wolf, altered cumulatively by climate and habits through the inheritance of acquired characteristics. Charles Lyell, whose work on geology greatly influenced Charles Darwin, reflected contemporary opinion when he argued in his 1835 *Principles of Geology* that the abilities of different races of dogs were modifications of instincts: "The fixed and deliberate stand of the pointer has with propriety been regarded as a mere modification of a habit, which may have been useful to a wild race accustomed to wind game, and steal upon it by surprise, first pausing for an instant in order to spring with unerring aim." He added that "when such remarkable habits appear in races of this species, we may reasonably conjecture that they were given [by the Creator] with no other view than for the use of man and the preservation of the dog which thus obtains protection."[57]

Setters were the most popular dogs for the older style of shooting party, being hardy and bred for different environments. When considering their history in 1867, Walsh concluded that setters had been around for four hundred years and had developed from pointers or spaniels. They had been fashioned originally to rouse birds for hawks to chase, before the gun became the dominant mode of hunting game. In Britain there were English, Gordon (the Scottish type named after the strain of the Fourth Duke of Gordon), and Irish setters; Youatt in 1845 recognized other national types: Spanish, Portuguese, French, and Russian. Walsh speculated that the dogs had learned to set from imitating pointers or had acquired that ability from crossbreeding. Their main advantage over pointers was endurance; they could work in rough country, in wet and cold conditions, and for many hours. Thus, when Walsh endeavored to define a breed type in 1867, he observed, "If we collected together twenty of the best setters in England, we should observe marked differences in their structure, coat and texture, and representing in their conformation the idiosyncrasies of their respective owners." Edward Laverack, a Manchester breeder whose dogs became the basis for the English setter, identified seven important strains associated with estates (Naworth and Featherstone Castles) or individuals (Lord Lovat, the Earls of Southesk, Seafield and Tankerville, and Mr. Lort).[58]

Spaniels, as the name indicates, were assumed to have originated in Spain and been imported to Britain, though exactly when was uncertain. They were mostly slower than pointers and setters, and trained to flush game from cover for men with guns; hence, their numbers had declined with the growth of battue shooting. They were most valued for their courage and fidelity. Their distinctive physical features were pendant ears and a thick coat, but otherwise there was variation between and within kinds, principally in size and color. The primary division was between land and water spaniels, with the latter being used in wildfowling. Irish water spaniels were much prized, and in the first half of the nineteenth century, one authority claimed that "Irish manufactures" were displacing the English variety, with Mr. M'Carthy's strain particularly prized.[59]

Land spaniels were described as either springers or cockers, according to size, but there were many variants—indeed, so many with the cocker that John Meyrick in 1861 stated that "it is impossible to give any specific description of it." Cockers were said to be found mostly in Devon and Wales, where shooting woodcocks was popular. Before the invention of the category of toy dogs, the following small dogs were regarded as types of cocker: the King Charles spaniel, the pyrame dog (the alleged ancestor of the Blenheim spaniel), the lion dog, and the shock dog. Form and function were intertwined in a two-way process. The retrieving characteristics of the King Charles spaniel worsened as the dog's size shrunk, and the Blenheim spaniel, another aristocratic creation by the Duke of Marlborough, found more favor, but that too was seen to have degenerated as it gave up the covert for the cushion.[60]

The heavier springer spaniels were employed in pheasant shoots, with three main types deriving from the counties of Sussex and Norfolk and the estate of the Duke of Newcastle at Clumber Park in north Nottinghamshire. There were mixed views on the usefulness of the Clumber for gun hunting. Some critics claimed that, being single estate and hence inbred, they must be weak, while others claimed that their aristocratic, pure blood had been improved by outcrossing. They were valued because they did not "give tongue"; that is, they did not bark. Silence aided their work and gave them the air of aloof, upper-class superiority. The Sussex spaniel allegedly had its origins in Hastings and had been bred to work in dense scrub, barking to reveal the whereabouts of birds. The Norfolk, sometimes oddly given the name of the West Midlands Shropshire spaniel, was the most common land spaniel, which again made it impossible to give it "a particular description." It was also alleged to have aristocratic origins on the estates of the Duke of Norfolk around Arundel in Sussex, though this was never substantiated.[61]

Pointers, spaniels, and setters would all retrieve game, but the shift to battue shooting led to dogs just being used to bring back shot birds. To serve this function a new type of dog was developed—the retriever, the main requirement of which was a soft mouth. In his 1845 book *The Sportsman's Library,* John Mills wrote, "It is not necessary for me to say much about retrievers, as any dog can be taught to fetch game." Dinks's *The Sportsman's Vade Mecum* agreed: "There is no true type of them. Every person has a peculiar fancy regarding them." Despite Mills's belief that "any dog" could be trained to retrieve, he recommended the large Newfoundland because its size, rough coat, and love of water made it useful in all terrains. On this understanding, retrieving dogs were a type associated with behavioral characteristics and not form. As such, there were no published pedigrees for these dogs, and it was left to individual choice how to breed and train them.[62]

### The Fancy: Ratting and Showing

The best-organized dog culture of the early Victorian period was among the working and lower middle classes involved in urban "blood sports," though such events were attended by all classes and included boxing.[63] Indeed, aficionados of these exclusively male sports had a name: the Fancy. In the eighteenth century, animal sports had included bullbaiting, badger baiting, and dogfighting, but campaigns against them led to their control and repression. While these campaigns were putatively against cruelty to animals, they were also motivated by fear of an aspect of plebeian culture that was seen as threatening to public and moral order. The 1835 Animal Cruelty Act, which made dogfighting and bearbaiting illegal, led to the development of new sports held in public houses. The most popular was rat killing, organized as matches between two dogs; the winner was the dog that killed the most rats in a set time. Some matches were made in advance, with high stakes and fixed odds, but others were arranged on the night of the event. Owners also brought their dogs along to event nights to show them off and invite challengers. Over time, such displays became events in their own right and were pejoratively termed canine "beauty shows." Needless to say, beauty was in the eye of the beholder, and typically prizes went to the most aggressive-looking animal.[64] Betting on the outcomes of ratting and beauty contests was for many people the main reason for attending these events.

The Fancy was centered in London, though there were similar groups in provincial towns and cities, especially Birmingham and Manchester. Its activities took place mostly in public houses, with events organized by the publicans and orchestrated by committees. The leading London establishments where the Fancy operated were owned by Jemmy Shaw: from 1845 to 1854, the Blue Anchor in St. Luke's and, from 1854, the Queen's Head in Haymarket.[65] Shaw developed a legendary status. He had been a boxer in the 1830s, becoming a publican and Fancy entrepreneur when he retired in 1841. He bred many champion ratting dogs, including Jacko the Wonder, as well as prizewinning rat-catching cats, ferrets, and a mongoose. By the late 1840s, a number of London public houses held separate ratting events and beauty shows, as well as evenings of mixed events. In July 1849, in one week at the Blue Anchor, there was a meeting of the Terrier, Spaniel, and Small Toy Dog Club, various ratting contests with dogs, and displays of ratting by Shaw's cat and mongoose. Some ratting and beauty events ran on the same night as boxing contests, providing varied entertainment and further opportunities for wagers.[66]

A dog show at the Queen's Head was the subject of an 1855 painting by R. Marshall now called "An Early Canine Meeting" (see plate 3). Many histories have used this painting to demonstrate that there were dog shows prior to 1859. However, a close look at the painting reveals more differences than similarities with the shows that came later. The artist represented a meeting of the Toy Dog Club, where bulldogs, bull terriers, black-and-tan terriers, and toy spaniels were judged on their beauty rather than against any defined standard. The walls of the room are shown lined with prints that record and celebrate the rowdy plebeian cultures of ratting and boxing. Shaw is standing to the left of the fireplace in bow-tie and waistcoat, as befits the host, while his guests wear stovepipe hats and black suits that suggest social pretensions, if not established wealth and status. The stakes in competitive events at Shaw's suggest that some participants were moneyed. Typical stakes in ratting contests were £2–5, but could be as high as £25; thus, the Fancy could be a meeting place for high and low life. William Secord, a historian of dogs in art, has noted the difficulty of identifying the breeds in this painting, concluding that it reflects how breeds have been changed. We contend that this interpretation is presentist. Our view is that no modern breeds can be identified because they had yet to be invented conceptually or produced materially.[67]

An account of a meeting at the Queen's Head, perhaps the one depicted in Marshall's painting, was given in *Bell's London Life and Sporting Chronicle* in

April 1855, where the dogs were identified with their owners, not as types. "The canine exhibitions at Jemmy Shaw's, last Wednesday evening, also on the 15th, surpassed anything of the kind for years, the beautiful studs of stock dogs of different breeds belonging to those well-known fanciers, Messrs Perks, Sabin, Bladen, Wollmington, Shaw, Mansfield, &c. were greatly admired. Many fresh members were enrolled, making this association of canine fanciers 61 in number. They met again next Wednesday evening, to propose, &c.—entrance free, open to all. Public ratting next Tuesday evening, at nine o'clock."[68]

The "Canine Fancy" column in *Bell's Life* covered beauty shows and ratting events in other London pubs, as well as the dog racing that took place in Manchester over two hundred yards at the Bellevue Course. Reflecting their origins in dog fighting, ratting, and horse racing, ratting events and shows were arranged as prize contests; that is, two dogs competed for stakes put up by their owners, with the winner taking all. For example, in 1845, *Bell's Life* reported that Shaw had "two terriers under 6 lb. weight he will show for beauty, or kill 12 rats, against anything else of the same weight." There was also side betting on the outcome by spectators, with odds based on previous results and "look"—size, weight and attitude.[69]

The earliest reports of beauty-only shows are from the 1830s for "fancy spaniels," typically the King Charles; most prizes went to the weavers of Spitalfields. There was a show of spaniels in August 1836 at the Portland Arms in St. John's Wood, during which a silver-chased collar was awarded to the owner of "the handsomest dog or bitch . . . for all properties." The term *properties* was widely used for what later became *points,* but it also embraced temperament and attitude.[70] The decisions of beauty show judges were challenged, and one response was that complainers needed education. Thus, in January 1846 at the Horse and Groom in Leicester Square, Messrs. Mansfield and Holborn lectured on the "many properties a bulldog has, to prevent so many disputes occurring." Anticipating problems in judging at new dog shows after 1859, Shaw sought to improve the Fancy by publishing rules for ratting and beauty contests.[71]

By the 1850s there were numerous Fancy clubs in London; for example, the East End Toy Dog Club, the West London Spaniel Club, the South London United Canine Association, and the Terrier, Spaniel, and Small Toy Dog Club. Prince Albert was reported to be the patron of one club, and the Southampton Spaniel Club invited Queen Victoria, who regularly passed through the port on her way to Osborne House on the Isle of Wight, to be its patron. At shows or afterward, dogs could be bought and sold, and stud visits arranged. The only person to make public use of pedigrees in the Fancy was Shaw himself. On the walls of his pub were engravings of his top ratting dogs, Tiny and Jem, along with details of their pedigree and performances.[72]

There were well-established studs among members of the Fancy, and these breeders would offer puppies for sale to any buyer. One of the best-known men in the Fancy was Bill George, alleged to be the inspiration for Charles Dickens's character Bill Sikes in the novel *Oliver Twist* (1837–39). George's establishment, known locally as the "Canine Castle," was carica-tured in a *Punch* cartoon in 1846 showing mostly bulldogs, but he also bred and supplied other types, including mastiffs.[73] Some fancy breeders adver-tised their dogs for sale in the classified pages of newspapers, as did pet shop owners.[74]

Working-class toy-dog ownership and breeding was not only for sport. The rat-catching abilities of small dogs had been cultivated in the industrial cities of northern England, where millworkers had bred small terriers to catch vermin in their homes and in factories. Hugh Dalziel, arguably the most influential journalist and author on the subject of dogs of the last quarter of the nineteenth century, claimed that the Skye terrier had been brought to the West Riding of Yorkshire by Scottish migrants seeking employment in the textile industry, and had been improved for pest control. Its long coat and affable temperament, which made it such a popular pet, seem to have been initially incidental, rather than the intentional results of selection. A second reason the working class preferred toy dogs was economy. Small dogs were cheaper to feed and easier to keep in crowded homes. In the 1840s, some so-cial investigators and sanitary reformers even claimed that in the homes of the poor, men would feed and treat their dogs better than their wives or children, comments intended to depict these men as irresponsible and igno-rant, and to show their poverty as self-inflicted. As well as providing vermin control, toy dogs were said to be very good domestic watchdogs. They reacted to "the approach of improper persons" by barking and, though unable to at-tack, continued their warning yelps as they retreated under furniture or into holes. The otherwise timid King Charles and Blenheim spaniels were known to be tenacious in keeping up "their sharp shrill note of defiance." Unfortu-nately, outside of the home these dogs were "always ready to be fondled by a stranger" and "liable to be stolen."[75]

*Lapdogs*

Historians have carried forward into the nineteenth century Keith Thomas's characterization of pet keeping before 1800 as a pursuit largely of upper- and middle-class women who cared for their dogs obsessively, fed them better than their servants, dressed them in fancy clothes, and included them in family portraits.[76] Thomas emphasized the social and psychological aspects of the relationship between owner and toy dog, which was "a symbol of fidelity, domesticity and completeness" in a period of rapid urbanization and other social changes.[77] It is not surprising that the lapdog was the favored victim of dognappers, who made their living stealing pets and demanding money for their return, exploiting the emotional value ladies invested in their favorites. The most well-known incident of dognapping was the repeated theft and ransom in the 1840s of Flush, the poet Elizabeth Barrett Browning's red spaniel. Although the association between lapdogs and refined femininity persisted throughout the century, mid-Victorian advice literature increasingly promoted dogs as suitable pets for every bourgeois home. Middle-class interiors were not commodious, and smaller dogs were less obtrusive in parlors packed with furniture and objects.[78]

There was some organized breeding of lapdogs through the Spaniel Clubs in the 1840s and 1850s, but in most cases the upper and middle classes acquired their pets from family and friends, or received them as gifts, as had Barrett Browning. Beauty and size mattered, but temperament and character were more important to forming an affectionate bond with the owner, tolerating a life spent mostly indoors, and proving suitable companionship to small children. Toy dogs were seen as unique individuals in their form, too, with crossbreeding between black-and-white King Charles and red-and-white Blenheim spaniels giving distinctive and individually unique colors. The upper and middle classes also carried into dog ownership their concern that new members of the family came from good stock or had pure blood. However, they were aware of the danger of close interbreeding, allegedly seen in some King Charles spaniels assuming "an expression of stupidity with which the character of the dog too accurately corresponds."[79]

The best-known upper-class female owner of a lapdog in the early Victorian era was Queen Victoria, whose favorite pet was a King Charles spaniel named Dash. He had been a present from the ambitious Sir John Conroy, her mother's former equerry and confidante. The Queen's age, just eighteen, not to mention her sex and Germanic background, meant that at accession she was not well regarded, and her affection for Dash was seen to confirm her immaturity. Lytton Strachey, one of Victoria's first biographers, wrote that even after the pomp of her coronation, "youth and simplicity" were reasserted as the young queen spent that evening giving Dash his customary bath. The King Charles spaniel was the exemplary lapdog: affectionate, elegant, and tolerant of a confined domestic life. While its long ears and silky black and white coat were defining characteristics, however, descriptions tended to be as much about character as form. Landseer's many paintings of Dash were, of course, portraits, and show that he was hardly toy sized and had few of the altered features that Youatt had condemned in such spaniels.[80]

The nature of the bond between lapdog and owner reflected and also redefined contemporary views of women's place in society, especially among the upper classes. Lapdogs were often portrayed as spoiled children and their owners as indulgent mothers whose emotional investment in their pets was foolish. Nowhere was this better seen than in an expanding consumer culture of luxury goods: lapdogs were dressed in fine clothes, wore jewelry, and were fed special foods, and many had their own plates and cutlery. The association between women and lapdogs rested on the perception that women, intrinsically emotional and maternal, needed an object for their love. As a trader in dog goods noted in Henry Mayhew's survey of London life, "it's them as has no children has dogs."[81] Lapdog varieties were largely, then, oriented around their prettiness and tolerance of being stroked and petted.

### Curs and Mongrels

The word *cur* was recognized as "the most ancient and general name of the dog in Europe," but by the Victorian period it was reserved for feral or street dogs. The earlier use was evident in Charles Hamilton Smith's volume on dogs in *The Naturalist's Library* series published in 1840, in which "The Cur Dogs" was the heading for domestic dogs, including terriers, mastiffs, and other mixed types, such as the bull terrier. Youatt in 1845 had the cur as a distinct mongrel type, "the sheep dog crossed with the terrier," which had "somewhat deservedly obtained a very bad name, as a bully and a coward; and certainly his habit of barking at everything that passes ... renders him often a very dangerous nuisance." Nonetheless, he described the cur as valuable to the cottager, for whom "he is a faithful defender of his humble dwelling ... and he likewise a useful and effectual defender over the clothes and scanty provisions of the labourer."[82]

By the last quarter of the nineteenth century, with the ascendancy of breed, curs were being lumped together with mongrels, but until then the term *mongrel* did not always have the same pejorative connotation. Among sportsmen the term was applied to any mixed-variety dog, and as we have shown, planned crossing was commonplace on estates and farms, and used by fanciers to acquire particular features for their strain. Indeed, some "mongrels" became established varieties, like the harrier—a beagle-foxhound cross of diminished size—and the bull terrier, in which the aggression of the bulldog was combined with the tenacity of the terrier. According to Youatt, the lurcher had been produced by a series of crosses: greyhound (for speed and quietness) with the shepherd's dog (for intelligence and willingness), and then with the spaniel (for finding game). Usually a mongrel was produced for just a single generation to obtain a specific feature, with the resulting dog bred back to type of form or function. This practice was often tried with "high-bred" sporting dogs, where the aim was to introduce intelligence, a faculty lost by inbreeding.

Curs were a motley-looking collection, being not just mongrels but "mixed mongrels" of every variety—and none. They were described as ill behaved, being the progeny of lascivious parents, and, like mixed-race human beings, were said (perhaps after a generation or two of hybrid vigor) to be weak and liable to poor health. If curs bred with other curs, and did so over and over again, the result would be the very worst sort of dog—"a degenerate mongrel race."[83]

If the indolent poor were the lowest of the low in the Victorian human hierarchy, curs held the same status among dogs. The association of curs with the urban and rural poor extended to the idea that they shared physical and mental attributes: both were unruly and impulsive, dirty, disease carrying, and threatening. Some canine experts claimed that because curs were typically the result of "miscegenation," combined with "illegitimacy," they inherited constitutional and moral weaknesses that made them prone to the canine plagues. There was a contrary view: "The mass of London dogs . . . are of no particular breed, but consist of a mixed multitude of mongrels. This deteriorates from their value in the eyes of breeders and fanciers; but it is likely that dogs are all the more intelligent from the mixture of blood, and much suited to city life than animals of a pure breed." The anonymous author then set out the class divisions among the capital's canines: "those who live within the doors and those who live without—the patricians and plebs of their races." The poor and working class did not confine their dogs indoors.

They were let out every day to roam the streets, sometimes joining packs and spending their days scavenging and fighting. In his book on domestic animals, James Rogers claimed that "the currish race" of mongrels was most likely to develop and spread rabies, being weaker and more aggressive. The size of the urban cur population was revealed in rabies scares, when street dogs were rounded up or killed by the hundreds and sometimes thousands. Such preemptive and violent actions were justified by curs' feral existence and the threat they posed, especially to children and family pets.[84]

## Conclusion

In 1859 a rival volume to Youatt's *The Dog* appeared when John Henry Walsh followed his book on the greyhound with the comprehensive *The Dog in Health and Disease*. The book did not just cover veterinary advice; it also dealt with the history of the dog, its varieties and management, all of which contributed to a holistic approach to health. The book began with descriptions of the varieties of dog, which were said to be "extremely numerous, and, indeed, as they are apparently produced by crossing, which is still had recourse to, there is scarcely any limit to the numbers which may be described."[85] In his classifications, his principal term was "varieties," but he also uses "sort," "strain," "type," "race," "tribe," and, of course, "breed."[86] He offered descriptions of the different varieties, which were qualitative and allowed latitude in the physical features, as with the mastiff, where comparisons seemed to imply some origin in crossbreeding with other types.

> A head of large size between that of the bloodhound and bulldog in shape, having the volume of muscle of the latter, with the flews and muzzle of the former; the ear being of middle size but drooping, like that of the hound. The teeth generally meet, but if anything there is a slight protuberance of the lower jaw, never being uncovered by the upper lip like those of the bulldog. Eye small. In shape there is a considerable similarity to the hound, but much heavier in all its lines. Loin compact and powerful, and limbs strong. Tail very slightly rough, and carried high over the back when excited. Voice very deep and sonorous. Coat smooth. Colour red or fawn with black muzzle, or brindled, or black; or black, red, or fawn and white. Height about 25 to 28 inches; sometimes, but rarely, rather more.[87]

There was no expectation that mastiff breeders would or should aspire to produce dogs with a uniform look.[88]

Form was just one dimension of a variety. Walsh gave equal importance to the temper and behavior of the mastiff, describing the dog as "being extremely docile and companionable, though possessed of high courage." It was "a most useful watch dog," but he wrote that, if purebred, mastiffs were perhaps too noble and mild. Hence, for duty as yard dogs, crossing with Newfoundlands and bloodhounds was recommended to achieve "a savage nature," and when used by gamekeepers against poachers, crossing with bulldogs was suggested "to give extra courage." [89] Thus, mastiffs were a resource for producing dogs for specific roles. Individuality in all varieties was expected and wanted. Owners of dogs for sporting competitions, from greyhounds to ratting bull terriers, did not want standard animals; they wanted exceptional ones that would win trophies and earn stud fees. Sportsmen wanted dogs that suited their estates and style of hunting, while pet owners wanted unique family members. Paradoxically, it was the extent of crossbreeding, to produce dogs tailored for particular roles, that led to modern conformation shows, as we discuss in the next chapter.

# CHAPTER TWO

# Adopting Breed, 1860–1867

Most histories of dog shows maintain that the first modern conformation show was held at the Corn Exchange in Newcastle upon Tyne on 29–30 June 1859.[1] There are reasons to doubt this claim, given that the Fancy held dog shows from the 1830s. (As we show in this chapter, these crossed over with the new shows.)[2] Also, the Newcastle upon Tyne Sporting Dog Show was restricted to pointers and setters and was an adjunct to the local agricultural show. If any event deserves to be regarded as the first modern, conformation show, it is the First Exhibition of Sporting and Other Dogs, held in Birmingham in December 1860, because it allowed entries for all types of dog.[3] The pioneer status given to the Newcastle upon Tyne show comes from its results being those first listed in the Kennel Club stud book; hence, its status represents a Kennel Club–centric narrative, to which this chapter offers an alternative.[4]

Here we chart the development of the new genre of dog shows, beginning with the initial domination of game-shooting interests and association with agricultural shows, and leading to the dog-only shows, where nonsporting dogs gained equal, if not higher status. Our narrative explores the social history of these events in the context of Victorian exhibition and competition cultures, the clash of values between "gentlemen" and "dog dealers," and contradictions of breeders and owners seeking to both improve and preserve types at the same time.

We begin, not with the Newcastle upon Tyne show itself but with the debate that led to its creation, which focused on the merits of dogs being judged on their conformation in competitions modeled on livestock and poultry shows. We set this in the context of the development of British rural sports as illustrated by the writings and pursuits of its leading figure, John

Henry Walsh. From 1857 to 1888, he edited the *Field: The Country Gentlemen's Newspaper,* and under the pseudonym of Stonehenge authored over twenty books on sports and other topics.[5] We then consider the mixed reactions to shows as they became more popular and commercial in the early 1860s, with growing tensions between those who sought to keep to the original aim of competitions, that is, to improve working and sporting dogs, and those who prioritized entertainment and profit by including dogs of all shapes and sizes. In 1865 sporting interests established field trials, competitions that simulated estate shoots, to determine the best dogs at finding (pointing or setting) game. Conformation shows and field trials followed the custom of agricultural shows in having experienced and knowledgeable men, preferably gentlemen, to judge. But as the number of events increased and those participating came from wider social backgrounds, shows became sites for rancor and disputes, as well as polite competition. There were complaints—sometimes backed up by legal actions—of bias, favoritism, bribery, corruption, and sharp practice.

One answer to the problem of judging was to agree on national standards for different types of dog, and in September 1865, the *Field* published the first conformation standard for a dog breed—the pointer. Written by Walsh, it took as an ideal a dog named Major that we contend was the first modern dog.[6] In succeeding months, conformation breed standards of other types were published in the *Field,* and Walsh published these in a collected volume, *The Dogs of the British Islands,* in 1867. He hoped that this volume would become the reference book for show judges across the country and would also guide breeders. Such efforts at standardization were common in other sports in the 1860s as they moved away from local competitions. The best known was the development of national rules for association football beginning in 1862.[7] As we shall see, Walsh's drive toward national conformation standards initiated the dominance of *breed* as a way of thinking about, producing, and engaging with dogs.

### A National Show of Sporting Dogs

The proposal for a show of sporting dogs was first made in February 1858, in a letter to the *Field* by Richard Brailsford, who signed himself "An Admirer of the Pointer and Setter." He was a well-known dog breaker and worked as a gamekeeper at the Knowsley estate of Prime Minister Edward Stanley, the fourteenth Earl of Derby. Brailsford maintained that current dogs were "a lot of weeds, wastrils [sic] and mongrels" and that they needed to be improved and bred truer to type. He wanted sportsmen to lead and take advantage of living in an era of "show mania." However, it had to be the right kind of show—one based on competition with the goal of improvement, as in agricultural shows—not an occasion for "Old Maids [to have] an exhibition of their tabbies."[8]

His proposal, and the spirited responses it provoked, anticipated many of the issues that occupied the dog-show fraternity over the next decade and beyond, particularly the tension between the cultures of commercial exhibitions for profit and gentlemanly competition for the improvement of dogs. The particular concern he aired was the degeneration of pointers and setters, which he maintained were being widely crossbred to suit battue shooting. Thus, the true type was becoming extinct and as a consequence would no longer be available for crossbreeding. It is not without irony, therefore, that the improvement of pure types of setters and pointers was not an end in itself but a means to secure the future of crossbreeding.

Brailsford made two propositions: a show for sporting dogs on the "same principles as Cattle, Sheep, Horse and Poultry Shows," and an association to improve the quality of dogs. The proposal drew many letters to the *Field;* all those published were in favor and argued that the success of cattle, horse, and poultry competitions augured well for the future of dog shows. A favored figurehead for the new shows and association was Grantley F. Berkeley, a colorful, not to say infamous, politician, who had been a reactionary during the reform era of the 1830s and 1840s. Since losing his parliamentary seat in 1852, he had devoted himself to writing on travels and rural sports. Unsurprisingly, given that the prevalence of crossbreeding was one reason for the show, some correspondents suggested classes for mongrels. "R. K. P." wrote that the term "mongrel" was becoming one of reproach, yet in agriculture and horticulture, hybridization and grafting were the main methods of improving stock. He argued that the "science" of improved breeds should be the "object of the association" and wanted "Mixed breeds and new varieties" as the main class, followed by "thoroughbreds" that were either "Sporting," "Utility," or "Fancy" dogs. Almost everyone thought that field trials were impractical in terms of logistics, cost, and ensuring fairness. One correspondent observed that poultry shows were not followed by egg-laying trials, nor livestock shows by slaughtering pigs and tasting the bacon, nor were horse shows followed by races. At agricultural and horse shows, it was well established that external form was a good indicator of physiology (putting on flesh) or action (movement of limbs).[9]

The term *breed* was used by many writers to the *Field,* but so too were *kind, variety,* and *strain,* and many ways of dividing dogs for judging were suggested. However, there were major disagreements over how such events ought to be organized, what dogs should be entered, and how they should be judged. In the correspondence printed, only a fraction of that received, twelve out of nineteen letters were in favor of a general, open show, and seven wanted to restrict it to sporting dogs. Brailsford supported the latter idea, arguing that "pets, mastiffs, bulldogs and the like . . . would considerably lower the character of the undertaking" or, more likely, their owners would.[10] He wrote that it was "painful to perceive" how his original goal of rescuing pointers and setters had been overtaken by advocates of all types of dog.

Open events for all "strains and varieties" were thought likely to attract large audiences, but it was important to deter the "wrong sort." A common theme was that shows should be endorsed, if not run, by gentlemen and that they should be organized by an association dedicated to the "improvement" of British dogs and not as entertainment spectacles run for profit. The amateur/professional divide was common in many sports in Britain at this time. More than that, the preference for "gentlemen"—that is, men of independent means (some of whom might also pursue a profession)—indicated a deep-seated suspicion of the entrepreneur. Political economists' deliberations on the social value of entrepreneurialism notwithstanding, those claiming to be gentlemen, who linked "fair play" with moral integrity, hesitated to associate with the profit-driven capitalist or the aspirational opportunist who typified the London-based fancy.[11]

Brailsford's initiative stalled, despite the support of some "influential names." He abandoned his initiative when he heard—wrongly, as it turned out—that the *Field* was organizing such an association. However, Walsh, through his editorship of the *Field,* continued to support a sporting-dog show; indeed, he was a judge at Newcastle upon Tyne in July 1859, and his newspaper postponed its gun trial by a week to avoid a clash.[12]

### John Henry Walsh, aka "Stonehenge," and the *Field*

Before becoming editor of the *Field,* Walsh had been a general practitioner in Worcester and early editor of the *Association Medical Journal,* the forerunner of the *British Medical Journal* (fig. 2.1). After losing part of his left hand in a shooting accident, he turned to journalism, writing and publishing on a range of sports and medical subjects. Under the pseudonym "Stonehenge," he published *The Greyhound* (1853) and *A Manual of British Rural Sports* (1856,

**Figure 2.1.** John Henry Walsh. Stonehenge, *Manual of British Rural Sports* (London: Routledge, Warne & Routledge, 1875), frontispiece

in its fourth edition in 1859), and, under his own name, edited *The English Cookery Book* (1858). In 1859 he published *The Dog in Health and Disease,* which he intended as the successor to William Youatt's *The Dog.*[13]

The *Field* was the leading newspaper for sports of all kinds and each week published columns on angling, archery, chess, coursing, cricket, falconry, farming, gardening, hunting, natural history, poultry, recipes, rowing, shooting, sporting, the turf (horseracing), veterinary matters, and yachting. Its proprietors claimed it was the largest newspaper in Europe. The articles, letters, and advertisements were aimed predominantly at those who lived and worked on landed estates, in great houses, and in public schools, and at those generally in a similar social milieu; however, it featured articles and reports on an eclectic range of subjects. The paper carried editorials on current affairs; encouraged technological innovation in farming, shooting, and transport; and was in the forefront of moves to establish national rules and competitions in many sports. Walsh was a founding member and for many years honorary secretary of the All England Croquet Club, which was inaugurated at the offices of the *Field*. When the club added lawn tennis to its annual Wimbledon croquet championship in 1877, entrants submitted their entries to Walsh at his home address in Putney.[14]

Until the 1860s, the *Field*'s coverage of dogs was focused on coursing (greyhounds), hunting (foxhounds), and shooting (pointers, setters, and retrievers). When accounts of dog shows began to be included, they appeared in a new section, "The Country House," presuming, it seems, that those exhibiting had kennels. (The word referred to both buildings and a pack or group of dogs.) Despite the *Field*'s rural orientation, Walsh favored the inclusion in shows of nonsporting dogs, what he called "man's domestic friends." He described such dogs in sentimental terms as companion animals, which were often a person's "solace in affliction, in poverty, or in disgrace": "When all the world deserts the outcast or the beggar, a poor, half-starved dog sticks to him to the death, and in many instances after he is laid in his lowly grave."[15] However, he was clear that only faithful house pets "of pure breed" should be allowed to compete, as these were the only dogs on which "improvement" could be based. He thought that an "open show" was likely to be more successful because its scale and diversity would attract entrants and a paying audience. On the question of judging, he was for nonsporting dogs to be assessed on conformation alone but felt that the winners of sporting-dog classes should be tested in the field for function, to ensure that good-looking winners were effective workers.

### Show Cultures: Newcastle upon Tyne, Birmingham, and Leeds, 1859–1861

The first new-style dog shows were held in the industrial cities of the north of England and the Midlands, in conjunction with agricultural shows that drew exhibitors and traders from rural hinterlands and the cities themselves. The numbers of horse carriages, urban dairies, piggeries, and henhouses, not to mention cats, dogs, and vermin, gave animals a presence in Victorian towns and cities.[16] Landed interests continued to hold economic, political, and cultural influence in urban areas, and local companies provided goods and services to farms and estates. The Newcastle upon Tyne Show in June 1859 reflected these dynamic social relations. It was organized by commercial interests—John Shorthose, a brewer's agent, and William Pape, a shotgun manufacturer. The exhibition was billed as an additional attraction to the poultry show, which had been inaugurated in 1847. The entries for the dog show came from across the north of England and the Scottish borders, but it became a national event because of publicity given in the *Field*. Following the debate the previous year, controversy was inevitable. Brailsford, who was a judge, wrote that he had to endure the taunts of huntsman, though it is unclear if they reproached him for simply taking part or because of favoritism, as his son's pointer won a first prize. The reporter for the local *Newcastle Courant* highlighted the seeming contradiction of judging on "shape, symmetry, and apparent pureness of breed," when, "although a dog may possess all these points, he may not, for want of proper training, scent, or other defects, be the best in the field for the sportsman." Inspired by the sporting-dog show, the annual Cleveland Agricultural Show in Redcar that September added a foxhound class to its schedule. Such classes were initially small affairs but grew in importance, and a national foxhound show was first held in Peterborough in 1878.[17]

In November 1859, Richard Brailsford coorganized a sporting-dog show in Birmingham, held in association with the city's annual agricultural, poultry, and horticultural shows. The *Field*'s correspondent wrote that nearly one hundred dogs and thousands of visitors were able "to witness the splendid forms displayed of English hounds, pointers, setters and spaniels." Again there were misgivings concerning the value of conformation without due consideration of function. A local Worcestershire newspaper observed, "Of course the prizes could only be awarded for appearances, as no trial of their valuable points as sporting dogs took place. In a sporting point of view, therefore, the exhibition must be completely useless, as any sportsman knows that a handsome dog, as far as figure and flesh go, may be the veriest cur in the field, while many an ugly looking brute may be 'worth his weight in gold.'" Nonetheless, the show was deemed a success and led to plans to hold another the following year. This event in December 1860 has the best claim to being the first modern dog show, as both sporting and nonsporting dogs were included. Ahead of the show, at a meeting in April 1860, it was agreed to form a society "for the purpose of establishing a national exhibition of sporting and other dogs."[18] Though its goal was to organize a national show, the society took a local title, the Birmingham Dog Show Society. Its first president was Viscount Curzon, who was joined on the committee by Midlands landed aristocrats: the Earl of Lichfield, Earl Spencer, the Earl of Aylesbury, Lord Bewick, and Lord Leigh.[19] The "modest" membership fee of 1 guinea allowed "other gentlemen well-known in sporting circles" to join but excluded lower-class fanciers.

Brailsford, among others, had hopes that the new society might start a public stud book for sporting dogs. William Lort, a member of the Birmingham Society and well-known local sportsman, raised the matter in the *Field*. His father had been one of the founders of the Midland Bank, and the son enjoyed the life of a wealthy "land proprietor" living in King's Norton, Birmingham. He moved from breeding and exhibiting poultry and livestock to doing so with dogs, and in time, his breadth of knowledge and genial character led to his being called upon to judge all sorts of animals at shows. Lort supported stud books, arguing that they were needed to ensure "careful breeding and proper change of blood."[20] Brailsford agreed and hoped that a register could be started in time for the November show. However, he had misgivings.

> Nothing in my opinion is more necessary than changes of blood. I have seen instances where sportsmen have been so opinionated with certain breeds that they have bred in and in until they could produce nothing but woods. We are all much too prejudiced to fashionable blood, for if Mr So-and-so had some of the finest and best pointers in England he could make little of them unless he could produce a pedigree from the late Sir Henry Goodriche, Mr Edge of Headley, Cpt. White, Mr Moore of Appleby or some such noted breeders, the first two of which gained their celebrated blood by crossing with the foxhound.[21]

As with cattle and sheep breeders, dog breeders were keeping private stud books, but the name of the breeder and trust in his stock counted as much as the "look" or lineage of an individual dog or bitch. Also, there was a tension between the practical men, such as Brailsford, for whom a dog's working ability was important (even if judged from conformation), and those for whom pedigree and lineage were most important.

The second Birmingham dog show was again staged in conjunction with the city's agricultural show in November 1860. It was promoted as the "National Dog Show" and received press coverage across the country. Its newsworthiness became clear when the *Illustrated London News* put a picture of the winners on its front page (fig. 2.2).[22] *Bell's Life* celebrated the event as another innovation from the city whose achievements went beyond industry: "The historian will note a new era in the annals of Birmingham. It will be no longer merely for the curious handicraft of the artificer, or suggestive only of the fiery furnace that encompasseth it. There is metal more attractive.

Here was it that the elegant Cochin China first forced himself into fashion and the 'buff' and 'fluff' rose to a paragon of excellence. Under the like auspices, and in the same iron-hearted city, the merits of the 'farmer's friend,' the Shropshire sheep, came to be asserted."[23] The correspondent of the *Birmingham Journal* mused on the contrasting displays of cattle and canines, tacitly recognizing the features of conformation breeds: "Indeed, the two tribes, although very dissimilar in character, have a property in common with each other, and that is that the beauty and symmetry in their forms, the variety of their colour, together with the domestic associations connected with them, are calculated to command the attention and admiration of the amateur as well as the connoisseur." The correspondent of the London *Times* was also impressed: "Viewing the strange diversities in form, capability, and disposition of the dogs in this whining, growling and barking menagerie, we can scarcely admit the doctrine that the animals are merely varieties of one species, and that all have been developed by differences of food, circumstance, and training from a single original pair; or, as some say, are collectively a tamed derivation from the lean and savage wolf. In this gathering of all descriptions of hounds, for instance, what extraordinary differences are observable in the nature and use of several breeds."[24] The reference to the dogs' origin in "a single original pair" might have been a reference to the Bible or perhaps to the ideas in Darwin's *On the Origin of Species,* published the previous autumn.

The judging of the dogs was in forty-three classes: twenty-nine for sporting and twenty-four for nonsporting, including classes for toy dogs and "other general varieties not used in sport," which were judged by Walsh. Among the nonsporting dogs, the largest group were Scotch terriers, followed by mastiffs, Newfoundlands, and toy terriers. The London Fancy stayed away, arguing that better dogs were to be seen in "the club rooms of any of the London public-houses." Bias among judges was an issue, in terms both of fairness and of what defined a prizewinner. To meet this concern, teams of judges were used; teams were also necessary because it proved impossible to find men who would act alone, as their decisions were likely to be challenged and their social standing to suffer. There was, according to *Bell's Life,* something for everyone at the show: "the luxurious 'King Charles' for fine ladies to fondle, and the three-seasoned 'hunter' for the sportsman to dwell over." Indeed, the event was so crowded that visitors were in danger of losing "the calf from your leg" or "the tail from your coat" from the attentions of bulldogs, pointers or the fearsome Cuban mastiff.[25]

**Figure 2.2.** National Exhibition of Dogs at Birmingham. *Illustrated London News,* 15 Dec. 1860, 1. © British Library Board

Echoing the London Fancy's disdain of the new exhibitions, the local Fancy announced a competing event in a public house—"A second Birmingham dog show," to be held at Mr. Burbidge's The Punchbowl in the city center on 1 January 1861. On display were to be the "finest bulldogs, black and tan terriers, [and] Maltese and Italian greyhounds," with "the celebrated J Brown of London" chairing the proceedings. Two days before, there had been a ratting meeting at the Hop Pole Inn, also in the city center. And the same week, at the King's Arms, "a large and commodious room" for ratting sports was opened, decorated with likenesses of prize dogs and a picture of the inaugural world-title boxing match, an illegal fight between the American John C. Heenan and England's Tom Sayers that had taken place in April 1860. This decor showed the continuing links between the Fancy, pugilism, and plebeian culture.[26]

The next grand exhibition that attracted national attention was held over three days at Leeds, Yorkshire, in July 1861. It was promoted as "The North of England Exhibition of Sporting and other Dogs" and held in association with the city hosting the annual Royal Agricultural Show. A significant feature was the large proportion of nonsporting dogs, 224 as opposed to 315 sporting. Foxhounds were again absent. Allegedly, huntsmen would only attend "a show of 'hounds,'" and one was coming up nearby in Cleveland, in the North Riding of Yorkshire. Among some sportsmen "hounds" were a separate and superior type of canine and were even counted differently—in couples. The nonsporting "second" division at Leeds attracted the ladies, who were patronizingly reported to be pleased by the exhibition of their "dear little loves," but entries from London did best in these classes.[27] The show was a private, commercial venture, and unlike in Birmingham, its main goal seemed to be entertainment and profit.

The show's commercial underpinning led to questions about fair play in judging, and there were calls for the proprietors to publish "the 'points' upon which they awarded the various prizes." The aim was to guide those planning to enter their dogs in future, as well as to "settle at once the unpleasantness there appears to exist over the late show." The account in the *Field* gave insight into the issues in its discussion of greyhounds, the largest class. Judges were supposed only to assess conformation, but they and exhibitors were conscious of bloodlines and previous performances. The National Coursing Club had been founded in 1858 to regulate the sport, but unlike the Jockey Club it had not published a public stud book; hence, pedigrees remained private. Nonetheless, the top greyhound at Leeds was Mr. J. Bell's Ringleader, which had the best possible pedigree, his grandsire having been Bedlamite, the top racing and stud dog of his generation. The report in the *Field* drew a distinction between greyhounds that pleased the judges by their conformation and symmetry, and those that would be good breeding stock, to which good blood would give "bone and substance." However, in practical breeding this consideration was nuanced by sex: for example, a fast greyhound male with "light understandings" (legs), might be bred with a female that had more "substance," to ensure that the offspring had strength as well as speed.[28]

Across all classes, judges looked for "noble expression," "substance," and "quality," as well as specific features or points, looking too for overall "symmetry." Color seems not to have mattered because "like a horse, a good dog can hardly be of a bad colour." While particular points were discussed, it seems that judging was primarily based on the impression made on the eye. In this sense, experience and tacit knowledge were all important in judging, which would have only been available to those of high social rank and gentlemanly by birth, marriage, occupation, or wealth.[29] Overall, commentators accepted that judges at different shows, in different parts of the country, rightly had their individual preferences, some of which might be idiosyncratic, but as long as there was fair play this variation was acceptable.

### Exhibition Cultures: London, 1862 and 1863

Dog shows built upon the social success of poultry and pigeon shows in the previous decades. Pigeon breeding and showing had a longer history and had been associated with the Fancy. Poultry keeping became popular in the 1840s and 1850s, producing what contemporaries called "poultry mania." In the middle and upper classes, there was a vogue to own and breed fancy poultry, and in turn to show birds in exhibitions, a fashion said to be started by the building of a vast aviary at Windsor. An article in Charles Dickens's *Household Words* compared the fervor at the 1851 Birmingham Poultry Show to "the old tulipomania" of the mid-seventeenth century. Two familiar issues had dominated discussions of poultry shows: the merits of judging on beauty rather than utility, and whether breeders were amateurs interested in the improvement of birds or dealers seeking profit. Newspaper coverage of the Manchester Poultry Show in 1855 noted that most exhibitors were "amateurs . . . ladies or country gentlemen, or . . . farmers" and that there were "few dealers." The 2,500 birds on display were said to show "high condition, beauty of plumage and purity of race." Poultry shows reached their peak popularity at the end of the 1850s. In 1860 there were two poultry shows at the Crystal Palace: in February, three to four thousand birds were exhibited, while pigeons and rabbits were included in the summer show in August, perhaps to boost interest and attendance.[30]

Prompted by the establishment of the Birmingham Dog Show Society, in the summer of 1860 the Duke of Richmond proposed creating a London-based national canine association, modeled on the Jockey Club, a move supported by the nobility and sportsmen. However, the duke died in October, and the prospectus never appeared. The need for such an organization was reflected upon in the *Observer* the following spring: had the duke's idea been visionary or absurd?

We have schemes for almost impractical railways, bridges, tunnels, leviathan ships, & c; for opening of new mines which for years have been considered worn out; for electric telegraphs which are to connect every part of the world; and even for a direct railway to India. There are new hospitals for the cure of disease, and hospitals also for incurables. Hydropathic, homoeo-pathic, and even mesmeric institutions are in existence, and receiving public support. Idiots are now cared for, a refuge has been provided even for houseless dogs. Some of these may appear ridiculous at first sight, yet it is impossible to say in this age of progress what eventually may turn out to be absurd; and what may be of great service to the public.[31]

It was hoped that the planned association would create on a permanent site an institution to oversee breeding and training, serve as a center for buying and selling dogs, and house a hospital and kennels for owners traveling over-seas. The association was also to hold its own annual show. The Zoological Society of London was involved, as varieties of dogs had been among the ex-hibits in the early days of its zoo in Regent's Park. Furthermore, "some of the highest persons of the realm" pledged their patronage to what promised to be a popular and profitable show.[32]

The first new-style dog show in the capital was in October 1861: "The Great London Exhibition of Sporting and Other Dogs." Held in a small venue in Holborn, it was reported to have been a disappointing affair, with few dogs of quality on display. A *Bell's Life* correspondent was critical, claiming that there was a great deal of "'weedy' stuff" sent up by "country people," and bet-ter dogs were to be seen that week by the London Fancy at Jem Ferriman's in the City Road. Encouraged by successes in other towns and cities, two new London shows were promoted in 1862—at Holborn on 26–30 May and Isling-ton on 24–28 June. The Holborn "Exhibition of Fancy and other Dogs" was significant for being the first modern show that did not include sporting dogs. The event was said to have been a great success, and the *Field,* for the first time, printed a full-page engraving of the prizewinners. The show was perhaps the first crossover of old and new shows, with the London Fancy entering their dogs. The *Field* correspondent hoped it had been educational, trusting that "dog-dealers" and their "back-parlour practices" learned the benefits of putting on shows in "a clean, airy and light place, where ladies and gentlemen can venture to inspect them."[33]

The event that really raised the profile of dog shows in the capital, but not in the manner hoped for, was held in June 1862. To the surprise of everyone it was billed as "The North of England Second Exhibition of Sporting and Other Dogs." It was the creation of T. Dawkins Appleby, a provisions dealer from Leeds who had organized the show there the previous year. Appleby promoted his venture as the "Monster Dog Show," suggesting entertainment as well as elevation. It was the first event held at the New Agricultural Hall, a venue built on Islington Green for livestock shows and unfinished at the time. Part of the roof was still open, which some wits commented was a great advantage, as it allowed noise and dog odor to escape. Advance publicity boasted that the show was to be the first with over one thousand entries— including the elite pack of foxhounds from the Duke of Beaufort's estate at Badminton—and would be graced with celebrities, such as Sir Edwin Land-seer, who served as a judge. Such was the confidence of the promoters that they planned a weeklong event, open to different social classes on different days. The entrance fee for an exclusive opening day was set at 5s.; for a middle-class second day, one paid 2s. 6d., and the three public days each cost 1s.[34]

It was a great success socially, drawing fashionable visitors and positive reviews, at least in the short term. On the first day the roads of Islington were jammed with the carriages of the aristocracy and members from both houses of parliament, all making the "expedition" from the West End to the East End or, as one commentator put it, from "Belgravia to Whitechapel." Over its five-day run, there were more than sixty thousand visitors. The principal attraction was the sheer variety of dogs, the report in the *London Review* claimed that in exhibitions that year, there was "none in which the plastic powers of art applied to nature were more strikingly shown" (fig. 2.3). There were celebrity dogs too: the bloodhound Druid, valued at 250 guineas; Canaradzo and Judge, two greyhounds that had won the Waterloo Cup; and Newfound-lands belonging to the famous England cricketer Nicholas Felix. Only a few complaints about the judging arose, even though in some classes no awards were made because the dogs were not up to standard. The list of prizewin-ning owners, who typically won a cup and either five or ten guineas, was published. There was some buying and selling, and many dogs had reserve prices; the highest prices seemingly were for deerhounds, with the most expensive being a bitch named Brenda, valued at £1,000. Most complaints were about logistics. Many owners did not accompany their dogs to shows, sending them by rail to be collected at London terminus stations by agents. Unsurprisingly, a few were lost in transit or escaped (or perhaps were stolen) from the hall. Common concerns among exhibitors were the emotional

*Figure 2.3.* Prize Dogs from the National Exhibition at the Agricultural Hall Islington. *Illustrated London News,* 5 July 1862, 21. © British Library Board

consequences for their dogs of separation, confinement, crowding, noise, and poor feeding, not to mention the danger of their dogs catching mange or distemper.[35]

The weekly press mused again on the meanings to be drawn from the dog shows in relation to English national character, and relationships with nature. The report in the *Daily Telegraph* saw the ranks of dogs as the "epitome of humanity," stressing character, not conformation. Indeed, English character was evident in the mastiffs, with their "immense strength, indomitable courage, unconquerable fidelity, but imperturbable solidity and placability." There was in fact a cross-section of society on display: "a clerical dog and here a fine lady, and here a parasite, and here a thief and here a sailor and member of the Royal Humane Society, and here a dog who is a blackguard and nothing more." In all, it was regarded as a remarkable, even "brilliant" event and more edifying than a visit to the zoo to see "the disgusting monkeys . . . 'our poor relations.'"[36]

Because he pitched his exhibition to the public at large, Appleby was accused of departing too far from the business of improving the nation's dogs. Calling this event the "Monster Dog Show" invited comparisons with the circus or freak show, where the corporeal spectacle and the audience's emotional responses were at the forefront. In some ways the show perpetuated, on a much more ambitious scale, the elaborate menagerie shows by the likes of George Wombwell.[37] There was no coincidence that a month before the Monster Dog Show, the showman P. T. Barnum had staged his first Great Dog Show in New York at his American Museum. Barnum's dog show followed a baby show, and those visiting were also able to see permanent attractions, such as "That very little shrimp of humanity, COM. NUTT, and that ponderous human biped, THE BELGIAN GIANT, the smallest and largest men ever known."[38]

Appleby's exhibition went from boom to bust in a fortnight. On 11 July, at a meeting of his creditors in Leeds, it emerged that he was insolvent. There had already been complaints that no prize monies had been paid out, nor trophies sent to winners. The cup maker, who had taken the trophies back for engraving, had kept them. Appleby owed £1,912 on his venture and £3,600 to his trade creditors; his assets were only £2,905. He claimed his problems were caused by having neglected his provisions business in Leeds for what he thought would be "a grand and paying affair."[39]

The Monster Dog Show was more widely condemned as a failure and became an opportunity for sportsmen and others to vent their feeling about the new shows. An editorial in the *Field,* entitled "Going to the Dogs," took against the show on three counts: (1) the lack of concern for animal welfare—the painter Edwin Landseer complained that many dogs had cropped ears; (2) the preponderance of nonsporting dogs, which meant that two-thirds of exhibits were "useless" canines to many of Walsh's readers; and (3) the private, profit-making nature of the event.[40] The editorial described Appleby as a "speculator" and reminded readers that he was a provisions merchants, or maybe just a "cheesemonger." What Walsh had hoped would develop as "a spontaneous movement on the part of the great dog-breeders of England" had been usurped and turned into "a great auction." However, there was also criticism for the titled and leading figures in the dog world who had lent their names to Appleby's venture. Perhaps he had abused their support by not constituting a committee or securing financial guarantees, but then why, the editorial wondered, had gentlemen allowed their names to be used?

In the wake of the Monster Dog Show, many correspondents to the sporting press objected to the whole project of new dog shows. "Old Towler" complained that too many owners were "breeding up to defects." Bloodhounds, those noble creatures, now had "long pendulous ears and lips—the drooping eye, the shambling gait, and the slovenly way of dropping and eating its food—all defects which judges deem beauties." Bulldogs had been bred with "a ridiculously short nose, and a great projection of the under jaw." He urged breeders to "get rid of that absurd 'stop' between the eyes; it is a very great defect, and injures the scent. We might as well breed a dog with one eye as with no nose." Old Towler extended his criticisms of modern bulldogs to other exaggerations of conformation: "Dogs of no sort should be subjected to such freaks, nor should judges countenance the abortions with mere fashion in breeding produces. If dog shows are to be established for the purpose of canine improvement, the excesses of in breeding [*sic*] must be attacked." With bulldogs, he went on to argue, breeders had neglected their "inestimable attributes of courage, pertinacity, stamina and nervous energy," which were vital to preserve, as they were invaluable in crossbreeding so as to bring such characteristics to other breeds. His views were echoed by "Aberfeldy," who asked, "What can be the use of breeding these goggle-eyed abortions called toy dogs, which are fit for nothing but to breed fleas and spoil carpets." The only exception seemed to be Queen Victoria's Pekinese Looty, which had been rescued from the Chinese Royal Palace in 1860 and had been featured in the *Queen* magazine the previous week, though he added that he "would not give such specimens house room." An article on Looty in the *Field* was far from flattering. She was said to go from "a state of perfectly calm and happy repose" quickly "to the greatest rage, amply expressed by bristling hair, startling eyes, grinning teeth, and a tail going through the vibrations peculiar to *fighting* dogs." She was no beauty, having "a crippled nose," which made her eyes water and gave her "a lolling out tongue."[41]

Over the summer of 1862 dog shows attracted criticism beyond the sporting press. In Dickens's magazine *All the Year Round,* a writer reflected on the dog show at Islington and the show at the Home for Lost and Starving Dogs in Holloway. There has been uncertainty over the authorship of this article.

Recently it has been attributed to John Hollingshead, but there is a stronger case for Dickens himself being the author. John Colam, who was secretary of the Royal Society for the Prevention of Cruelty to Animals (RSPCA) at the time, later wrote appreciatively that the article had turned around the fortunes of the home, stating that Dickens's words were "at the time worth any sum of money to our cause."[42]

Dickens contrasted the Islington event, which was "all prosperity," and the event a mile away that "was as complete a contrast to the first as can well be imagined," one where "all is adversity." He observed, "For this second dog show is nothing more nor less than the show of the . . . poor vagrant homeless curs that one sees looking out for a dinner in the gutter or curled up in a doorway taking refuge from their troubles in sleep." His musings were mainly on the similarities between the personalities of humans and dogs. Dickens, a known lover of dogs, wrote that he had discerned a moral difference between the two sites, which echoed individual and social-class characteristics: "I must confess that it did appear to me that there was in those more prosperous dogs at the 'show' a slight occasional tendency to 'give themselves airs.' They seemed to regard themselves as public characters who really could not be bored by introductions to private individuals. . . . As to any feeling for, or interest in, each other, the prosperous dogs were utterly devoid of both. Among the unappreciated and lost dogs of Holloway, on the other hand, there seemed to be a sort of fellowship of misery, whilst their urbane and sociable qualities were perfectly irresistible." Dickens thought the Holloway asylum, which many contemporaries regarded as ridiculous and likely to divert charity from more worthy causes, was "a remarkable monument of the affection with which English people regard the race of dogs, as evidence of that hidden fund of feeling which survives in some hearts even [through] the rough ordeal of London life in the nineteenth century."[43]

Dickens was less sanguine about the Islington show, claiming that it was "a great comfort *not* to understand" the preferences of judges.[44] Like Old Towler he wondered why "the bloodhound's skin should hang in ghastly folds about his throat and jaws," or why bulldogs had to be "bandy, blear-eyed, pink-nosed, blotchy under-hung, and utterly disreputable." He was typically humorous—suggesting that the small intellect of the St. Bernard was what "might be expected of a race living on top of a mountain with only monks for company"—and sharp politically, asking why a medal had been given to a Cuban hound, a dog notorious "for chasing down fugitive slaves." Dickens's comments, although coming from different premises, moved toward similar conclusions as those reached by the rural sporting lobby—disapproval of the conformation and character of the newly invented "show dogs."

On sporting dogs Dickens was silent, making no reference to the scandal that had overtaken the Islington show, which rural sportsmen were using to reinvigorate their calls for a national society for improvement of dogs. Their hopes lay with the National Society of Sporting and Other Dogs in Birmingham. In November 1862, the society's third show was again run alongside the city's agricultural show, with dogs competing for public attention and allegedly winning it over the attractions of cattle, sheep, pigs, rabbits, canaries, crops, and flowers. Newspapers contrasted the "mere speculation" of the recent Islington show with the thrice-proven Midlands event backed by a wealthy committee of "noblemen and gentlemen." It did not disappoint. Over six hundred dogs were exhibited and over twenty-five thousand people paid a total of £1,376 to attend. Unsurprisingly, the sporting dogs were more numerous, 441, and allegedly of better quality than the 199 nonsporting. Once again the bloodhound Druid led the prizewinners.[45]

The following year, 1863, saw two shows succeed in London from unpromising beginnings. The specter of Appleby's "speculation" was raised again early in the year when Mr. Edward Tyrrell Smith, a theatrical entrepreneur and new proprietor of Cremorne Gardens on the Chelsea Embankment, announced he was organizing a dog show in March. Cremorne Gardens had been developed as a place of "alfresco amusements," offering concerts, dancing, fireworks, marionettes, balloon ascents, and equestrian exhibitions, as well as ballets, farces, and other indoor events in the site's Ashburnham Hall. The plan for a dog show was welcomed in the London popular press, with the hope that the lessons of the previous year had been learned. However, an editorial in the *Field,* entitled "Give a Dog a Bad Name" and probably penned by Walsh, took against the scheme with great vehemence. It complained that it seemed London was to have another show for profit, rather than improvement. The same day the editorial was published, the *Field* office received a letter from Smith's solicitors threatening litigation if an apology was not forthcoming. An editorial in the following issue gave details of the exchanges that had followed the previous week's tirade and offered a cynical apology. The context was a report in the *Sporting Life* of a meeting of the London Fancy, which Walsh read as confirming his fears that Smith was in their pocket. The Fancy held several of its beauty contests in public houses during the same week as the Cremorne Show, a week that also saw the All England Ratting Sweepstake held at the Crown Court in Windmill Street. Mr. Ives, who ran the "Over the River" Mammoth Canine Establishment in Lambeth, across the Thames, looked forward to meeting old friends at Cremorne and making available his stud dogs.[46]

Any association with the London Fancy led the *Field*'s editorial to argue that "what Mr Smith designs to carry out is not a Dog Show, but a 'Dawg' Show" (referring to a Cockney Londoner's pronuniciation). The distinction was explained at length.

> The difference may be slight in appearance, but it is, in reality very vast: there is a wide distinction between a dog and a "dawg" as there is between a sportsman and a sporting man. A dog is the faithful and intelligent animal which the sportsman makes the companion in his pursuits; a "dawg" is the creature which the dog-dealers use to hunt money out of the pockets of cynomaniacs, pet loving old dowagers, and other persons having more money than brains. A dog is a genuine Newfoundland, a retriever, a foxhound, a pointer, a setter, a terrier, a spaniel, or a mastiff; a "dawg" may be any of these or none of these, for he may be a monstrosity or a mongrel, whose only use is to bring money to his master's pocket. The dealers in "dawgs" term themselves in the language of the report "the fancy," but the meaning of the term is not easy to understand; for what it is they fancy, except the money of their customers, it passes to conceive. At any rate, we are quite willing to admit, if it be any satisfaction to Mr E. T. Smith, that any show organised and patronised by "the London fancy" does not interfere with our ideas of a dog show, or with our notions of what it is proper sportsmen should take part in.[47]

Smith dropped his threat of litigation and took steps to improve the entry and public profile of his show. He enrolled experienced doggy people and elite supporters, hiring Richard Brailsford's son Frederick, who was gamekeeper at Shugborough Hall, to find more sporting entries.

Visiting a week before the opening on 21 March, Walsh did not find his worries assuaged: "The decorations smack a little too much of the Drury-lane scene-painter." The *Field*'s report the following week was scathing on all aspects. The major complaint was poor organization: twelve hundred dogs were on the benches in a space suitable for just six hundred. They were chained, with no water, and on display from 8 a.m. to 11 p.m. (fig. 2.4). Sporting dogs were outnumbered two to one by nonsporting ones. Many dogs had been placed in the wrong class and did not match the catalogue, suggesting that organizers, owners, and judges did not know what they were doing: "On the whole, therefore, it may be gathered that the show of 1,200 dogs at

Cremorne was made up of a large mass of wretched brutes collected from the 'highways and byeways' in which were mixed a few superior animals of their respective classes, and that, but for the gross mismanagement of those who had the control and arrangement of the dogs, we should been enabled to announce that a success had been achieved, the more wonderful, because of the total unfitness for the task undertaken." Perhaps the future prospects of all dog shows had been damaged. Correspondents to the *Field* complained of their dogs being lost, catching diseases, and—inevitably—being judged wrongly.[48]

There was, however, an alternative narrative. The verdict of the London press was that Cremorne was "brilliantly successful" socially. The Prince of Wales and Princess Alice had visited on the opening morning, along with the Duke and Duchess of Wellington and other members of the nobility. Smith had followed the Islington show's practice of having graded entrance fees on different days, and over the week there were 100,000 visitors. The *Morning Post* commented that the "extraordinary sensation" excited in certain circles against the event had proved unfounded and that public confidence in Mr. Smith had been rewarded. The weekly press was also positive. The *London Review* was in no doubt that "an exhibition of this kind may be repeated

VIEW OF ASHBURNHAM HALL, THE SCENE OF THE GREAT DOG SHOW.

*Figure 2.4.* "View of the Ashburnham Hall, The Scene of The Great Dog Show." *Illustrated Sporting News,* 28 Mar. 1863, 21. © British Library Board

every year with increasing popularity and success." The *Times* celebrated the nation's canines, claiming that, like the country's horses, they had the highest reputation abroad: "From the days of the Romans, who took our rare breed of mastiffs to fight in their amphitheatres down to the present year of grace, English dogs of distinct breeds have always been considered preeminent above similar breeds in other countries, and no other country as ever produced more distinct varieties than England." The reporter contended that dog shows had moved upmarket from "the dirty purlieus of some small 'public,' kept by an ex-champion . . . frequented by the lowest of 'the fancy' . . . [where] no gentleman ever sent his dog." The report in the *Era* confirmed that "the visitors . . . have not been confined to the middle classes, for hundreds of the 'upper ten' were viewed in the building on the first two days." The popular press clearly had different criteria than sportsmen for measuring success.[49]

In June 1863 a second, now "international" dog show was held at Islington, which many thought had to be an improvement on the Cremorne jamboree and the Islington show's first incarnation. Royal endorsement from the Prince and Princess of Wales on the first morning set the tone. The *Field* thought the improvement was due to the event's organization by a sportsman rather than a showman. No doubt driven by the need to increase income, there was also a reluctant change of mind with regard to the exhibition of toy dogs: "We therefore beg the pardon of the fair sex, and the proprietors of nondescripts of every class, and are only too happy to have them . . . at the Islington Hall."[50]

Over the next two years, the Islington International Show became an annual event in the society calendar that also had wider appeal. The same phenomenon was seen across the country through the early 1860s. By 1865 the National Dog Show Society in Birmingham had raised the funds to construct its own venue, the Curzon Hall, which allowed a greater number of entries. The Manchester Dog Show was successful too, helped by being staged with the city's poultry and rabbit show at the popular entertainment venue the Belle Vue Zoological Gardens over Christmas each year. The number of dog shows around the country proliferated, with events in Bradford, Brighton, Cheltenham, Darlington, Doncaster, Dublin, Falmouth, Newcastle upon Tyne, Nottingham, and Stoke-on-Trent, and an increasing number were put on separately from agricultural and poultry shows. Within shows, the number of classes multiplied, encouraging more entries and guaranteeing more prizewinners. At Islington in 1865, classes were divided between dogs and bitches, large and small size, weight, and color, and there was a catchall class for "Toys of All Kinds."[51]

The ratting and beauty shows of the Fancy continued, and there was increasing crossover between the new and old fancies. This was in line with the wider growth of leisure and of organized sports by the urban middle classes, which saw the co-option of upper- and working-class support as spectators and participants. However, the alliances made were uneasy, and there were tensions. Some are familiar to historians of the Victorian era—rural/urban, amateur/professional, improvement/profit, sport/entertainment—and all complicated by class, gender, and regional identities. In 1865, "A. H. B." bemoaned the presence of the Fancy, the "dealers and stealers of the metropolis" at Islington. He wrote that they were "a class compounded of the roughest material, of back-street slum, and gutter manufacture, composed of low publicans, pugilists, cabmen, cobblers, journeymen tailors and mechanics of all sorts, their chief stock in trade consisting, in most instances, of a *drunken old woman, and a good breeding bitch.*" James Greenwood, a journalist and contemporary chronicler of London's popular culture, was scathing about the London Fancy in his *Unsentimental Journeys; or Byways of the Modern Babylon* (1867). His description of the city's contemporary venues and events suggested continuities with the early Victorian period. In Greenwood's eyes, people and dogs merged: "Apart from the bustle and the uproar sat two or three of the most miserable objects that could be imagined,—ragged, thin, and anxious looking, and each accompanied by a gaunt hollow-sided bulldog." Many men were alleged to spend more on their dogs than their families, and he thought that their infatuation would lead them to the workhouse. Even so, A. H. B. observed that "amidst the whirlpool of gin, beer, baccy, fisticuffs, and dirt," many of these men were "genuine dog breeders [who] possess a practical knowledge . . . that the wealthier can never obtain."[52]

## Judging and Standards

At and after shows there were always complaints about judging. In part, these were from disappointed exhibitors who had inflated ideas about their dogs, and the common response was that, in any competition, there had to be losers as well as winners. However, there were persistent issues. Who should judge and how? What criteria were dogs to be assessed against? And what was the purpose of shows?

As we have shown already, the best judges were said to be "gentlemen," whose independence and experience meant that they would make fair decisions without fear or favor. Ideally, they would be sportsmen who knew their dogs and were familiar with the etiquette of agricultural competitions. However, past shows had proven that individuals were known to have their preferences; hence, it was usual to announce the judges beforehand to allow exhibitors to pick the best events for their stock. Pairs of judges or committees were another way that the biases of individuals could be mitigated.

The central, and of course very Victorian, principle of dog shows was that competition would select the best individuals, and breeding from these winners would guarantee the progress of the nation's canines. In its coverage of the 1865 Islington dog show, the *Pall Mall Gazette* commented wryly that "Piccadilly contends with Hoxton, the PRINCE OF WALES with the publican; for the mania for competitive examinations increases so that not only are public servants chosen by competition, but dogs compete for valuable prizes, and there is even a competition to discover the greatest ass."[53]

However, laissez-faire competition was open to abuse; it even went as far as judges awarding prizes to their own dogs and being chosen in anticipation that they would reward the dogs of favored owners. As early as December 1863, the reporter from the *Field* had argued that while most judging was in good faith, there was a more fundamental problem. "But the misfortune is that, at present there is no fixed code of rules applying to all kinds of dog, so that exhibitors are at the mercy of the judges, who, moreover are not selected until the entries are made. Hence we find at one show the judges of retrievers preferring the smooth coated variety, and then Mr Hill's Wyndham gets his innings; while at the next, curly coats are in ascendant, and Mr Hill's Jet or Mr Riley's Royal is at the top of the tree."[54]

While everyone might agree that competition needed to be based on agreed standards, how were these to be set? After the Cremorne show in April 1864, there had been a call for a "standard committee" at each show. The idea was to invite "those gentlemen who possess a knowledge of any particular class of dog" to contribute their standard of perfection, and for the committee to consider all the contributions and produce an agreed standard of points and their worth. The application of the scheme was seemingly straightforward: "Each judge previous to judging, [is] to be furnished a card containing the committee standard for his class or classes . . . Dogs shall consist of so many points; for the superiority of each part they shall receive so many points on the card, and opposite that card take the same number away if defective in that part, or give as many points for or against as is deemed desirable, placing perfection and imperfection side by side; add up the numbers for and against, take one from the other and you have the result."[55] But it would be difficult for a standards committee to agree what constituted perfection and imperfection, as "what some think excellence other see as defects."

The *Field* endorsed the proposal for standards committees, maintaining that "we have long insisted on . . . laying down rules for judging dogs" and, with a hint at class cooperation, called for a "committee of gentlemen who might, if necessary, obtain the assistance of certain members of 'the Fancy.'" It accepted that there would always be difficulties in deciding if a dog met "the standard of his class" but charged that without guidance "judging is a farce" and, too often, a "delusion and a snare." The credibility of all shows was at stake, not least when the owners of the new Alexandra Palace introduced a third, large London show. This was seen principally as entertainment—and all the more so when the following week the venue was booked for the country's first mule and donkey show. Walsh hoped that the madness for shows would spread no further, as it might require "a public lunatic asylum for its prevention or cure."[56]

Correspondents to the *Field* all backed the call to improve judging by having national rules and standards. David Hume, a breeder from West Hartlepool, set out a system to "do away with the bungling work we have seen of late" and elaborated a metric system for gun dogs: "Let each pointer or setter (to be perfection) be supposed to possess 100 points; thus; head, 15; ears, 15; hip, 10; body, 10; tail, 10; legs, 10; size, 10; colour, 10." Such a system would also address the problem of entrants in the wrong class and introduce consistency into classes where there was a wide range of types. One such was the St. Bernard, for which it was "impossible to define the points . . . as there does not at present exist any distinct breed, all the dogs met with abroad being more or less crossed."[57]

Walsh was initially against such formal standardization, sticking to the old ways: "Select for your judges *gentlemen* and *sportsmen,* men who are in the eyes of the world capable of playing their part efficiently—and let the losers grumble as they will." However, by the end of 1864, he seems to have changed his mind and was advocating a scheme of numerical scoring for different conformation points. He had been persuaded by the leading pigeon breeder Frederick C. Esquilant, longtime honorary secretary of the Philoperisteron Society, whose numerical system was used by the National Columbarian Society. It was also used in working-class dog shows. Walsh acknowledged such methods had been common in "the bar-parlour; but, nevertheless worked so satisfactorily in keeping within bounds of the 'fancy' who patronised exhibitions held there, that we have long been of the opinion it ought, if possible, to be extended to the more important shows of horse, cattle, dogs and poultry."[58]

Walsh's own scheme for sporting dogs had five groups of points, together totaling one hundred: head and neck (30), legs and feet (24), body (20), stern (16), quality and coat (10). He tested it retrospectively on five bitches from the pointer class at the Birmingham show in December 1864. Two bitches were rejected as not matching the points, and with the remaining three, first and third places were reversed when scored with numerical points. The system even worked with toy dogs, though what to count as points was "the result of whim or caprice," so perhaps they could be assessed objectively by their market value.

Walsh hoped that either the Royal Agricultural Society or the National Dog Show Society in Birmingham would lead by drawing up "a series of definitions" for different breeds. The Birmingham judges discussed the proposal and turned it down. Walsh had not helped his cause with the National Dog Show Society when the *Field*'s report questioned the result of the pointer class at their show in December 1864. Judging at Birmingham that year had been particularly controversial, with many grievances aired afterward. In response, the *Field* asked again for breeders, exhibitors, and organizers to create a national body to regularize judging, thereby likely offending the members of the Birmingham society further, as they had constituted themselves as a national society.[59]

### Sporting Dogs and Field Trials

When initial plans for new dog shows developed in 1858, judging sporting dogs on their conformation alone was controversial. The new proposals on standards reopened the question. Standard points might be acceptable for nonsporting dogs, but sporting dogs were another matter. After the December 1863 Birmingham show, there were complaints that it had been taken over by urban breeders and owners, which threatened the interests and values of rural gentlemen and their sporting dogs. The specific issue was the judging of retrievers: prizes had been given to the best-looking dog, but critics asked if these canines were good in the field.[60] Also, while urban owners found curly-coated dogs more attractive, sportsmen thought them harder to train and less versatile. In shows, good smooth-coated outnumbered curly-coated dogs ten to one; hence splitting the class would favor the latter in terms of prizewinning. Also, improvement would be threatened, and poor-quality curly-coated dogs would win, increase in number, and go on to spread their inferior blood.

One possible answer was to have different classes for sporting dogs and those kept for exhibition or as pets. There was little support for this proposal. Sportsmen contended that outside the ring it would impossible to keep the two types apart and crossbreeding would lead to the loss or dilution of key features. For example, it was suggested that retrievers would lose "nose, patience, obedience and a soft mouth." Also, any separation would further compromise attempts to make retrievers a defined type. In *The Dog* in 1859, Walsh had stated that typical retrievers were crossbreeds—between a New-foundland and setter for work on land, and the terrier and water spaniel for work in water. Indeed, some correspondents suggested that further crossing resulted in better dogs. "Huzzlebee" recommended "crossing a pure St John's Newfoundland with the setter and highly-bred retriever," while "Experien-tia" considered "setter blood indispensable to produce a first class retriever." Clearly, such owners were breeding for work not show, each having views on the cocktail of blood needed to give the best abilities.[61]

Were shows inimical to such aims? Not at all, if they returned to Richard Brailsford's original aim that they should identify and encourage pure blood and recognized types. "Bow-Wow," in a letter to the *Field* in January 1864, was clear on the question of the purpose.

> My answer is that it is, or ought to be, to encourage the keeping up in their purity any breed of excellence, and prevent them falling into decay. I freely admit that in retrievers, as in other animals, the cross-bred is frequently the most valuable; but as the rule does not apply to the descendants of the mongrel, and that to obtain a first cross it is necessary some one should possess a pure sire, the only duty that, in my opinion devolves on the judges at a dog show is, to determine on whether the animal entered as a "retriever" be a pure-bred specimen of any well-known breed. To compass this object, authenticated pedigree is, in my opinion, absolutely essential.[62]

He drew attention to the long experience of the value of pedigrees to breed-ers producing stallions for racing and short-horn bulls on the farm but also argued that performance in competition, whether on the turf or at agricul-tural shows, was a better indicator of good blood than any pedigree.

A more radical approach to reconciling the merits of judging by conforma-tion with performance in the field was to hold game-finding competitions for pointing, setting, and retrieving that simulated work with gundogs (fig. 2.5). Such trials had been canvassed in the discussions back in 1858 and 1859 ahead of the first sporting-dog show and now arose again. The result was that John Douglas, the head gamekeeper at Clumber Park and the most favored show manager, organized a trial for pointers and setters to be entered in the International Dog Show at Islington in May 1865.[63] This event is now celebrated as the first-ever field trial. The trial, which required a large area well stocked with game, took place on 19 April 1865 at the estate of the brewer Samuel Whitbread at Southill in Bedfordshire, an established coursing venue. Walsh warmly welcomed the trial and, drawing on the way that coursing was judged, suggested a numerical valuation of the three key qualities: nose, 50; style of ranging, 30; and extent of breaking, 20, with "the last including the degree of intelligence exhibited." As in coursing competitions, there was a series of knockout heats, where two dogs were put to ground to flush out game and judged on their performance. The winners in each round went through, and the best two that survived the knockout rounds competed in the final. The first to find game did not always secure victory, as judges were looking for style in sweeping an area and staunchness in approaching the game. Two judges on horseback followed the dogs, assessing independently how they "ranged" and pointed, and then agreed on a winner.[64]

The brief notice in the *Field* was positive, even though Walsh was prevented by a bout of rheumatism from attending and acting as judge. The conditions were not ideal. There was insufficient cover for the birds, and the air was hot and still, so scent did not carry.[65] Nonetheless, good dogs were able to demonstrate their merits.

The pointers were divided into large and small sizes, the former including "Mr. W. R. Brockton's Bounce" and "Mr. W. G. Newton's Ranger," and the latter "Mr. J. H. Whitehouse 's Hamlet." In a maximum of 40 for nose, "Bounce" and "Hamlet" were accredited full marks, "Bounce" taking the highest compliment too in pace and range, and also for temperament. He was, therefore, estimated by the judges, the Rev. T. Pearce and Mr. Walker, of Halifax, to have been absolutely perfect. "Hamlet" was the same, both taking 90 in a hundred, but "Ranger" only got 30 for nose, and half marks for pace. This tallied much with his character at home, as although a good, steady, workmanlike dog, he yet was never quite brilliant, such as "Bounce" had the credit of being, and the late Mr. Whitehouse, a capital sportsman, would always contend that he never shot over a better dog than "Hamlet."[66]

*Figure 2.5.* Belle and Ranger at the Bala Trials. The illustration of a trial of stud pointers and setters shows two dogs setting out on their range, with a judge on horseback and spectators watching. *Field*, 30 Aug. 1873, 221

The value of the field trial was endorsed when Bounce and Hamlet were, respectively, large- and small-size winners in the stud pointers champion class at the Islington show.[67]

Not everyone was impressed with the field trial, either in principle or practice. Before the event, "Down Charger" warned that such events were artificial and likely to be no guide to the true worth of gundogs unused to working before an audience and without guns. Moreover, these competitions only told of ability on one day, and even then the number of birds and the conditions for scenting might vary from hour to hour. He also warned that dogs that were good in the field were not necessarily the best at stud. After the trial, the *Field* published many letters of complaint, which ran on through the summer until Walsh closed the matter down in October. There were grievances about the judging and the fact that a separate trial was run, in Yorkshire in May, for two dogs whose owner could not make it to Southill. Critics claimed this proceeding was highly irregular and unfair, and for some it confirmed that results were a lottery. Nonetheless, field trials became regular events, although their link to conformation shows was short-lived.[68]

## Breed

Walsh put forward his proposal for agreed conformation standards and numerical points in a series of articles in the *Field*. The first was on 9 September 1865 and featured "The Model Pointer." It described a particular dog, Mr. Smith's Major, that had won many prizes, most recently at Birmingham the previous December (fig. 2.6) If there was a moment when the modern dog was invented, this was it. The drawings, description, and enumeration of Major's points was the first attempt to define the conformation standard of a breed, which makes him the first modern dog.[69]

Walsh once more divided the dog's body into five parts, each with a points value: head and neck (30 points), frame and general symmetry (25), feet and legs (20), quality and stern (15), and color and coat (10). These parts were in turn divided. Thus, within "frame and general symmetry" scores were given for loins (7 possible points), hindquarters (6), shoulders (5), chest (4), and symmetry (3). In all, the pointer had sixteen parts or points to be scored. While framed quantitatively, points were given rich qualitative descriptions.

> The *head* should be of full size, wide across the ears and with a well developed forehead. Nose long, broad and square in its front outline: that is to say, even jawed, not pig-snouted. The lips should be well developed, but there should be no absolute flaws. The ears should be soft, long and thin, set on low down, and carried quite close to the cheeks. Eye of medium size, soft and intelligent, varying in colour with that of the skin. The neck should be set on with a convex line upwards, springing from the head with a full development of the occipulal [*sic*] bone, and coming out from between the shoulder-blades with a gentle sweep.[70]

An important factor in the choice of Major was that he was "an exceptionally-large dog of his strain," demonstrating how "by selecting all our modern improvements in domestic animals have been effected." With pointers, such a physique was better suited to work on the new sites of game shooting, in "the high heather of the north" and changing agricultural terrain elsewhere, seen in the "the luxuriant growth of southern turnips and mangolds." Major's pedigree on his father's side was from prize stock; his mother's was "never ascertained, but she was evidently of 'good blood.'" Walsh's idea was that the standards published in the *Field* were

FRONT VIEW OF "MAJOR'S" HEAD.

MR. SMITH'S POINTER "MAJOR."

**Figure 2.6.** Mr Smith's Pointer "Major." Stonehenge, *The Dogs of the British Islands* (London: Horace Cox, 1867), opp. p. 31

"first drafts," offered as the starting point for discussion, and this happened with Major.[71]

The second breed standard was published the following week for a Gordon setter, and in subsequent weeks and months, articles on other breeds followed, written by leading authorities. These pieces became a regular feature, followed by correspondence. By the end of the year, standards for deerhounds, Clumber spaniels, Norfolk spaniels, truffle dogs, and fox terriers had been published. The series continued into 1866, and the articles were published together in June 1867 in an edited collection under Walsh's pseudonym, *The Dogs of the British Islands*.[72] The book was intended as a reference guide for breeders, exhibitors, and judges, but it had a wider influence. When adopted, it advanced two other ideas and practices: applying to all varieties the idea of *breeds* as defined by conformation, and establishing the standard across the *breed* to create as far as possible a uniform-looking population. Walsh wrote in the preface that the volume offers "the principle points of several breeds," acknowledging the growing use of the term in show catalogues, sporting literature, and discussions of dogs. He hoped that the standards in the volume would be adopted nationally by judges and thus guide breeding toward giving a more consistent, even uniform look across breed populations. While the adoption of standards was certainly seen as leading to "improvement," some warned that dogs would likely become more uniform internally. Prizewinning dogs would be sought at stud, leading to higher degrees of inbreeding as breeders looked for the best blood in terms of purity and prepotency.

The new approach and goal can be seen in the differences between the descriptions of types in Walsh's *The Dog* (1859) eight years before *The Dogs of the British Islands* allowing for the different purposes of the two volumes. The seven divisions of 1859, largely based on natural history, became just four in 1867: "Dogs used with the Gun," "Companionable Dogs," "Hounds and their Allies," and "Toy Dogs."[73] There were significant differences in the style and detail between the two volumes. Consider the pointer once again. The illustrations of the types in *The Dog* showed the animals in action; in *The Dogs of the British Islands,* each was posed to display its conformation (compare fig. 1.3 and fig. 2.6) The descriptions in *The Dog* were more discursive, with discussion of Spanish, English, Portuguese, and French pointers, and speculation on origins, changing form, and adaptation to work with the gun, with the best dogs having "a large brain, and a well developed nose." In 1859 the English pointer was said to have been improved and possessed "the extreme development of the head and muzzle that is now sanctioned" and to be "a very fast galloper compared with the old fashioned dog." Descended from the Spanish pointer, the English variety had "a very different frame, legs and feet," having been altered to suit game shooting on landed estates. That such claims for "artificial selection" were published in 1859, the same year as Darwin's *On the Origin of Species,* was coincidence, but Walsh does write about the "descent" of the variety from ancestral dogs, and discusses how form and character had been changed over time by selection and crossbreeding. It seems that few pedigrees were available, and, indicating the modernity of the English pointer, he recommended that anyone wanting to go back more that a couple of generations should consult a naturalist.[74] Overall, the sense was that the English pointer had, and would continue, to change. The chapter in the 1867 book, significantly titled "The modern English Pointer," in contrast was short and briefly mentioned the types across the country before repeating, without changes, the long, point-by-point description of Major that had appeared in the *Field* in September 1865.

In *The Dog* in 1859, the bulldog was classed among the "Watch Dogs, House Dogs and Toy Dogs." In 1867, no doubt to the surprise of many, though not the London Fancy, it was listed among "Companionable Dogs." In the earlier book, Walsh dwelled on the bulldog's origins, tenacity, and manner of attack, and blamed any viciousness on ill treatment by ignorant owners. Its history was perhaps an object lesson on the dangers of show fancies, as after the banning of dogfights and the bulldog's entry into beauty contests, there had been a tendency to breed for exaggerated points, not least diminutive size. In *The Dog,* the description of its conformation was set out in a single twelve-and-half-line paragraph. The head took just over four lines: "The head should be round, the skull high, the eye of moderate size, the forehead well sunk between the eyes, the ears semi-erect and small, well-placed on the top of the head, rather close together than otherwise, the muzzle short, truncate and will furnished with chop." Eight years later in *The Dogs of the British Islands,* the description of the head alone occupied twenty-one lines.[75]

In both books, the greyhound enjoyed pride of place; unsurprisingly, perhaps, as the second edition of Walsh's *The Greyhound* had been published in 1864. In *The Dog* (1859) he began by explaining how the English or smooth greyhound had been changed in the first half of the nineteenth century. First, new game laws had opened up coursing to the middle classes and led to more and larger meetings. Second, the creation of a national sport had occasioned regular gatherings, on standardized courses, and a national market in buying and selling top dogs. These changes had led already to the disappearance of local and regional variations and the creation of a beautiful and elegant hound—the English greyhound. In *The Dogs of the British Islands,* Walsh argued that with respect to breeding, points mattered less than with other varieties, because performances and pedigrees were known for many generations; hence, "although the breeding may be guessed from the appearance of the individual, it is far better to depend on the evidence of the Coursing Calendar."[76] The description of the ideal greyhound was very detailed and

*Plate 1.* Edwin Landseer, *A Distinguished Member of the Humane Society,* 1838. © Tate
Images

*Plate 2.* The Bull Dog. Sydenham Edwards, *Cynographia Britannica: Consisting of Coloured Engravings of the Various Breeds of Dogs Existing in Great Britain; Drawn from the Life* (London: C. Whittingham, 1800), 28

***Plate 3.*** R. Marshall, *An Early Canine Meeting* (1855). From The Kennel Club Collection

*Plate 4.* Sheffield Dog Show Poster, British Kennel Association (1885). Courtesy of the UK National Archives

***Plate 5.*** Charles Burton Barber, "Yorkshire Terrier Dog, Italian Greyhound and Pug."
V. Shaw, *The Illustrated Book of the Dog* (London: Cassell, Petter, Galpin & Co., 1881),
opp. p. 177

***Plate 6.*** Bedlington Terrier "Geordie"—Dandie Dinmont Terrier "Doctor." V. Shaw, *The Illustrated Book of the Dog* (London: Cassell, Petter, Galpin & Co., 1881), opp. p. 121

Figure of I. Wolfhound . 35 niches high .
mouer stands 6.11 in :

**Plate 7.** George Augustus Graham and his "model" Irish Wolfhound. *The Irish Wolfhound* (1885), in pages added to first edition. www.irishwolfhoundarchives.ie /graham1885.htm. We would like to thank Elizabeth Murphy for this image.

*Plate 8.* New PETA "Freak Show" Ad Shines the Spotlight on Crufts. www.peta.org .uk/blog/new-peta-freak-show-ad-shines-the-spotlight-on-crufts/

highlighted qualitative features of form and character. However, attention to points was cursory, with just five general divisions offered: head and neck (20 points), frame and general symmetry (40), feet and legs (30), tail (5), and color and coat (5). The approach in 1859 had been quite different. Walsh had described the features of the greyhound in a more literary manner, based on lines of doggerel attributed to Wynkyn de Worde and first published in the 1490s.

> The head of a snake,
> The neck of drake,
> A back like a beam,
> A side like a bream,
> The tail of a rat,
> And the foot of a cat.[77]

In 1867, the rhyme was simply included as a coda, in a short, final paragraph.

In *The Dog,* Walsh had described "two very different kinds" of Newfoundland: "the large, loose-made and long-haired variety," and "the small compact and comparatively short haired," the latter included among the retrievers as being one origin of the English retriever. With the larger dog, its color could be "black, or black and white, or white with a little black, or liver colour, or a reddish dun, or sometimes, but rarely, a dark brindle not very well marked." *The Dogs of the British Islands* acknowledged just one type of Newfoundland dog, the larger animal. Moreover, in just eight years of dog shows it had changed: "The purest specimens are of an intense black colour . . . Any admixture of white is a defect."[78]

## Conclusion

It was only in the third edition of *Dogs of the British Islands,* published in 1878, that breed was included in its subtitle, "A Series of Articles on the Points of their Various Breeds." However, it was in the 1860s, with growth in the scale and popularity of dog shows, that *breed* was adopted for varieties of dog and other terms dropped away. This change was more than linguistic. It precipitated a major shift in the morphology and anatomy of dogs, as the range of variations within types was reduced, with more and more dogs approaching the breed standard to give greater uniformity across the population. Importantly, the preferred physical form for each breed remained subject to renegotiation at many levels, from the judging of classes at local shows, up to attempts, like that in *The Dogs of the British Islands,* to set out the principal points of all breeds to be used in all shows. With breed came an interest in ancestry and a preference for dogs with "good" or "pure" blood, which meant verifiable evidence in the form of pedigrees setting out inheritance from good stock. As with livestock and poultry, competition at shows should have selected the best conformations, and winners with strong pedigrees and good blood would not just bring home cups, rosettes, and ribbons but would command high prices at sales, for stud fees, and for their puppies. In *The Dogs of the British Islands,* Walsh recalled his travels around Scotland looking for good examples of Skye and Dandie Dinmont terriers, but the days when good dogs were discovered were past; now they were produced by fancy breeders to conformation standards.[79]

The adoption of breed for dogs was not without its problems. With live-stock, physical form was likely to be an accurate guide to the amount of meat or milk they would produce, and fancy poultry and pigeons were bred for ex-hibition. With sporting breeds, form was taken as a proxy for function in working or sporting abilities. Thus, skull size was linked to intelligence; muzzle length to scenting; coat to warmth, protection, or visibility; and so on. Many such linkages were imagined or based on historical texts and engravings that fanciers studied to discover ideals, though other owners were happy with the very modernity of their improved breed.

Given that the dog shows of the early 1860s were run, either in association with agricultural shows (as in Newcastle upon Tyne and Birmingham) or with poultry and pigeon shows (as in Manchester), they were shaped by the values and cultural practices of those shows' constituencies. Landowners dominated agricultural shows, though there was space for tenant farmers and smallholders at local events. As Harriet Ritvo has shown, breeding ped-igree livestock symbolized aristocrats' power to manipulate nature to their ends, while at the same time legitimating inherited position and wealth.[80] The middle and working classes had supported poultry and pigeon shows, and the Birmingham dog shows in the early 1860s established inclusive shows, for all dogs and socially for aficionados and the public. The London and other city Fancies became associated with the new shows, though they often defined their activities against what they saw as the snobby amateur-ism of latecomers to canine affairs. Class divisions and tensions were also evi-dent at the big exhibitions in London and other cities. These differences were starkest in the claims that the main results of the new, grand shows had been the resurgence of "toy-dog breeders, dog dealers and stealers."[81] But ten-sions were evident at every level, demonstrated by the different entrance charges to ensure different clienteles on different days; the question of trust in the integrity of judges, especially those who were not gentlemen; and concerns about the motives of organizers and exhibitors. In the decade after 1865, these issues dominated the worlds of breeding and exhibiting dogs, and a number of organizations vied to regularize the canine affairs of the country.

# CHAPTER THREE

# Showing Breed, 1867–1874

The dog shows of the 1860s were typically mid-Victorian enterprises in rationale, organization, and reach. Most were local affairs, an addition to middle-class leisure pursuits, but those in the cities became great exhibitions that hosted competitions to determine the national elite in each breed. The variety of canines and their antics on the bench and in the ring provided entertainment in what the more scientifically minded regarded as pop-up natural history museums. In theory, shows were open to all, but in practice entry and success was increasingly restricted to middle and upper class owners with the money and connections to acquire or breed top dogs. Within these groups, organizers had to balance the interests of "professionals" and "amateurs"—recognizing the gentlemanly ambition of improving the nation's dogs as well as the commercial need for financial viability—and provide appropriate rewards for winners and allow opportunities for trade in valuable animals. Dog shows became popular across the classes, driven by the participants' passion for dogs and their investment of time, effort, and money. However, they also provoked anxieties in some observers, who wondered if competitions, rather then being wholly respectable, meritocratic ventures, were being distorted by speculation, profit seeking, and dishonesty.

In the developing culture of the new fancy dog shows, different groups claimed to represent the interests of the nation's canines, that is, those who organized shows, exhibited, judged, spectated, and followed fancy dog shows. It was the social and economic relations between these groups that produced the new human-dog relations that developed breeds materially and culturally.

The key questions for the show fancy were who should judge, how they should judge, and by what standards should a dog's performance (the term

used in stud books) at shows be assessed? First, we consider the continuing complaints about the organization of shows, from accusations of maladministration to uncertainties about shows' very purpose. Doggy people
agreed that their activities (like other sporting and leisure activities)
needed regulation, but there was a struggle over who should govern the
nation's canine affairs and to what ends. After 1869, several organizations
claimed national authority to regularize shows; they all maintained that
fair competition needed rules and their enforcement. The shows had proved
to be open to fraud, favoritism, and speculation, and perhaps the degeneration rather than improvement of the country's dogs. We consider in turn the
organizations that vied for control, exploring their social makeup, models
of governance, and ideas on improvement. Regional rivalries, based on different social and economic relations, were important among the four competitors: in Birmingham the National Dog Show Society (sometimes the
Birmingham Dog Show Society), in Liverpool the National Dog Club (NDC),
in Nottingham the National Canine Society (NCS), and in London the Kennel
Club. The Kennel Club won, though this was only fully secured in 1885 with
a rapprochement with the National Dog Show Society. (We discuss this
matter in chapter 4.)

### Ignorance, Favoritism, and Fraud

The breed standards set out in Stonehenge's *The Dogs of the British Islands*
(1867) had an uncertain status. On the one hand, they were based on the articles written by "some of the most distinguished breeders of the day," including Walsh himself and Thomas Pearce (who wrote under the pseudonym
"Idstone").[1] Many fanciers regarded the descriptions as definitive, giving the
much-needed delineation of breed standards that judges could use to achieve
consistency across shows and over time. On the other hand, the standards
were offered as "first drafts" and open to revision; indeed, the book included
some of the letters published in the *Field* from critics who offered alternative
views. The book's standards were not used at the National Dog Show in Birmingham at the end of 1867, even though Walsh was its referee. The Birmingham Committee also persisted in the practice of judging in private before
the show, which exhibitors had long criticized, calling instead for open, public judging. The Birmingham Show that year was the least successful to
date, with entries down. Numbers recovered the following year, but there
were still protests about prizewinners and accusations that the show revealed "ignorance, favouritism and fraud" among the judges.[2]

Ignorance had been claimed for several years, particularly with the Mount St. Bernard Class. In 1867, the judges refused to award second prize to a smooth-coated dog on the grounds the breed was characterized by a rough coat.[3] The Rev. John Cumming Macdona had one of his St. Bernards disqualified. At this time, Macdona was rector of Cheadle in Cheshire, having graduated from Trinity College Dublin and served in parishes at Sefton and Sandringham. He gave up the cloth to become a barrister, remaining a leading figure in canine affairs until he was elected Conservative Member of Parliament for Southwark in London.[4] He requested the judge, Matthias Smith, to explain his decision to readers of the *Field*. This Smith did, prompting a long correspondence highlighting the vicissitudes of judging. The dispute reopened the question about the "true" form of the dogs kept by the monks in the hospice on the St. Bernard Pass. This drew upon travelers' tales, interpretations of paintings, and exhibitors' knowledge and experience. The rescue work at the hospice had come to prominence in the early nineteenth century through stories of the heroism of the dog Barry.[5] His taxidermied body remains in a Bern museum, altered in 1923 after complaints that he did not look like a St. Bernard (fig. 3.1).[6] Barry, was a smooth-coated dog, with a different conformation than the larger, heavier St. Bernards favored at British shows—the best known being Macdona's Tell (fig. 3.2).

In *The Dogs of the British Islands,* the St. Bernard standard was actually based on Tell, who was allegedly descended from Barry.[7] The smooth variety was mentioned as an afterthought. However, in the columns of the *Field,* the majority opinion was that both coats were acceptable, with the smooth dog being the original type. Either way, the decision at the Birmingham Show in 1867 was said by many correspondents to the *Field* to reveal the inexperience and lack of knowledge among the local judges. Those correspondents who had visited the hospice reported seeing only the smooth-coated variety. They said the monks had tried to improve their dogs by crossing with a Newfoundland to produce what some called mongrel dogs with thicker and hopefully warmer coats. However, the new features were a liability; a longer thicker coat held ice and snow, weighing down the dogs in drifts. Thus, the monks were said to have given these dogs away to lowland villagers, and, it seemed, these dogs had been bought by less adventurous travelers. Only Victorian Alpinists who had reached the hospice had seen the true Barry type. The possibility of another dispute between Macdona and Smith was avoided the following year by having two classes, rough and smooth. Tell, reputed to be worth £10,000, won the rough-coated class. In the smooth-

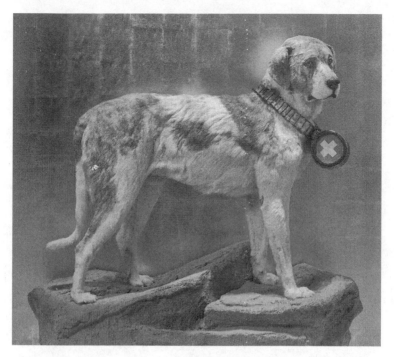

**Figure 3.1.** Barry (1800–1814) © Natural History Museum of Bern.

coated class, no dog was found to be of sufficient merit for first prize, but a second prize was given to another of Macdona's dogs, Bernard.[8]

The St. Bernard controversy exposed the uneven levels of knowledge among judges, but other grievances centered on allegations of favoritism. In the mastiff classes in 1867, the *Field's* report claimed that the winners, Mr. Bamford's Wolf and Mr. Clive's Prince, were "less striking" than others in the class. Wolf had not won top prizes before, as he was said to be poorly proportioned, with "weak slack loins, a contracted body, and a bad coat." His one redeeming feature was "a fine head," and breeders now wondered if this point was going to take precedence over others in competitions. Perhaps the judges, the already much-maligned Matthias Smith and John Percival, were biased as well as ignorant?[9]

Would controversy have been avoided had Walsh's weighted points been adopted? The Birmingham Committee had been resistant to the idea of points and stuck with the established method of "the critical eye" of the judge and the overall "look" of the dog. The committee also resisted calls for open judging, claiming that privacy avoided overt and covert pressures from exhibitors. The issue of favoritism blew up in the sporting press in the summer of

THE REV. J. C. MACDONA'S "TELL."

*Figure 3.2.* The Rev. J. C. Macdona's "Tell." The drawing shows Tell against the backdrop of the Alps and the St. Bernard Hospice. Tell was, of course, kept in Britain, in Cheadle Hulme just south of Manchester, where Macdona was the vicar. Stonehenge, *The Dogs of the British Islands* (London: Horace Cox, 1867), opp. p. 76

1868, when Pearce wrote, "year by year the prizes go to the same men and same dogs," and "I can tell who will have the prizes from my armchair, and will write the list!" Accusations of favoritism and collusion spilled over into charges of bribery. A correspondent from Lancashire, who wrote in November 1868 under the pseudonym of "Vox," claimed that some judges accepted money for "the bestowing of favours."[10]

Fraud was alleged to occur with entries and in judging. Accusations of "trickery" were made at Birmingham in 1867, when fox terriers with allegedly "doctored" ears were awarded prizes. More common were complaints of "irregularities" in the catalog over pedigree, ownership, and even the identity of the dogs entered. These factors were inextricably linked around trust. Furthermore, the market in the selling and buying of prize-winning dogs developed, along with income from stud fees. So intertwined were the relations that judges might be the breeders of the dogs they were assessing. The opening of shows to nonsporting dogs meant entries no longer came solely from gentlemen of independent means. Indeed, the success of shows had further opened the door to "dog dealers," those for whom the fancy was an occupation rather than a vocation. The uneasy crossover between the two spheres of canine enthusiasm was again evident in December 1867, when winners from the Birmingham Dog Show were presented at the distinctly plebeian Monster Meeting of the Fanciers of Birmingham at the Bowling Green Tavern in the city.[11]

### The National Dog Club

The tensions over the practices and results at Birmingham in 1867 rumbled on in the pages of the *Field,* which printed grievances about the judging of classes in other breeds: mastiffs, Dandie Dinmont terriers, pugs, fox terriers, bloodhounds, and black-and-tan terriers. A new objection was raised about the consistency of judging between different shows and the future of such events nationally. Edwin Nichols, a breeder and exhibitor of bloodhounds, mastiffs, and Newfoundlands, complained of the different results at the Birmingham and Manchester shows, just four weeks apart. The bloodhound winner at Birmingham came third in Manchester, where the class was won by a dog that had been previously unplaced. Nichols highlighted similar "differences of judgment" with deerhounds, bulldogs, and fox terriers. Walsh, of course, was unsurprised at the disparities, and he had the answer: the agreement of standard and openness in judging. In his newspaper he had "repeatedly insisted, there never will be any certainty in judging until a book of points is placed in the judges' hands, and they are required to fill it up."[12]

Hope was soon at hand. By the end of 1868 the National Dog Club had been founded, one aim of which was to publish a book of points to be used at its shows. The NDC had begun as the Shooting Dog Club (SDC) in 1867 to support owners of gun dogs. Walsh endorsed the plan in the *Field,* though an editorial worried about the double meaning of the club's name. Several meetings of the SDC took place in 1867 and 1868. Its leading figures were G. Russell Rogerson, a Liverpool solicitor, and Richard Lloyd Price, owner of the 64,000-acre Rhiwlas estate in North Wales.[13] During the Birmingham Show on 1 December 1868, the SDC changed its name to the National Dog Club, expanding its remit to conformation events. The new committee contained many men who had taken a leading role in shows: William Lort, John Cumming Macdona, Edward Laverack, and Francis Brailsford.[14] They were self-appointed gentlemen from the Northwest and West Midlands who promised to adopt breed points, prepare a stud book, and hold "an exhibition of all kinds of dogs" the following year. This event was planned for London in early June 1869, to run alongside the Metropolitan Horse Show. It was to be the first London dog show for three years. Walsh expressed optimism about the venture at several levels. First, shows were becoming more commercially viable, being attended by the middle and lower classes as public entertainment, for whom they provided "innocent" recreation and "healthy amusement." Second, the NDC promised to use the standards set out in *The Dogs of the British Isles* as the basis for its own code of points, which were to be published ahead of the show and required to be used by judges. Indeed, judging was to be public and the results based on the averages of the numerical scores given. Walsh's hope was that transparency and standardization would secure public trust and ensure propriety.[15]

The NDC's London show in June 1869 was an immediate success. Nearly a thousand dogs were exhibited, half sporting and half nonsporting, and several classes were "excellent." The *Illustrated London News* carried a full-page illustration of the prizewinners (fig. 3.3). The report in the *Field* conflated the canine and social progress made since the Birmingham show a decade earlier: conformations had been improved to an "incalculable extent," not least because "a superior class of men were doing their best to produce first-class animals." The Prince of Wales attended, and there were good crowds. The value of published standards was clear in a dispute over the judging of Newfoundlands, where the black-and-white-coated type made famous by Landseer was suggested as the standard. But the judges deemed dual coloration a "defect"; Landseer in all his paintings was said to have misled the public—Newfoundlands had to be black.[16]

There was also an objection in the classes for Dandie Dinmont terriers, when the Rev. S. Tenison Mosse claimed that awards went to overweight dogs. The breed was named after the terrier of the character Dandie Dinmont in Walter Scott's novel *Guy Mannering* (1815). It had one of the longest entries in the second edition of *The Dogs of the British Islands* (1872), due to the number of additional letters Walsh included expressing differing views on the standard. In the months before the NDC show, there had been a long correspondence on the breed, in which owners debated the merits of the various proposed standards. What was the Dandie Dinmont terrier? Was the standard to be based on fact or fiction? Was it to be drawn from Scott's own pet pepper-and-mustard terrier, the terrier described in the novel, or the

***Figure 3.3.*** Prize Dogs in the National Dog Show at Islington. *Illustrated London News,* 12 June 1869, 20. © British Library Board

dogs now found in the Scottish Borders? From our perspective it is clear that there was never a single true type; rather, terriers varied continuously across the Scottish Borders in look, coloring, and size. Indeed, there was likely a continuum with terrier types to the north and south; hence, there was a larger process of breed invention happening, differentiating terriers associated with Bedlington in Northumberland and Skye, and perhaps otterhounds kept across the north of England.[17]

The NDC show, like Appleby's seven years earlier, also went from boom to bust. It lost money and was tainted as a failed "speculation." In his editorial in the *Field,* Walsh was perplexed and wondered why London shows repeatedly failed financially, while those in the rest of the country covered their costs or made a surplus. He suggested that promoters might learn from the success of the horse show that had run at the same time and with similar attendances. This event had provided "melodrama," in the form of "show jumping," where the entertainment came not in seeing the skills of horse and rider in clearing fences but in witnessing their inexperience, with riders falling off and horses bolting. In a more serious vein, he mused critically for the first time on the value of the new shows: "In discussing these exhibitions, it must not be supposed that either of them have any great beneficial effect. Both are extremely interesting as objects of curiosity, but neither can be regarded as absolutely useful. In the case of the horse, it cannot be pretended that his breeding is improved by the prizes offered; and in that of the dog, though we believe in many instances their exhibition has greatly altered for the better some of the breeds, yet as they are, with few exceptions, objects of luxury rather than of use, no real good has accrued." He was speaking, of course, from the elite and masculine world of rural sports, hence his skepticism toward urban dogs and their owners, many of whom were women.[18]

The committee of the NDC kept a low profile in succeeding months, allowing others to mull over its future and that of dog shows. Some exhibitors and breeders still hoped that it would set standards that would be adopted at all shows. Others argued that it should return to its original aim, an organization for landowners and gentlemen, dedicated to field trials and the improvement of gun dogs. Another lobby wanted the NDC to develop as a social institution, with a London "club" for meeting, dining, and drinking. None of these hopes were realized. The NDC's book of points continued to circulate and was used at some shows, but the committee never recovered from the reputational damage of the failed London event and ceased to meet in 1870, though they held a pointer and setter trial in the spring of that year.[19]

### A Committee of Gentlemen

Despite the disappointment of the 1869 show, a number of former NDC activists were determined to establish a successful and profitable London event. They formed a "Committee of Gentlemen" to plan for 1870. The group included Lord Calendon, Lord Holmesdale, and Thomas Meyrick MP, but the principal organizers were Sewallis Evelyn Shirley MP and J. H. (John Henry) Murchison. The pair represented, respectively, old and new money. Shirley was a descendant of a baronetcy created in 1611, which took the title Earl Ferrers in 1711 (fig. 3.4).[20] He owned estates in Warwickshire and Ireland. In 1868 he had become Conservative Member of Parliament for the Southern Division of Monaghan in Ireland, a seat he held until 1880. He split his time between Ireland, London, and Warwickshire, where he kept a large kennel at Ettington Park. He acquired a large library and natural history collection, including a Great Auk (a flightless seabird extinct since 1844) and a supposed dog-fox hybrid. A keen sportsman and an excellent shot, he had kept a pack of beagles and hunted regularly while at Eton. Shirley

*Figure 3.4.* Sewallis Shirley (1844–1904). E. W. Jaquet, *The Kennel Club: A History and Record of Its Work* (London: The Kennel Gazette, 1905), 2

entered the dog show world at Birmingham in December 1867, winning the fox-terrier class with Jack. In August 1869, he won the bull-terrier class at the South Durham and North Yorkshire Horse and Dog Show in Darlington. At the same event, Murchison won the fox-terrier class. If they did not meet then, they almost certainly did that December at the Birmingham show, when both won prizes for their fox terriers. They may have already known each other through politics, as Murchison had stood as a Conservative in Truro in 1859.[21]

J. H. Murchison was the eldest son of Dr. Alexander Murchison, a medical practitioner and plantation owner in Jamaica. His younger brother was

Charles Murchison, a leading authority on fevers and liver disease, and Roderick Impey Murchison, Britain's leading geologist of the day, was a second cousin. J. H. was a "mining engineer and promoter of mining enterprises," first in Devon and Cornwall, and then, from offices in the City of London, as the chair and secretary of many mining companies across Britain and the world. As well as marketing companies in the financial press, he published on geology, mining, and mines, and was elected a fellow of the Geological Society of London in 1852. While he was unsuccessful with his parliamentary ambitions, he remained active in politics, publishing pamphlets, as well as tracts in the reform debates in the mid-1860s. He first appears in reports of shows in 1869, winning at Exeter, Bingley, Darlington, and Birmingham. By this time, he had considerable wealth and seems to have speculated in stud dogs for breeding, just as he did in mining stocks. He lived in Surbiton, Surrey, southwest of London, but kept a kennel of over two hundred dogs in Thrapston, Northamptonshire, one hundred miles north of the capital, where he employed several kennel staff, in the manner of a landed gentleman. While fox terriers remained his favorite, he branched out with St. Bernards, greyhounds, and Dandie Dinmont terriers, becoming a respected judge of many breeds.[22]

The "Committee of Gentlemen" organizing the 1870 Crystal Palace Show issued a prospectus stating that their object was "to collect together all the most valuable and perfect dogs of every description." They sought to disassociate themselves from commercial interests, stating that they had "no motives activating them beyond the encouragement of breed and quality." Not all enthusiasts were convinced, as letters to the sporting press indicated. One correspondent, "Worstead," wrote to *Bell's Life* expressing the hope that the show would be "entirely free from what is now known as the common dog show taint" and that its promoters were not from a class interested only in "merely company speculations" for the improvement of their bank balances. The writer was also concerned that there appeared to be "symptoms of . . . 'the orthodox dog show smack' of prizes given to the organising clique" and that "the clerical element" dominated the "old sportsmen." The promoters promised open judging to published standards, but this did not reassure Worstead. His letter ended by observing that too often "our real judges must have lost their written instructions from head-quarters," as they had been giving prizes to dogs "having corrugated pendulosities, and with anything but well-knit loins, besides having very questionable olfactories." Other correspondents to *Bell's Life* complained about the entry fees, modest prizes, and location, and worried that the names of the judges had not been published. In a neat summary of the manifold complaints dogging shows, one suggested that the event be titled "The National, Clerical, Legal and Know-nothing-at-all Dog Show" and claimed that it had been "got up by a nice little society for mutual admiration."[23]

The place and timing chosen by the committee was significant. By 1870 the Crystal Palace Company had established its unique site as *the* London venue for public exhibitions and rational recreation. It was in a green and newly landscaped park in fashionable Upper Sydenham, a setting very different from the vulgar commercialism of Cremorne and "the cowshed" at Islington. What its promoters called the Great National Exhibition of Sporting and Other Dogs was also independent of any agricultural event. Entry was certainly expensive, in part to avoid financial failure and in part to move it upmarket, as the report in the *Western Mail* explained: "The Committee had laboured to keep out inferior specimens, one of the expedients being to limit the exhibitors to subscribers of a guinea, and to have an entry fee of 7s for large and 5s for small dogs. The low fancier with his 'leetle dawg' has therefore been virtually put out of court, and the kennels of the wealthy in consequence have largely contributed to the wonderful display." A respectable show required breed and quality among exhibitors as well as dogs.[24]

The show ran from 21 to 24 June and drew 956 entries, with breeders and exhibitors attracted by the promise of £1,000 in prizes. The event shared its first day with several other public and private events, including drill exercises, orchestral and organ recitals, and swimming by more than three thousand boys from the capital's philanthropic schools, reviewed by His Serene Highness the Prince of Teck and Princess Mary. Visitors could also visit the fine arts collections and picture gallery, and wander the gardens to admire statues, flowers, birds, and the fountains. As one newspaper noted, the dog show offered the "best amusement" for a paying public, even those who failed to distinguish a spaniel from a terrier, in what was an otherwise dull month. In this incarnation, shows took their place in the miscellany of cultured entertainments and rational recreations put on in the new exhibition halls that had been built to continue the spirit of the Great Exhibition of 1851.[25]

There were mixed reviews. The *Western Mail*'s reporter wrote that it was "one of the best dog exhibitions ever known," and the *Times and Sporting Gazette* regarded it as successful. The *Illustrated London News* and the *Graphic* printed large illustrations of the prizewinners. Not unexpectedly, the *Birmingham Daily Post* saw the show as inferior to their city's event. *Bell's Life* agreed, though noted that exhibitors from the West Midlands carried off many prizes.[26] The committee members were styled "gentlemen dog dealers," a wounding dig at men who claimed superiority over other fanciers. There was also scepticism about the new champion classes, elite groupings exclusive to dogs that had won previously.

> We know for a fact that shows, generally speaking, have done a vast amount
> of mischief amongst dogs used for field sports, as prizes have been given so
> frequently to "mongrels," more especially what have been called Champion
> pointers and setters, which have been "cracked up" by their owners as
> extraordinary animals in the field that gentlemen have been induced to breed
> from them, and thereby fill their kennels with rubbishing stock; conse-
> quently, very few of the fine old English setters, or the beautiful pure-bred
> setters are now to be seen unless in a country that has not been tainted with
> dog shows.[27]

*Bell's Life* was inundated with more letters of complaint, all of which attempted to expose the show as corrupt, with the organizers filling their own pockets. Many observed that the committee had promised a reformed show but that "the old firm" of the NDC had dominated it. "A Looker On" calculated that the committee had won £199, a sum that was soon corrected to £263, plus eight medals. Judging had been open but also open to abuse. It was alleged that two judges, Handley and Barrow, had bought a black-and-tan terrier bitch, Queen II, just before the show, sold her to an exhibitor for £120, and then awarded her first prize.[28]

To crown the complaints, a relative of Jemmy Shaw, still a leading figure in the London Fancy, complained of injustices at the "mock dog show," as the "so-called judges . . . did not know half the properties and points of most of the class of dogs they were supposed to judge." Shaw entered some dogs, indicating that high entry fees had not had the desired effect. The extent of the crossover between old and new fancies was evident when prize toy dog winners from the Crystal Palace were shown at Jemmy Shaw's on the following Saturday, and there was a discussion of the results "with the oldest living fanciers" at the Old Red Lion, in Clerkenwell, on Sunday. The incursion of the old fancy into the new, respectable, and, supposedly, improved show world, revealed the still-contested nature of middle- and upper-class claims about having the knowledge, experience, and probity to lead the country in the improvement of its dogs.[29]

A member of the "Committee of Gentlemen" replied to all the malcontents in a letter published in *Bell's Life*. He made the familiar observation that those most interested in breeding would in all likelihood be the most knowledgeable, having the best dogs with the best blood, and that the judges at the Crystal Palace had been "a judicious mingling of 'amateur and professional' talent." He appealed to the sporting spirit and asked that those "fairly and honestly beaten" should accept defeat with good grace, otherwise "every man possessing a mongrel cur will be quite happy in the conviction that he is the undoubted possessor of the best 'dog in England.'" This appeal did not assuage the critics, who had a new line of attack that undermined the group's fundamental claim to integrity: it had been a commercial rather than gentlemanly venture. "R. H. S.," writing to the *Sporting Gazette,* negatively described it as "a gate money meeting," a term applied to the recently introduced practice of charging for entry to horse-race meetings. Another correspondent declared that it had been a show for speculators, where judging had been by "dog dealing divines" and that, as at other shows, there had been "tricks, jobbery and chicanery." When it came to economics, respectability could be quite fragile: anyone losing money was damned as a speculator; anyone making money was seen as profiteering. Supporters maintained that shows run by "gentlemen," who did not have to think about money, would be inevitably better and that any profit would be reinvested in the improvement of the nation's dogs.[30]

In the wake of the latest London "difficulty," all eyes turned to the next big event, the eleventh national show at Birmingham in late November 1870. Could this remain the blue-ribbon show, and its committee be accepted as honest, honorable, and principled men? Many breeders and exhibitors were not sure. They wanted to see an end to judging on benches by individuals in private and a move to open, public judging by teams in the ring. In fact, Birmingham judging was in two stages—initial selection on the bench and final judging in the ring—but some fanciers wanted all dogs to have their chance in the ring, lest the show become one of curiosities. Advocates of public judging argued that openness was the best guarantee of fairness, both in terms of scrutiny of the process and of all the evidence being seen. Direct comparisons with court proceedings were drawn. Against this, defenders argued that open, public judging was only favored by the wealthy, who had keepers they could afford to send to shows and walk their dogs to the best effect. Privacy also avoided the unedifying spectacle of owners, breeders, and handlers attempting to influence the judges by nods, winks, and backchatting.[31]

The Birmingham committee was unyielding and continued with their established practices. A group of thirty-seven former exhibitors—including Lord Combermere, the president of the cattle show that ran concurrently—stayed away. The *Birmingham Daily Post* called them "secessionists," but despite their absence the show was successful, with one hundred more entries than in 1869 and a financial surplus of £200. Admission was one shilling, with the working classes admitted for 6d after 5 p.m. on the final two days. Compared to previous years, there were fewer protests about judging decisions, except that the correct conformation of the mastiff was yet again contested. Where there were disputes, the point standards in Walsh's *The Dogs of the British Islands* were cited as authoritative.[32]

In early June 1871, the "Committee of Gentlemen," undaunted by the reaction to its first meeting, organized a second Great National Exhibition at the Crystal Palace. The event enjoyed better weather and more generous space, with the judging of sporting dogs on the terraces. There were over eight hundred entries, benched in 107 classes. The press deemed it a success. The reporter in the *Gentleman's Magazine* commented wryly on the social profile of exhibitors. There was the man just hired to lead the dog; the owner "vainly trying to look unconcerned, yet inwardly so full of excitement"; "the three-shows-a-month man, strolling round with great confidence"; and the nervous novice. The popular press enjoyed the exhibition of people as well as dogs. An editorial in the *Manchester Guardian* noted, "Pretty girls and stately dowagers, cold beauties and laughing damsels, the frank openhearted fellow and the cynical abomination, the swell and the snob, the well-dressed gentleman and the Whitechapel rough, meet on common ground where dogs are concerned." Only the report in *Bell's Life* accentuated the negative, noting that the number of entries was down and that the organizers had lost the support of "some very influential and respected gentlemen" since the previous year.[33]

An editorial in the *Field* reflected on the show. Walsh expressed frustration over the absence of standards when correspondents started to dispute the correct color of retrievers' toes. Reform was needed; otherwise, the future of shows was at stake: "We have refrained from discussing this much-vexed subject for some time, owing to our almost losing hope of much improvement in the principles on which judging dogs for exhibition is now conducted. Private judging, public judging, semi-public judging, judging by 'rule of thumb,' and judging by points have all been tested without giving general satisfaction. . . . There seems to be something radically opposed to consistency in the mere fact of contact with the dog." He was exasperated that after each important show the *Field* was inundated with letters from disappointed exhibitors; hoping to deter correspondence, he observed that he was very selective about which ones to publish. Walsh bemoaned the collapse of the NDC and with that the loss of its code of points. Acknowledging that the NDC was beyond resuscitation, Walsh argued that what was needed more than ever was a "court of appeal" or "tribunal" to set rules and monitor standards. His editorial ended with a hopeful call for leadership: "We do not wish to enlist anyone in the undertaking without opening his eyes to the consequences, but would any exhibitor be possessed of the courage assigned to the bulldog, the mastiff, or the Bedlington terrier by the several votaries of these breeds, we shall be happy to hear from him and assist him to the full extent of our powers."[34] In time, Shirley, who was on the Crystal Palace Committee in both years, took on the mantle of leadership by founding the Kennel Club, but that was two years hence. In the meantime a new provincial group, the National Canine Society, tried to assume control of canine affairs.

### The National Canine Society

In 1872 there were fifty dog shows across the country, from Liskeard in the south to Edinburgh in the north, and from Dublin in the west to Boston in the east. The year ended with the Manchester Show in Belle Vue Gardens, the city's premier zoological and entertainment venue, over Christmas. The events were popular with all classes, who paid variable fees, either to compete with their dogs or to be entertained by the spectacle of the sheer variety of forms and sizes, and cacophony of yapping, snarling animals. Dog shows had joined the many Victorian ventures of animal exhibitions that combined competition and entertainment. Presented as living museums, these events promised "improvement" of the animals exhibited and of the public who visited.[35]

The next Crystal Palace Show in June 1872 was hailed in the *Times* as the "best ever held in London," as there was good weather and judging on the terraces (fig. 3.5).[36] Judging was in public but by "the old routine school," with predictable results and the usual complaints, including the accusation that it was run by "a clique" for its own benefit. This criticism was pursued over the rest of the summer by the newly constituted, Nottingham-based National Canine Society, ahead of their October show, timed to be part of the town's annual Goose Fair. The committee of the NCS was headed by John Chaworth Musters, a landowner of Wiverton Hall, near Tithby in Nottinghamshire,

MR. SHORTHOSE'S NEWFOUNDLAND VICTOR.    MR. MURCHISON'S FLEETFOOT.    MRS. MONCK'S MOPSY.    MR. M'DONALD'S MOLLY.
MR. P. BULLOCK'S SUSSEX SPANIEL GEORGE.    MAJOR ALLISON'S RETRIEVER VICTOR.

DOGS AT THE CRYSTAL PALACE SHOW.

**Figure 3.5.** Dogs at the Crystal Palace Show. The dogs are shown on the terrace and include those belonging to Kennel Club founding members John Shorthose (Newfoundland, Victor, back left), J. H. Murchison (Greyhound, Fleetfoot), and Major Allison (Retriever, Victor). The other dogs are those of Mr. P. Bullock (Sussex spaniel, George), Mrs. Monck (Maltese terrier, Mopsy), and Mr. McDonald (Italian greyhound, Molly). *Field,* 26 June 1872, 579

with J. G. V. Wakerley acting as secretary. They decided to upgrade their annual event to be the Grand National Dog Show and to set it against those in Birmingham and the Crystal Palace. They drew up a prospectus that was bound to offend. It began, "This show is not raised as a speculation, like many others. The efforts of the promoters are disinterested, and aim solely at the establishment of a truly national society of the preservation of our national dogs." In addition, no committee member or judge was permitted to exhibit, and judges were to be elected. The aim was to exclude the usual clique of "circuit judges," the men who gave prizes to the same dogs and owners at every show. The men behind the Crystal Palace shows were not mentioned by name, but it was clear who Wakerley had in mind. There was a swift reaction in the sporting press to these East Midlands upstarts, not least as Wakerley ratcheted up the tension by responding to every comment at length.[37]

The NCS show attracted nearly one thousand entries in ninety-five classes, including dogs billed as celebrities by the popular press (the bloodhounds Draco and Dingle among them). It ran at only a small loss and had mixed reviews. The local press was positive, saying that it was well run and attracted a high-class entry, and *Bell's Life* welcomed its freedom from the "regular dog firm." But there was a different story in the sporting press, with many owners claiming that their dogs returned home starved or injured; apparently, bulldogs and mastiffs were tethered too close together and fought each other. Despite the pugnacious tone adopted in the NCS prospectus, the "Committee of Gentlemen" was invited. To the surprise of many, they not only attended on the final day but brought with them and awarded special prizes. The group—Sewallis Shirley, J. H. Murchison, Captain Montressor, Rev. T. O'Grady, and Mr. John Douglas—were said to have acted with "spirited tact." Why did they go? We can only speculate. Most likely it was to try and coopt the NCS to the club they were planning.[38]

## The Kennel Club

The foundation day for the Kennel Club has been set retrospectively as 4 April 1873. In fact, this was actually the day of a planning meeting for the Fourth Great National Exhibition of Sporting and Other Dogs at the Crystal Palace, to be held two months later. The name "Kennel Club" was significant. It was an organization for those with the wealth to have kennels; the word referred both to buildings and a group of dogs. Unlike two of its rivals, it was not a society but a club, a term associated with exclusivity and high social status. The club's membership records reveal that members were signed up as early as February 1873.[39]

The first member was Paynton Pigott, a barrister living at Tong Castle, Shifnal, in Shropshire, the home of the Wolverhampton ironmaster John Hartley. Four others were elected that month: Thomas Haselhurst (a general practitioner from near Bridgnorth), Lord Lurgan (Liberal government whip), Capt. John Platt (son of the owner of a large textile machinery company in Oldham), and Edward Wingfield Stratford (soldier, son of owners of estates in County Wexford, and high sheriff of Kent that year). The April planning meeting was chaired by Shirley and attended by twelve men.[40] Like the membership of the NDC, they were professionals, landowners, and farmers, and included established doggy people, such as John Cumming Macdona and William Lort.

The prospectus for the Fourth Great National Exhibition of Sporting and Other Dogs at the Crystal Palace in June 1873 announced that it was being run by the newly formed Kennel Club. Although the number of entries was down from previous years, the event was hailed as a success on canine and human social terms.[41] However, it was not long before criticisms began to be aired, in what soon became a rerun of earlier controversies. Seemingly unmoved by the visit of Shirley and friends to the Nottingham show the previous autumn, Wakerley led the assault on behalf of the NCS, backed by two correspondents to the *Field*. "Fixed Idea" had first written at the end of May concerning the first spring dog show in Manchester, which had been held at the Free Trade Hall and organized by Samuel Handley and John Douglas, members of the "Crystal Palace clique," soon to be the Kennel Club. Fixed Idea's main complaint was familiar: the show was a speculation, where the committee acted as judges and won most prizes. "O. R. K." was just as direct. He divided dog shows into three types: those, like Birmingham, "raised by disinterested promoters for the encouragement of the purity of breed"; those, like the recent Manchester show, "raised by speculators"; and lastly, those, like the Kennel Club event, "raised by breeders and exhibitors for the purpose of securing prizes and commendations for . . . their own property." "Fixed Idea" wrote again and was equally damning: "I visited the Crystal Palace Dog Show held under the management of the Kennel Club; and if I drank vinegar at Manchester, I can truly say I had gall for my last dose." He went on: "Mr Shirley doffs his wig and joins his friends Mr Murchison, Mr Hemming, and Mr Whitehouse; Mr Lort and Mr Handley kindly retain their posts; and presto, the honours of the same are divided amongst the clique, whilst the bulk of the public exhibitors stand like country geese watching for the crumbs that may fall from the feast over which the game is played."[42]

Other correspondents pointed out that Shirley won prizes for nearly three-quarters of the dogs he entered, with the five other committee members averaging seven prizes each. The other 450 exhibitors took just one-third of the prizes. An editorial in the *Field* defended the distribution, stating that few of the awards were undeserved and that the winners were likely to be owned by those most committed to shows and the improvement of breeds. These men were, of course, the best judges. However, he professed that, while he had accepted an invitation to join the club, he maintained his independence and "was not in any way responsible for the doings of the Kennel Club." In particular, he objected to how judges had been selected, arguing that the process should be more open to avoid the suspicion that the judges were self-serving. Other correspondents also defended the Kennel Club and wondered if O. R. K. and Fixed Idea belonged to the "Nottingham clique."[43]

Wakerley and NCS made waves again later in the year, announcing their October show as a "national" event and thus challenging the Birmingham Society and the Kennel Club, which used the same term for their shows. The "Country Correspondent" of *Bell's Life,* who had previously sympathized with the NCS, turned against them. He complained that it had adopted a "party spirit" against other shows and that their "assertions about corrupt practices had been much exaggerated." Moreover, their pejorative use of "circuit judges had given great offence to a very large section and it was unnecessary." Needless to say, Wakerley replied at length, and his position hardened against the Kennel Club. He wrote that "until other societies, however powerful they may be, raise shows upon proper principles, a 'spirit of opposition and warfare' to them will be evinced by the NCS." An exchange followed in which Wakerley continued to defend the principle "that the committee of management who elect judges ought not to exhibit." Giving further offence, at one point he likened the Kennel Club committee to "wolves that would feed on their flock."[44]

Walsh denied Wakerley the opportunity of any more "trumpet sounding" for the NCS in the columns of the *Field* in the weeks leading up to its next Nottingham show, and Kennel Club members stayed away. Further revenge was gained in reports. The Country Correspondent of *Bell's Life* acknowledged that the hall was well appointed and judges of repute had been found, but he observed that even "at a reform meeting, the judges cannot get away from the same merit that has been adjudged first at Birmingham, Manchester, &c, by those poor gentlemen who have been called anything but pleasant names." He went on to say that "the show has not taken a particle of right to the title of national" and that the claims of the NCS to be the group "to regulate the management of dog affairs, and stand as an authority" had been found wanting. The report ended with the Kennel Club receiving a ringing endorsement. It "was formed in the first place by gentlemen who have made the culture of the various dog families quite a study, devoting an immense amount of time and fortune to the subject; and to their instrumentality we owe the first establishment of shows and the improvement of dogs." The National Canine Society's show the following September was held, no doubt reluctantly, under Kennel Club rules.[45]

At the National Dog Show Society show in Birmingham in December 1873, rivalry with the new Kennel Club was evident, though this did not prevent Shirley and Lort acting as judges. Afterward there were no complaints about organization, only the typical protests about judging, with particular bile reserved for the awards made in the mastiff class. The judge under fire was the Rev. Malcolm Bush Wynn, who hailed from Scalford, near Melton Mowbray, and had first competed with mastiffs in 1870. He quickly became a leading figure with the breed and in the summer of 1873 founded a Mastiff Club, which, according to his critics had one aim, to ensure that only its members judged and won. There were immediate objections to creating what, once again, looked to be a clique that went against the view that dogs shows should be open and meritocratic. It was predictable, therefore, that Wynn's awards at Birmingham were challenged. "The Critic" wrote to the *Field* that he was "disgusted" at the turn of events: "Last year Mr Hanbury was judge and he awarded the cream of the prizes to Mr Wynne [*sic*]. This year Mr Wynne judges, and gives all the first prizes and cups to Mr Hanbury." He concluded that "it is a perfect farce for an outsider to send a mastiff to Birmingham at all." There were accusations of abuses in other classes, which led "Caractactus" to ask for a body like the Jockey Club, such that, "gentlemen and sportsmen of known position and honour, like Messrs Shirley, Price, Lort, Laverack, Stonehenge, Idstone, &c form a committee that would furnish impartial judges in all classes, investigate charges of corruption and rules of shows . . . and purify it from all its abuses." Arguably, the new Kennel Club had made matters worse, as there were now three organizations claiming to rule "the canine world." Could these groups come together and one organization be formed? In Walsh's view there was as much chance of the Birmingham committee amalgamating with the Kennel Club as of "oil mixing with water."[46]

Through the winter and spring of 1874, views on how to reform "the canine world" continued to be aired, with the claims of the Kennel Club to be the "Canine Court of Appeal" gaining support. In its favor was the commonsense, efficient, and courteous organization of its annual show, especially the practice of announcing the names of judges months in advance. Then there was the standing of Shirley and his "Committee of Gentlemen."[47] Indeed, one report on the Crystal Palace Dog Show in June 1874 maintained that the club had "already established for itself no little prestige as the Jockey Club of the canine world."[48] Its standing also came from the respect enjoyed by show manager John Douglas, who had been head of the aviaries on the

Duke of Newcastle's Clumber Park estate and brought experience of poultry shows to canine affairs. Yet there were still complaints that the club operated to private rather than public advantage, as its small number of members still won the majority of prizes. As long as organizers exhibited their own dogs and chose who was to judge them, there would be questions over impartiality. The canine correspondent for *Bell's Life* in London noted with pleasure the improvements that the Kennel Club had made and praised how the "the new movement" had dissociated itself from the London Fancy and made the dog show "the resort of fashion and the *élite* of society."[49] An important measure of success for the club, then, had nothing to do with dogs; rather, it was considered successful because it had moved its shows up the social scale by excluding lower-class "dog dealers" and their base practices.

## Conclusion

Paradoxically, the different factions of doggy people struggling for dominance of the nation's dog affairs from 1869 to 1874 highlighted a set of shared priorities, but there were differences over how these should be addressed and by whom. The travails of promoters and organizers also brought acknowledgement that shows needed to be financially viable as well as socially respectable. It was important that profit was not an end in itself but a means to secure a stable future for shows and the improvement of the nation's dogs. The Kennel Club was the most recent claimant to the position of being the canine Jockey Club or the "Canine Court of Appeal," and it had key advantages, being based in London and boasting an elite, expert membership spread across the country. Its membership terms effectively excluded the Fancy and "dog dealers," while its rules and regulations sought to ensure shows were respectable and well ordered. The emergence of the Kennel Club as the ruling body of the nation's canine affairs has often been portrayed as inevitable; however, we have shown that it was born among rivals. In the next chapter, we discuss the continuing struggles for power and influence within the new dog show fancy, the growing social profile of shows and fancy dog culture, and the effect these changes had on the form and character of the nation's dogs.

# PART TWO

# 1873–1901

# CHAPTER FOUR

# Governing Breed

From 1873 the Kennel Club sought to become the "Jockey Club of the canine world." Although its remit was only Great Britain, the club was soon emulated across the world.[1] Gentlemen of knowledge and experience served on its committees, and membership was restricted as much as possible to the "right sort." The pitch for ascendency depended upon a mutually beneficial relationship: the organizers of show and field trials who adopted Kennel Club rules would enjoy legitimacy and some protection, while the club gained power and influence. While the leadership of the Kennel Club sought to bring unity, there remained divergent interests and conflicts, and its dominant position was not secured until the mid-1880s.

As rivals fell away or were coopted into the club's ambit, alternative power bases developed—a Fourth Estate and specialist breed clubs. The success of shows led to the growth of a specialist doggy press, and canine affairs found a larger place in national and local newspapers and in journals. Many journalists became authors and consolidated their influence in landmark books that displaced the reference works of Youatt, Stonehenge, and Idstone.[2] The Fourth Estate, particularly new men such as Hugh Dalziel and Vero Shaw, was consistently disparaging of the Kennel Club's actions—or rather lack of them—when it came to the improvement of the nation's dogs. The Kennel Club was said to be preoccupied with governance and was deemed negligent in allowing improvement to be taken over by specialist breed clubs, which, while claiming expertise, had narrow concerns. These clubs proliferated in the 1880s. They were established by enthusiasts for a particular breed, who set standards, became judges, and in time organized their own single breed shows. Many specialist clubs represented geographic areas and cultural identities. For example, there were national, regional, and local clubs, respectively, for Scotch terriers, Yorkshire terriers, and Bedlington terriers. The success of these groups initiated a new politics in the canine world, with tussles for authority between the Kennel Club and breed clubs, and within and between breed clubs. From the 1870s, the activities and profile of the working-class Fancy fell away with the decline of public house–based sports. Increasingly, the institutions and culture of shows were termed the (new) *dog fancy* or *canine fancy*, and its principally middle- and upper-class participants and followers were called fanciers.

## The Ruling Authority in Dog Affairs

The Kennel Club's rules, set out in its first stud book published in the spring of 1874, ensured exclusivity by allowing no more than one hundred members. It went without saying that it was only for men. Membership could be increased if a general meeting agreed, but the intention was to restrict membership to the elite of the dog world; joining was by invitation only. It cost five guineas to become a member, with an annual subscription of the same amount. The club's first home was a flat in Victoria Street, but in April 1877 it moved to 29A Pall Mall, opposite the Carlton Club, an establishment favored by Tory politicians and whose members included Prime Minister Benjamin Disraeli. The premises were next door to the Junior Carlton Club, which had the same political allegiance and which a number of Kennel Club members used as their London address. The Prince of Wales was the club's first patron. It was not a gentlemen's club in the sense of having social facilities for meeting, dining, and drinking; rather, its offices were an administrative center dealing with correspondence around membership, shows, and the stud book. However, the club's proximity to London's sites of elite association indicates that it was an institution for gentlemen with wealth or connections or both.[3]

While the leader of the club, Sewallis Shirley, was aristocratic and Tory, its committees were mixed politically, socially, and geographically. At the time of the first general meeting, held at the Great Western Hotel in Birmingham in December 1874, the committee consisted of four landed aristocrats (Shirley, the Marquis of Huntly, Lord Onslow, and the Hon. William F. Byng, son of the Earl of Strafford), three professional men (John Cumming Macdona, clergyman; Frank Adcock, solicitor; John H. Murchison, mining engineer and speculator), two landowners (John H. Whitehouse, farmer, and Richard Lloyd Price, squire of the Rhiwlas Estate), and three sons of successful industrial entrepreneurs (John Harold Platt, son of John Platt textile machinery company; Joshua Horton Dawes, son of William H. Dawes, "Gentleman Ironmaker"; and Sam Lang, son of Samuel Lang, a leading Midland iron-master). The Marquis of Huntly was a Liberal, who had been Lord-in-Waiting to Queen Victoria, while Platt's father had been the Liberal MP for Oldham from 1865 to 1872. New money had adopted the ways of old. Platt and Dawes shot on the estates bought by the industrial wealth of their fathers and grandfathers. The committee was also drawn from across England and Wales: Adcock (Watford), Byng (London), Dawes (Birmingham), Huntly (Northamptonshire), Lang (Bristol), Macdona (Cheshire), Onslow (Guildford), Murchison (London), Platt (Bangor, North Wales), Price (Bala, North Wales), Shirley (Stratford-upon-Avon), Whitehouse (Worcestershire).

The club's activists were young, with an average age of thirty-four years. Murchison was the oldest at forty-eight, and Platt was just nineteen. They were also reformers. The club's first rule was to "endeavour in every way possible to promote the improvement of dogs, dog shows, and dog trials." Most club rules set out procedures for the operation of shows and trials, assuming a role similar to that of the Jockey Club and also adopting its model of an open stud book. There were, however, no specific requirements on the vexed questions of judging and standards. The latter was particularly significant, as it indicated that the club saw its role primarily as regulating shows, with the assumption that competition was the best way to improve the nation's canines. The club's policy could be cast as Darwinian, or rather as based on Social Spencerism, after Herbert Spencer, who systematically popularized competitive evolutionism as the guarantor of progress. It assumed that judges' decisions (selection) would identify the best (fittest) dogs that would go on to reproduce (survival), yet allow for conformations to change (evolve) over time.[4]

Another route to improvement was the provision of information on pedigrees and performances in the club's stud book. Its compiler and editor was Frank Pearce, the son of Thomas Pearce. A large volume of over six hundred pages, it was essentially a reference work. The first part of the book recorded the results from the major shows and trials up to 1873, beginning with the Newcastle upon Tyne show in 1859 and the Southill Trial in 1865. Included were the top prizes awarded at all the Birmingham, London, and Manchester shows, and there were detailed reports of field trials. The second part of the stud book presented the pedigrees of dogs that had been successful in shows or trials, cross-referenced with the results of performances. Each named dog was given a number, and, where known, details of its breeder were listed, along with the dog's year of birth, pedigree, and chief performances. Most entries were very short, but a few dogs had their family tree set out like a royal or aristocratic lineage. Mr. Edwin Brough's bloodhound Rufus had both a short and long entry (figs. 4.1 and 4.2).

The breeder's name was important because the value of each entry depended on the trust that could be placed in the veracity of the pedigree. Gentlemen were, of course, held to be more trustworthy than "dog dealers." The entry for Macdona's St. Bernard Tell was typical in having a pedigree for just a single generation, though claiming distant descent from the famous Barry.[5] The club expected, of course, that in successive editions pedigrees would become longer and more useful. Breeders would then be able to identify the most successful sires and dams, and those with good bloodlines, as had long been practiced with Thoroughbred racehorses. The most elaborate pedigree in the first stud book was for a bulldog, Abbess, owned by Frank Adcock, which went back twelve generations and recorded 105 ancestors. This documentation relied substantially upon pedigrees that had been kept by the London Fancy, including dogs owned by Jemmy Shaw and Bill George, revealing the crossover between old and new fancies.

The hope of the committee was that, by providing open information on a dog's performances and pedigree, shows and trials would become fairer and there would be greater consistency in results between events. Owners and breeders would be able to sell, buy, and breed with more confidence, and a public record would give authenticity and build trust in shows. What was implicit in the codes and rules was that shows had seen sharp practices and deception, if not outright dishonesty and fraud. A key aim was to bring respectability: to stop dogs being entered under false names with made-up pedigrees, to ensure dogs were entered in the correct class in competitions, to deter favoritism and corruption in the awarding of prizes, and to reduce the number of frivolous complaints about awards by providing robust appeals procedures.

The value of the pedigrees in the stud book was disputed. For some people they were "a mere string of names" that, even if accurate, revealed nothing of value about an individual dog.[6] At shows it was hoped that judges would make decisions based on the conformation of the dogs on the benches and not on their pedigrees. Nonetheless, breeders looked for assurance that the *qual-*

*Pedigree:* By Gelert out of *Norma;* Gelert by *Raglan* out of *Vengeance;* Raglan by Old Raglan out of *Welbeck;* Norma by *Thane* out of Becker's *Duchess;* Thane by Oscar out of *Lady;* Welbeck by Romewood out of Countess.

*Chief Performance:* Edinburgh, 2nd prize, 1873; (shown in Mr. Kerr's name).

57. ROSAMOND.—Mr. Fraser Cobham's, The Firs, Amwell, Ware; breeder, Mr. Musters; born 1870.
*Pedigree:* By Holford's *Regent* (No. 50) out of Muster's *Hilda;* Hilda was by Cowen's *Druid* (No. 16) out of Bird's *Duchess.*
*Chief Performance:* Crystal Palace, 1st prize, 1871.

58. ROSWELL.—Mr. Reynolds Ray's, Dulwich, Surrey, S.E.; breeder, Mr. E. Bamford; born 1866 (formerly belonging to Mr. J. K. Field and Mr. J. Bird).
*Pedigree:* By Duke of Beaufort's *Warrior* out of sister to Field's *Rufus.* (For *Rufus's* pedigree see No. 61).
*Chief Performance:* (When Mr. Field's) Birmingham, 2nd prize, 1870; Crystal Palace, 1st prize, 1871; (when belonging to Mr. Bird, J.P.) Birmingham, 1st prize, 1871; Crystal Palace, 1st prize, 1872; (when Mr. Ray's) Birmingham, champion prize, 1872; champion prize, 1873; Crystal Palace, 1st prize, 1873; Nottingham, 2nd prize, 1873; Paris, 2nd prize, 1873; Maldon, 1st prize, 1873.

59. RUBY.—Lord Bagot's, Blithfield Hall, near Tamworth; breeder owner (bitch); born 1858.
*Pedigree:* By *Forester* out of *Venus* (see *Gamester,* No. 31).
*Chief Performances:* Birmingham, 2nd prize, 1860; 1st prize, 1861.

60. ROYALTY.—Mr. C. E. Holford's, High Oak House, Ware; breeder, owner; born March, 1870.
*Pedigree:* By Holford's *Regent* out of Becker's *Brenda* (see Nos. 50 and 5).
*Chief Performances:* Birmingham, 1st prize, 1870 (when eight months old); Manchester, Belle Vue, 3rd prize, 1870.

61. RUFUS (see table on p. 210).—NOTE. According to one authority, *Mona* should be *Nana* by Garrard's *Bluff* out of *Blancka,* but it is believed the given pedigree is correct. *Tarquin* and *Grafton* were stone-coloured hounds.
*Chief Performances:* Birmingham, 1st prize, 1867; 1st prize, 1868; Crystal Palace, 1st prize, champion class, 1872; Manchester, Belle Vue, 3rd prize, 1867; 1st prize, 1873; Chester, 1st prize, 1873; and other prizes at local shows.

62. RUFUS.—Mr. Alfred S. Boom's, H.M. 15th Regiment; breeder, Lord Faversham; born 1856.
*Pedigree:* By Lord Faversham's *Stranger* out of a bitch of Lord Ossulton's.
*Chief Performances:* London, 2nd prize, 1861; Islington, 2nd prize, 1862; 2nd prize, champion class, 1863; Ashburnham Hall, 1st prize, 1863 and 1864.

P

*Figure 4.1.* Mr. Edwin Brough's Bloodhound Dog Rufus. Frank C. S. Pearce, ed., *The Kennel Club Stud Book: A Record of Dog Shows and Field Trials* (London: The Field, 1874), 209

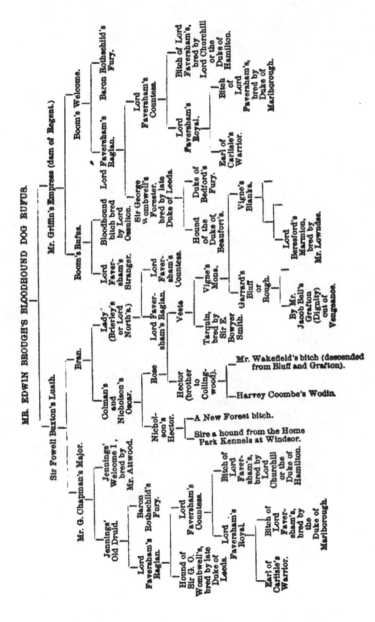

MR. EDWIN BROUGH'S BLOODHOUND DOG RUFUS.

*Figure 4.2.* Pedigrees. Frank C. S. Pearce, ed., *The Kennel Club Stud Book: A Record of Dog Shows and Field Trials* (London: The Field, 1874), 210

*ity* of conformation would be reproduced, trusting that pedigrees revealed pure blood and hence prepotency. While doubts persisted on its usefulness in breeding, the stud book helped to stabilize dog shows as reliable venues for trade, becoming a prospectus for sellers and a guide to buyers and breeders on past performances at shows, if not prepotency.

The club continued to face accusations that it was a clique, and it remained the case that its leaders won many prizes at their own shows. This situation was said not only to be unfair or corrupt but to give committee members the power to remodel breeds to standards that were at odds with those favored by the majority of aficionados. This accusation was made most strongly against Frank Adcock over the award of prizes to large bulldogs at the club's June show in 1877. In fact, bulldogs had been shrunk as well as enlarged, a situation recognized at the show, where there were classes for large, medium, and small dogs. The *Sporting Gazette* sarcastically described the class of large bulldogs as containing "five animals so equivocal as to be utterly beyond criticism, creatures of the Adcock-Toro creation for which a special niche must be newly arranged in natural or *un*natural history" (emphasis in original). Walsh declared Toro to be "barely half bred" and not a true bulldog.[7]

Adcock's "theory" was that since the end of bullbaiting, breeders had reduced the size of British bulldogs, and his mission was to return them to their original form.[8] The result, according to Dalziel, who spoke for many aficionados, had been the creation of a "gigantic mongrel race" and "the obliteration of the characteristic type." This complaint was not just about size but blood, too. A report on the first bulldog show remarked that bulldogs had suffered from "the mistaken efforts of its professing friends, who, with the intention of reviving and improving (?) the breed have crossed it with others and introduced impure blood." This example of a Kennel Club insider seeking to remodel a standard was one factor that led to the foundation of a new Bulldog Club. Its principal aim was to address the "contradictory decisions at various shows" by "publishing a description of . . . the true type of that breed, and to urge on show committees its adoption and uniform application." However, it was important that the campaign against dogs over fifty pounds did not go too far. Just as worrying was the trend to produce dogs that were too small. Shrinkage had developed after "the bulldog lost its peculiar occupation" and became fashionable as a companion, leading to toy-sized dogs from "crossing it with the terrier," which had resulted in another "travestie [*sic*] of the true breed."[9]

### Jockey Club of the Canine World

The Kennel Club's ambition to become "the ruling authority in dog affairs" faced continued opposition, with its regional rivals keen to protect their independence and authority. The Birmingham Dog Show Committee continued to claim that their rules were superior, having been tried and tested over many more years. Walsh's suggestion for a confederation of the existing societies still had supporters. However, the Kennel Club was never in the mood for compromise, and, with the sense of entitlement that members' social position and assumed expertise gave them, they continued to act as natural rulers. The club enjoyed credit for making shows more respectable. A report in *Bell's Life* on the club's show at the Alexandra Palace in December 1877 remarked upon "an almost entire absence of the rough element, and hardly any but well-dressed ladies and gentlemen parading the long dog avenues." The move upmarket was reflected in the prices paid for prizewinners: one hundred guineas for a bloodhound puppy and a black-and-tan Manchester terrier, ninety guineas for a St. Bernard puppy, and many others over forty guineas. The report applauded the democratization of the commerce of breeding.[10]

> There is no reason why such a rise in value should be deplored, for some excellent people breed dogs, show them, and sell them, including nobility, country gentlemen of the highest standing, and dignitaries of the Church; and if it is a source of pleasure and profit to those in a still humbler station of life, is there anything to be gainsaid of the subject? For such persons might be certainly doing worse, and there may be as much satisfaction to a man of moderate means in being successful at a show with a team of fox terriers as a stud of race horses can afford to a patron of the Turf.[11]

There was, perhaps, no longer a tension between respectability and profit, as shows had become sites for class interactions for mutual benefit, with gentlemen learning that making money was necessary and dog dealers finding that fair trade paid. However, the blurring of the line between dog dealers and dog lovers meant that the latter were seen to have an unfair commercial advantage; they were the "gentlemen amateurs" who ran the shows, chose the judges, and influenced the results.[12]

Unsurprisingly, given Walsh's antipathy to aspects of the club's operations, the *Field* continued to carry many letters from exhibitors who claimed maladministration and poor judging at shows held under Kennel Club rules.

Others were unhappy, too. In December 1879, Richard Lloyd Price resigned over the continued endorsement of William Lort as a judge. It was not that Lort did not know his breeds but rather that his temperament was flawed; under pressure in the show ring he was too readily swayed. Lort was experienced and expert but seemingly not enough of a gentleman to be reliably independent. Soon two of Walsh's protégés, Hugh Dalziel and Vero Shaw, entered the fray. Dalziel developed a telling critique of the club, arguing that it had done little to improve the nation's dogs. The proliferation of shows, and trials for that matter, without agreed standards of judging had spread confusion, leading to results continuing to be inconsistent between events. Dalziel accused the club's officials of being "a mere association of showmen." He made the familiar observation that shows organized under club rules were biased toward their promoters, who appointed the judges and won the most prizes. Dalziel wanted democracy. He argued for judges to be elected by exhibitors and characterized the Kennel Club's position as being "these are the judges we have appointed, and if you don't like them you need not exhibit."[13]

Dalziel's attack was answered by the club's supporters, one of whom claimed that its shows were "never more scientific and never more popular." Shirley replied, too, beginning by stating that the election of judges had been tried and found wanting, especially with the NCS at its Nottingham shows. It was "sheer nonsense" to say that judges did not know what they were doing and that breeds had not been improved. The expansion of shows under club rules demonstrated public confidence. He ended with a thinly veiled sleight of journalists like Dalziel and Walsh, observing that "as a rule judges are much more consistent in their opinions than many of their critics, the reporters." The letters published in the *Field* in subsequent weeks were evenly split between supporters and critics, and went over familiar ground. Defenders of the club, such as William Allison, contended that its committee had "character" and were "a body of honourable gentlemen." He went on to state that judges had to be independent and of a social position to take on an often unenviable role, in which they must face the censure of disappointed exhibitors. On whether judges should be appointed or elected, Allison observed that the problem with election was that "a clique of exhibitors" might organize to have their own special judge appointed. He did not elaborate, but his meaning was clear when he added that "anyone who knows anything about the lower stratum of dog-show life will be able to understand what I mean."[14]

In the spring of 1880, having learned from experience and hoping to assuage critics, the club published new rules for shows and field trials. The main changes with shows were to strengthen registration and identification in order to counter outright deception. Walsh broadly welcomed the rules, though he objected strongly to Rule 3, which required all dogs entered in shows to have their name registered with the club at the cost of one shilling. This level of fee was justified as covering costs and deterring poor-quality dogs, but it was also designed to exclude "the lower stratum of dog-show life." Walsh had an additional objection. He calculated that the annual total of fees received would be £5,000, potentially a tidy subsidy to the club's new venture—the monthly *Kennel Gazette,* a journal that promised to outshine the *Field* in its coverage of canine affairs. Shirley politely refuted each of Walsh's criticisms.[15]

The *Kennel Gazette,* first published privately by Shirley before he gave it to the club, reported on committee meetings, show results, and registrations, as well as carrying news and features. It revealed that the committees dealt mainly with membership, the annual stud book, show organization, and appeals on procedures at shows. At most committee meetings there were less than ten members present, while general meetings had higher attendance, facilitated by holding them in association with major shows in London or Birmingham. Day-to-day club business seems mostly to have been transacted by Sewallis Shirley and the secretary, George Low. At shows there seems to have been good compliance with the rules, as few judgments were appealed to the committee, and those were mostly dismissed due to lack of evidence, thus tacitly endorsing the judges' decisions.[16]

Many shows adopted the club's new rules, but the major events in Birmingham, Manchester, and Darlington stuck to their own procedures. In succeeding years the Kennel Club escalated pressure on these shows to adopt its rules and secure its place as the single national authority. It continued to meet resistance, and hostilities broke out in the summer of 1880. Shirley alienated the fancy in the northeast by attacking the laxity of registrations at the Darlington Dog Show and, with snobbish condescension, ridiculing its acceptance of dogs with names such as "Wag, Roy, Bobs, Bangs, Jets, Nettles, Vics &.c," whose identity was impossible to validate. The following year club members were advised to abstain from exhibiting or judging at shows not held under its rules. Opponents called this a "Boycott," a sensitive term for Shirley at this time. It had been coined the previous year following events on the estate of a fellow absentee Irish landowner, Lord Erne, whose agent, Capt.

Charles Boycott, was ostracized over the eviction of tenants in a protest organized by the Irish Land League.[17]

In the summer of 1882, the club moved from advising members to abstain from noncompliant shows to intimating that stronger measures would follow if their participation in these continued.[18] An indication of what might happen was given in an anonymous letter received by the Darlington secretary and assumed to come from a club insider. It threatened the disqualification in future of all dogs and exhibitors that took part "in any show not under Kennel Club rules." Thus, action was to be taken against noncompliant shows and only indirectly on noncompliant club members. The Darlington secretary reacted angrily to "the persecution of provincial shows" by a small clique and stated that his society would not be "bullied by a secret society in the Kennel Club, nor submit to its dictation." Familiar insults aimed at undermining the integrity of the club's committee resurfaced: it was no more than "an association of 'entrepreneurs' of a couple of London shows." In the eyes of one critic, the club had "purified shows" but had become self-serving and had been "too limited" in its goals with regard to canine improvement, a gap that was being filled by specialist breed clubs.[19]

In October 1882, the Kennel Club stepped up pressure on Birmingham, announcing that the results of its December show that year would not, as previously, be included in its stud book nor would the winners of their champion classes be recorded. J. H. Murchison claimed that he had "long regarded the Birmingham display as a Barnum exhibition, carried out as a commercial undertaking and not tending to the improvement of dogs." This seemed rich coming from an organization whose London shows were by this time managed by Charles Cruft, of Spratt's Patent dog biscuits, who became known as "the British Barnum."[20] There were again calls for an alternative ruling body, and, ahead of the Birmingham show, a meeting was held of representatives from shows that had not adopted Kennel Club rules. Walsh was elected chair, and he backed the proposal for a new association and stud book, offering the pages of the *Field* to aid the project. Former Kennel Club stalwarts John Shorthose and Cumming Macdona supported the new venture, in part because it might force a compromise with the Kennel Club over its attitude to northern shows. Macdona was fed up with the "hectoring mandates and whining appeals of the Kennel Club" and their condescension toward provincials and their events. The Birmingham show in 1882 was reported to be to its usual standard, though *Bell's Life* thought its organizers were wrong to resist the Kennel Club's authority.[21]

The next big show after Birmingham took place in January 1883 at the Crystal Palace, where a gathering of dissenters resolved to found a new National Dog Club (NDC) to rival the Kennel Club. Three Kennel Club members spoke against the proposal, arguing for unity, but the motion was carried. Walsh still hoped for a compromise and maintained that "the Kennel Club had been the aggressors" by attacking the Birmingham and Darlington shows, which "were only acting in self defence." He went on to say that "if any compromise were proposed, it was for the Kennel Club to come to the new association, not for the new association to go to the Kennel Club." Vero Shaw, who was kennel editor of the *Field,* attended the next meeting of the Kennel Club Committee in March 1883 on behalf of the new NDC. His mandate was to seek a compromise, but he was rebuffed. An editorial in the *Field* backed the northern alliance and looked forward to the new NDC's success as a more representative organization. The Darlington and Birmingham shows that year were held under new NDC rules, and Shaw promised to edit an alternative stud book. The new NDC's rules were published in the *Field* on 31 March 1883.[22]

Another national dog organization was formed at this time—the British Kennel Association (BKA). It was founded in February 1882 with the sole aim of promoting dog shows around the country, and one of its first acts was to meet with railway companies to negotiate reduced fares. The general secretary was Walter K. Taunton, a Hatton Garden jeweler, who was well known for exhibiting mastiffs and foreign dogs, especially Eskimo dogs and a dingo. He took part in the trials to use bloodhounds to track criminals.[23] The BKA had offices in Fleet Street, London, and seems not to have been short of money. It was no challenger to the Kennel Club; indeed, it may have been a covert operation on its behalf. Its first show, held under club rules with Shirley as a judge, was a grand affair held at Aston in Birmingham and had a larger entry than the city's annual December show the previous year. The association offered prizes at other local and breed shows before its second (and last) show in Sheffield in October 1885. After that, nothing more was heard of the BKA.[24] The poster advertising this show suggests it was for the genteel upper and middle classes and their well-heeled dogs (see plate 4).

Some Kennel Club members, notably John Henry Salter, wanted a compromise and proposed a meeting of a "doggie parliament."[25] Salter, an Essex general practitioner and prodigious shot, had a growing influence in the Kennel Club, having redrafted its field-trial rules in 1880.[26] Many men associated with the new NDC recognized the problems of having two stud books and also wanted an accommodation. In 1883, the Kennel Club boycott affected the Durham and Birmingham shows, and their entries went down. The latter show was featured in the *Graphic,* accompanied by a not completely flattering illustration of show practices rather than the usual gallery of prizewinners (fig. 4.3). There were drawings of weighing, grooming, walking a dog in the ring, and judging, and of characters, canine and human: the arrogant prizewinning dog and the dainty pet, the "loving mistress combing out" her pet, and "one of those obnoxious haunters of shows, 'the umbrella cad,' who makes it his special mission never to let sleeping dogs lie." The

**Figure 4.3.** Notes at a Dog Show. *Graphic,* 8 Dec. 1883, 7. © British Library Board

report in the *Times* confirmed that the standoff with the Kennel Club had affected the quality as well as the number of entries. The correspondent in *Bell's Life* felt Birmingham had suffered from joining "the weaker side"; hence, it was time to end "the deplorable quarrelling and the endeavours to obstruct authority that have long been a marked feature of the canine world."[27]

The Darlington Show Committee and the Scottish Kennel Club yielded to London pressure in 1884, and the following year the Birmingham committee reached an accord, though the *Times* called it "a judicious surrender." Birmingham adopted Kennel Club rules, except that judging remained in private, and, in return, two Birmingham representatives were invited to join the club's committee for relevant business. The new NDC faded away, though its spirit was maintained by Birmingham's National Dog Show Society, whose winter event in 1885 was a grand affair, with entries up and a visit by the Prince and Princess of Wales. The *Kennel Gazette*'s report on the event put a gloss on the controversy, stating that now "the misunderstanding . . . has been entirely cleared away and all the greatest shows of the kingdom will be held under one set of rules." At the show, Shirley won two prizes with his retrievers, and it was probably no coincidence that the rapprochement with Birmingham coincided with the end of the British Kennel Association.[28]

The Kennel Club's attempt to gain total control of the nation's dog shows still had uneven success. Jaquet's official history of the club pointed to the fact that between 1874 and 1881, the stud book, which also became an annual calendar of results, recorded the results of 123 shows, of which 77, over half, were held under its rules. The number increased year by year, such that from 1882 the club was confident enough only to publish reports from their licensed shows—twenty-eight of forty-seven covered in the *Field* that year. However, the number was increasing faster than Kennel Club influence, though many of the noncompliant shows were local affairs in towns and villages or held in suburban parks and halls. In 1885, fifty-five shows were held under club rules, but the *Field* reported on a total of 144. There were similar figures for 1886. The rise in numbers reflected more comprehensive coverage by the *Field* and the increase in local and specialist breed club shows, testifying to the rapidly growing popularity of breeding, owning, and exhibiting dogs.[29]

The Kennel Club leaders had appointed themselves rulers of the nation's canine affairs because they had the experience of running successful shows, and the social standing to be above quarreling and act as a fair court of appeal. Although they had seen off or co-opted challenger organizations, in the late 1870s and the early 1880s new adversaries emerged: a Fourth Estate of journalists, authors, and lobbyists; and specialist breed clubs.

## A Fourth Estate

From the late 1870s, the canine press grew in number and circulation, and many new books on breeds and all aspects of breeding, feeding, keeping, treating, and exhibiting dogs were published. The two most influential new men were Hugh Dalziel and Vero Shaw, but many others found a niche in canine affairs, such as Gordon Stables, Henry Webb, George Jesse, and Harry Wyndham Carter, and they were joined by authorities who wrote on specific breeds.[30] The emergence of a Fourth Estate testifies to the economic growth of the new show fancy; both Dalziel and Shaw were able to turn their hobby of breeding and exhibiting into a full-time occupation. They did so without being Kennel Club insiders, though both were aided by Walsh's patronage.

In 1871, Dalziel gave his occupation as "Commercial traveller in drugs." Although living in Birmingham and regularly judging shows across the Midlands, he seems not to have been active with his home city's dog show society. By the 1880s he had moved to Kent and had begun styling himself a "Public Writer." He contributed to many journals (writing under the nom de plume of "Corsincon"), produced several books, and served as kennel editor of the *Bazaar*, a triweekly publication for the selling and buying of antiques and other collectibles, which now included dogs. Dalziel's first books were on veterinary medicine: *The Diseases of Dogs* (1874) and *The Diseases of the Horse* (1879).[31] His major work on dogs also appeared in 1879: *British Dogs: Their varieties, history, characteristics, breeding, management and exhibition*. The book had chapters on seventy-seven breeds, many more than in the Kennel Club stud book, reflecting the many classes at shows across the country and the interests of his potential readership. Dalziel wrote most of the chapters but was helped by unnamed "Eminent Fanciers" who contributed chapters on individual breeds. The chapter on the origin of the dog reflected the uncertainties raised in the evolution debates, concluding that the subject was still "waiting to be unravelled by a Darwin or a Wallace." In presenting the chapters on breeds, he followed the established division between sporting and nonsporting dogs. The first division was ordered by styles of hunting and the second by work or other roles. There was a third division, "House and Toy Dogs," which was disparaged as mainly about the dogs of ladies and ruled by fashion: "The fashion only changes in the selection of the reigning favourite, and caprice ordains that the bandy-legged dachshund, lolling in the lap of luxury yesterday, may, by the fickle goddess, be to-day dethroned in favour of that natty little dandy, the Yorkshire terrier, who, in his turn, struts his brief span of power upon the stage, most tyrannically governing the mistress who lavishes the exuberance of her affections upon him, till he again has to give place to some aspiring and successful rival." The nascent women's movement had yet to engage with the doggy world but did so in critical ways in the 1890s, as we discuss in chapter 7.[32]

Vero Shaw was born in India and qualified as a solicitor before becoming a journalist and writer. He played cricket for Cambridge University and a few games for Kent. Over his career his primary interest moved from dogs to horses, and in the twentieth century he became a promoter of horse shows. In the 1870s he was associated with the Birmingham fancy, editing its stud book, before serving as kennel editor of the *Field* and then the *Stock-Keeper and Fanciers' Chronicle*. Shaw bred, exhibited, and judged at shows, principally with bulldogs and bull terriers. His bulldog Smasher was a champion throughout the 1870s and early 1880s, and the dog's portrait was widely reproduced. Shaw was said by Dalziel, to be "the newest and most brilliant luminary in canine literature, before whom all past and present dealers in doggy lore must, sooner or later, pale their ineffectual fires." Shaw's principal work, *The Illustrated Book of the Dog* (1881), was the first publication consistently to give capital letters to breed names, making them proper nouns.[33]

Shaw also had assistance from "Leading Breeders of the Day," and his book ended with a one-hundred-page appendix titled "Canine Medicine and Surgery." This section was written by Gordon Stables, who had qualified in medicine in 1862 and served as a ship's surgeon until 1875.[34] After publishing *Medical Life in the Navy,* he began to produce veterinary handbooks, feline fiction, and numerous boys' adventure novels. He was prolific, publishing 130 books, and his range of topics was wide, as shown by a sampling of titles: *Cats, their points and characteristics, with curiosities of cat life, and a chapter on feline ailments* (1876); *The Domestic Cat* (1876); *Jungle, Peak and Plain: A boy's book of adventure* (1877); *Friends in Fur: True stories of cat life* (1877); *Dogs in their relation to the Public: Social, Sanitary and Legal* (1877); *Wild Adventures in Wild Places* (1881); *The Cruise of the Snowbird: A Story of Artic Adventure* (1882); and *Tea, the drink of pleasure and health* (1883). Stables also served as kennel editor of the *Livestock Journal and Fancier's Gazette,* which had merged two publications in 1875.

Dalziel and Shaw's books on dogs used new printing technologies to include high-quality illustrations, typically of recent champions, that exemplified breed ideals. Significantly, photography was largely eschewed—in part, no doubt, due to the quality of the prints but also, importantly, because photographs did not give sharp, conformation-defining lines. Walsh kept the same illustrations of breeds in action in all editions of *The Dog in Health and Disease* but used breed-defining drawings in *The Dogs of the British Islands*. For the later books, the best artists were sought, and care was taken with the quality of reproductions. The illustrations were all taken "from life" and most were in profile to show the ideal proportions of head, body, legs, tail, and other features. Dalziel's main illustrator was R. H. Moore, a wood engraver, whose drawings were used in the *Fancier's Gazette, Stock-Keeper and Fanciers' Chronicle,* and the Royal Society for the Prevention of Cruelty to Animals's (RSPCA) *Animal World.* Shaw was the first author to include color plates in a dog book since Sydenham Edwards in 1800. One painter whom Shaw commissioned to produce the color plates was Charles Burton Barber, who specialized in portraits of children and pets, enjoying commissions from royalty and painting the now-famous composition of Queen Victoria and her ghillie, John Brown. The importance of beauty in toy dogs was evident in Barber's portraits of the Yorkshire terrier, Italian greyhound, and pug (see plate 5). The other great dog illustrator of the era was Arthur Wardle, who was commissioned to paint portraits of the toy dogs of the wealthy but was best known for paintings of wild animals in their natural settings (though these were based on drawings made at the London Zoo). In the twentieth century, his dog portraits appeared on postcards and cigarette cards, and his drawings were used in government posters.[35]

Dalziel and Shaw began their doggy careers in journalism, joining a number of reporters on sporting and fancy affairs. The main sporting publications were *Bell's Life in London and Sporting Chronicle* (1822–), the *Field* (1853–), *Land and Water* (1866–), the *Illustrated Sporting and Dramatic News* (1874–), the *Live Stock Journal and Fancier's Gazette* (1875–1886), and the Sporting Gazette (1862–79), which became the *County Gentleman, Sporting Gazette and Agricultural Journal* (1880–1903). Another was the *Fanciers' Chronicle,* which started in 1879 in Newcastle upon Tyne, moved to London, and merged with the *Stock-Keeper* the following year. Then, of course, there was the *Kennel Gazette* (1881–). All of these publications covered other fancies, for example, the *Fanciers' Chronicle* was for "poultry, pigeon, dog, pet stock and bee fanciers." Dog-only publications, apart from the *Kennel*

*Gazette,* did not do well. In 1880 *Dogs of the Day* managed just one edition, while the *Kennel Review* lasted four years before folding when its owner and editor, Harry Wyndham Carter, fell into debt and was arrested for shooting a bailiff who was trying to seize his home.[36] The *Field* and the *Stock-Keeper and Fanciers' Chronicle* succeeded in part because they covered most domesticated breeds and enjoyed significant revenue from advertisements, from whole-page promotions of Spratt's dog biscuits to numerous small ads for sales and stud services.

All publications carried reviews and results, often lengthy, of dog shows, large and small. Many reports were syndicated nationally to local newspapers. Reviewers were, unsurprisingly, opinionated, and they commented on every aspect of shows, from the state of the venue to the judging of a single class. A lot was at stake in this public scrutiny, with the reputations of organizers and exhibitors and the value of dogs affected by adverse remarks. An indication of how much reviews mattered is indicated by a case in June 1880. Mr. E. Nicholls, a Kensington breeder of bloodhounds, sued the *Field* and the *Fanciers' Chronicle* for libel, demanding £1,000 and £700 damages, respectively. He was angry over comments made about his bloodhound Napier, which had been awarded second prize at the Alexandra Palace Show. A popular headline for the story in the press was "Alleged Libel on a Dog." The report in the *Field* had commented that Napier was "too slack loined altogether for second place" and that he was most probably "faked," his eyes having been tampered with. The *Fanciers' Chronicle* agreed, stating that Napier ought to have been disqualified. The hearing lasted two days. Nicholls's barrister called thirteen witnesses, including two leading veterinarians (Professor Pritchard and Mr. Hunting), while the defense relied on "the greatest dog authorities of the day," including Walsh, Dalziel, and Shaw. Nicholls lost but had a small victory, as the jury's verdict was that "the statements in *The Field* were not true in substance, but that they were made *bona fide* and without malice in both cases."[37]

The Kennel Club took umbrage at the case for two reasons. First, agreeing with a comment by the judge, they insisted that such matters should never have gone to court but should have been dealt with by the club, in the same way that the Jockey Club dealt with disputes in horseracing. Second, they objected that the reporters, by questioning the decision of the judges, were effectively challenging the club's authority, especially given that there had been no complaints about Napier's placing at or after the show. The report in the *Field* had gone further and had made the by-then regular criticism that there would have been no controversy if the Kennel Club had adopted standard points for breeds.[38]

## Breed Clubs

While there was agreement by the mid-1880s that the Kennel Club had largely met the goals of improving shows and trials, there were continuing complaints that it not done enough toward "the general improvement of dogs." The club's leadership assumed that, as in society and in Nature too, competition was the best guarantee of improvement. With fair rules, the best dogs would emerge and become the bloodstock from which improvement would accrue. Many breeders and exhibitors were unhappy about this laissez-faire attitude toward quality and standards, and took matters into their own hands by establishing specialist clubs. The emergence of breed clubs exposed the shifting character of the dog fancy, with the authority of the Kennel Club potentially pitched against demands from breed aficionados to be taken seriously and for their clubs to be given a formal role in canine affairs.

In 1877, Walsh editorialized in the *Field* on changes that had occurred since the 1865–66 series of articles that had first set out breed standards and been the basis for his *The Dogs of the British Islands*. This editorial was a prelude to a new series that would update all the essays for a third edition of *The Dogs of the British Islands*. Hounds were little changed, he noted, because "for two or three centuries . . . masters of foxhounds and harriers had brought them to perfection." However, among gun dogs, retrievers had replaced pointers and setters as the favorite, while the first modern dog, Major, had lost his place as the model pointer because he had proved a poor sire and deficient in field trials.[39]

As we have seen, a show-oriented, new Bulldog Club had been established in 1875 and soon moved into contentious territory when it directly challenged the Kennel Club's role by publishing its own stud book and a list of approved judges. However, its leaders sought a compromise, offering special prizes to winners at Kennel Club shows that met "the type agreed upon in the Bull Dog Club standard." This outcome set a precedent for the Kennel Club of not only accepting the standards set by specialist clubs but also agreeing to have their members as judges. The influence of the new breed clubs was further strengthened as their standards were disseminated in dog publications and newspaper features. Walsh grudgingly followed suit in the fourth edition of *The Dog in Health and Disease* (1887), commenting that he had included the descriptions of the specialist clubs, "even when I have not been able to fully agree with them."[40]

In the 1880s many more clubs were established as local and regional groups and held their own shows (see table 4.1.) The Irish Kennel Club (1878) and the Scottish Kennel Club (1881) were founded to promote and protect the breeds of each nation. The latter's aim was "to preserve the purity of the Scotch breed of dogs," but like its London counterpart, with which it had good relations and overlapping membership, it declined to set standards. Aficionados of other breeds emulated the new Bulldog Club. An early venture was the Dandie Dinmont Club, founded in the Scottish Borders, the type's home region. A breed standard was published in the early months of 1876 and started a "prolonged and acrimonious" controversy.[41] Many of the early clubs were for terriers, with breeders and owners seeking to differentiate the many varieties that had been nurtured locally and, in the new world of breeds, to give these dogs new identity and status. For example, the Newcastle and Gateshead Bedlington Terrier Club was founded in 1876. Bedlington was a small town north of Newcastle with iron works and several coalmines, with a large and vibrant working-class fancy culture. There was a long tradition of poultry and pigeon shows, to which dogs were added in the 1860s. Given their social background and geographical remoteness, Bedlington terrier aficionados struggled to have their dog accepted into the great national shows. They were ultimately successful, although the breed was seen as a new invention, as shown by correspondence in the *Field* titled "Manufacturing a Bedlington" (see plate 6).[42]

In 1879, inconsistent judging at shows caused "the pent up feelings of the Irish terrier breeders" to "burst forth," and the breeders created their own club. Walsh had refused to include this breed in *The Dogs of the British Islands* because it was a dog about which "no two seemed to agree." The members of the Scotch Terrier Club, founded in the same year, were revisionists, advancing the claim that the dogs exhibited on the English show circuit were not true Scotch terriers; they provided both a standard and the dogs to match. However, their attempts to establish the breed were not helped by changes of name, with Scottish, Scots, Highland, and Cairn proposed as alternatives, and confusion over the status of any of these vis-à-vis the Skye terrier.[43]

There were mixed views on specialist clubs. In November 1880, an editorial in the *Kennel Chronicle* argued that the "New Dog Clubs" had fostered a "rage for publishing ever-varying standards of excellence," and that "Confusion as to type rages now, and each additional set of arbitrary definitions

TABLE 4.1
*Dates of the Establishment of National Breed Clubs to 1885*

| Year of foundation | Club name |
| --- | --- |
| 1864 | Bulldog Club |
| 1873 | Mastiff Club |
| 1875 | Bulldog Club |
| | Bedlington Terrier Club |
| | Dandie Dinmont Club |
| 1877 | Fox Terrier Club |
| 1879 | Irish Terrier Club |
| | Scotch Terrier Club |
| 1881 | Collie Club |
| 1882 | Pug Club |
| | Irish Setter Club |
| | St. Bernard Club |
| | National Spaniel Club |
| | Dachshund Club |
| 1883 | Old English Mastiff Club |
| | Mastiff Club (New) |
| | Beagle Club |
| 1884 | Airedale Terrier Club |
| | Basset Hound Club |
| | Black and Tan Terrier Club |
| | Fylde Fox Terrier Club |
| 1885 | Cocker Spaniel Club |
| | Great Dane Club |
| | Irish Red Setter Club |
| | Toy Spaniel Club |

*Note:* The years cited are taken from notices and reports published in the *Field* and other publications. Rivalries meant that certain clubs were anxious to claim priority of establishment; hence, the years given are indicative rather than definitive. Also, we have not included regional breed clubs, such as the Sheffield Fox Terrier Club founded in 1885.

makes confusion worse confounded." On the other side, an editorial in the *Stock-Keeper and Fanciers' Chronicle* in April 1880 argued that breeds without a club, especially sporting breeds such as the Irish water spaniel, were in decline in quantity and quality. In October 1882, when launching the new *Kennel Review* as "The Organ of the Specialist Clubs," Harry Wyndham Carter was positive about their role. Writing in the same edition, "Hertfordshire Fancier" wanted clubs to "define accurately and by fixed sharp lines what type of animal we are trying to breed." He called for the "plainest pictures that pen and ink can produce of a typical animal," questioning the value of paintings, photographs and models. To consolidate the influence of the specialist clubs, he wanted them to find impartial judges and assure accurate registration and pedigree, but there was one reform he was against—the admission of women.[44]

The next step in the development of breed clubs was the proliferation of single breed shows. In 1883, a report in the *Graphic* set this in context: "This is an era of specialists—of specialist doctors, of specialist painters, of specialist journalists, and even of specialist dog shows." There had, of course, been spaniel shows in the London Fancy from the 1830s and foxhound shows since the 1860s, but neither were exclusively conformation shows. Modern single-breed shows began with the Bulldog Club in the mid-1870s and were generally seen as second rank, being run by and for cliques of enthusiasts, with narrowly drawn entries and fixed ideas in judging. Such comments were made about the first St. Bernard's Club show in 1882, held at the Duke of Wellington's Riding School in Kensington. It ran for three days, and John Cumming Macdona judged more than 250 dogs in twenty-two classes. Most reports questioned the quality of the entries. *Bell's Life* stated that "there was a terrible lot of rubbish," and many dogs looked to have pointer, collie and retriever features. There were similar comments about the "British" St. Bernard being crossbred or a mongrel, with its variability indicating that it was still a breed in the making, in terms of what the breed standard was and the number of dogs that conformed to it.[45]

By the early 1880s, specialist breed clubs were established in canine affairs and a growing influence in shows and the wider fancy culture. There were contrasting views on their impact. On the one hand they had popularized certain breeds "to an extent that would have seemed incredible but a few years ago," while on the other hand they tended to strive for needless novelty and disregarded "all sustained consideration of the duties in life for which their pets were originally produced." There was more comment in the sporting and fancy press about particular clubs and how they had become cliques, perhaps like the Kennel Club, and had chosen the wrong breed standard, as with the Spaniel Club in 1882–83.[46] Specialist club members claimed to be the experts on their breed, but they were said to be enthusiasts and, of course, were interested parties at shows as breeders, judges, and traders. Critics also painted them as self-appointed authorities that had acquired too much power, being able to reshape their breed physically, with a penchant for exaggerated fancy points. An unsigned commentary on the evolution of shows published in the *Stock-Keeper and Fanciers' Chronicle* mused upon the "universal change that is taking place in all breeds of show dogs." It noted that some changes were gradual, while others were happening in "leaps and bounds," with specialist clubs seeming to favor the "transmogrification of show dogs." The writer argued that most recent changes were improvements and that many of "the old dogs were all wrong"; hence, he was puzzled by a new trend—breeding to the conformation of "the old types" that were now seen to "represent the *beau ideals* of canine excellence."[47]

## Conclusion

The Kennel Club's rapprochement with the Birmingham National Dog Show Society in 1885 effectively secured its preeminent position. Its rules were increasingly adopted by dog shows at all levels, and having one's dog recorded in the Kennel Club's stud book became essential for every serious breeder and exhibitor. Its role with field trials was equally secure, though that had not been much challenged. The club was not a parliament, despite the admission of two Birmingham representatives; it was a self-appointed governing body and court of appeal. Its membership was drawn from across the country, but its leadership came from a narrow stratum of those with the wealth, land, or time to devote to canine affairs. In London, it had created a new geographical site of elite sociality. Gentlemen and ladies, the latter admitted as "affiliates" in 1883, could meet in polite rivalries and establish networks of exchange in trading and breeding their animals. At shows, gentlemanly behavior had been institutionalized, and if it had not proved possible to keep out "the lower stratum" entirely, at least they were subject to sanctions. Well-ordered and orderly events ensured that dogs and exhibitors were able to perform and be judged fairly, and that shows were respectable sites of entertainment and education.

By this time, the dog show fancy had spread across the country and had begun to influence how all dogs were seen. Daily and weekly papers carried stories of the antics of particular breeds and their owners, with poodles the favorite butt of jokes.[48] Also, in news stories about dogs, such as mad-dog chases and cruelty cases, it was increasingly common for the breed of dogs involved to be named.[49] In a mad-dog incident in Bradford in May 1886, four hundred savage strays were "'run in' and destroyed," but in the reports of specific dog-bite victims a white Pomeranian and fox terrier were identified.[50] Nonetheless, breeders and exhibitors in the new fancy saw a growing gap between their "improved" dogs and the rest. Hunting hounds and gun dogs were becoming distinct from their "show dog" peers, but there was hope that among the wider society those acquiring house dogs and pets would follow the fancy and seek particular breeds. There were now a variety of sizes, colors, and forms to suit all tastes.

The first rule of the Kennel Club was "in every way to promote the general improvement of dogs, dog shows, and dog trials." As we have demonstrated, in its first decade the club's leadership was preoccupied with governance: did this mean it neglected "the general improvement of dogs"? Walsh and other critics argued that it had, with the most serious failure being to set and promote breed standards.[51] A critical article in the *Field* in January 1888 claimed that breed clubs "appear to be usurping some portion of the work the parent club originally possessed."[52] The issues were familiar—who should define breed standards and whether these would hinder "improvement" by fixing current ideals. Past struggles, the continuing challenges of regulating shows, and its liberal philosophical antipathy to top-down decrees, meant that the Kennel Club remained happy to concede the setting of standards to specialist breed clubs and to their validation in competition. Improvement had become an internal affair for the new fancy, led by breed clubs. All the "good" specimens of particular types of dog abroad in the country had been found and inducted into fancy culture; high-class dogs with coveted qualities and pure blood were no longer at large. Instead, aficionados looked for "improvement" from within the ranks of their own dogs, by breeding with and for the best. To guide their endeavors, they relied first and foremost on the experience of practical breeders, and then from the 1880s on, they were encouraged to draw on science.

# CHAPTER FIVE

# Improving Breed I: Experience

At the heart of the new fancy and its dog shows was a paradox: breeders, exhibitors, and judges assumed that there was an ideal, standard conformation for each breed, yet they were all seeking "improvement," which meant changing the standard and the bodies of dogs. The goal of improvement was not just to have more dogs reach the ideal but to achieve greater uniformity and conformity across each breed, with less "mongrelism."[1] In the 1860s, exhibitors had largely sought improvement by finding good specimens. However, by the late 1870s, exhibitors had adopted the new approach of breeding with their own dogs, which necessitated having a kennel of top dogs, with pure blood, and making judicious mate selections. A small but growing number of breeders made a living from producing dogs for exhibitors and for sale as pets. Self-consciously distinct from the "dog dealers" of earlier decades, the new "genuine-breeders" supplied a market that wanted quality products in terms of conformation and blood. Competition at exhibitions and the market in stud fees and sales were the main drivers of breed differentiation. As well as splitting existing breeds, there was new interest in reviving old and extinct breeds, and in importing foreign dogs.

Breeding show dogs was based on ideas shared across the communities producing Thoroughbred horses, all types of livestock, and fancy animals. There were two main strategies breeders used to create and stabilize new forms: (1) selecting desirable variations and inbreeding to "fix" these in the blood, and (2) outcrossing or crossbreeding with other families, strains, and breeds to acquire desired features or to counter degeneration from excessive inbreeding by bringing new blood. The new approaches posed the question of how the value of a dog was to be assessed. Was value based on a dog's past, as revealed in its pedigree; or on its present, as evidenced in its conformation and performances at shows; or on its future, as promised by its prepotency at stud? The three approaches were, of course, linked, as pure blood, confirmed by pedigree and demonstrated by success at shows, was assumed to reproduce type most effectively.

Fanciers sought to change breeds by drawing upon their experience and that of the breeders of other domesticated animals. They argued that more varieties were being stabilized, standardized, and perfected—internally with purer blood and externally with an ideal conformation. Critics, however, lamented that most "improved" dogs were nonsporting and toys, and that fancying had become too fancy, with exaggerated and unnatural points, and the unnecessary proliferation of breeds. The standardization of all breeds, old and new, had to be negotiated. With some breeds consensus was quickly reached, but with others agreements took longer, and in some cases views were irreconcilable, leading to the splitting of a breed into two (and sometimes more) types. We illustrate these issues with discussions of Yorkshire terriers' differentiation from other terriers, bulldogs' return to their supposed fighting form, and Irish wolfhounds' "resuscitation" from extinction. These endeavors drew upon contemporary ideas on the "art" of breeding, which we consider next. Breeders engaged in the making and remaking of breeds used selective breeding, inbreeding, and crossbreeding. They also had to consider nonhereditary influences, especially "maternal impressions" and telegony—the notion that some "influence" from the father of the firstborn shows in subsequent offspring. This chapter demonstrates how the controversies that plagued dog shows on how to judge the top dog were mirrored in the energetic debates on the authenticity of breed identities and how to breed.

## The Proliferation of Breeds

The acceptance of foreign varieties was one reason why the number of breeds in the Kennel Club stud book rose from forty in 1874, to fifty a decade later, and to seventy-six by the turn of the century.[2] There were tacit criteria for the acceptance of a breed into the stud book: a sufficient number of dogs for regular appearances at shows, good pedigree records, and a breed standard. Most breed clubs claimed history and heritage for their dogs, but the leading authorities who wrote on canine affairs agreed that modern breeds were recent creations, a fact implicitly endorsed by stud book compilers, who accepted few pedigrees that went back before the 1860s. Imported dogs, such as Dachshunds, St. Bernards, Pomeranians, Newfoundlands, Maltese, and Italian greyhounds, were sufficiently longstanding immigrants to have been in the first stud book in 1874. By the early twentieth century, Great Danes, basset hounds, Borzois, Griffon Bruxellois, Poodles, Japanese, Pekingese, Chow Chows, schipperkes, and six others had joined the "Foreign Dogs" category.[3] However, the principal reason for the increase in the total number of recorded breeds was the splitting of existing British breeds. The Kennel Club was not proactive in this proliferation; rather, it reacted to the lobbying of breed clubs and followed the classes allowed by show committees across the country. The greater differentiation of British breeds was evident with the recognition of places (e.g., Welsh spaniels), colors (red and white spaniels), coats (flat- and curly-coated retrievers, rough and smooth collies), and other markers (prick- and drop-eared Skye terriers). In 1874 there were three breeds of spaniel; a quarter of a century later, there were ten. The overall number of terriers, in both the sporting and nonsporting divisions, rose from eleven to fifteen, including the newly accepted Airedale, Clydesdale, and Yorkshire breeds.[4] The designation of all these terriers in one sense recognized the variations of the pre-breed era, but there was a crucial difference: then, the variants formed a continuous series; in the post-breed era, the variants were separated and segmented.

In one sense, all breeds were new, owing to the novelty of the concept and its material productions. Considerable rhetorical effort was required to create breed standards, and it took a lot of work to produce conforming, uniform, and discrete breed populations. Some breeds were regarded as genuinely new and without history; these were often called "manufactures."[5] Novelty was sometimes an advantage. Vero Shaw argued that many established breeds had "been the subject of acrimonious and narrow-minded disputes and petty quarrels," whereas new breeds could be judged simply on their merits.[6] The accepted story of the Yorkshire terrier, for instance, was that Scottish weavers had brought their ratting dogs with them to the textile mills of Lancashire and the West Riding in the early nineteenth century and remade their farm dogs into ones suited to pest control in factories, urban pet keeping, and fancy shows.[7] Through selection and crosses with other terrier types, a variety with a long smooth coat, first known as the Halifax blue-tan terrier or Yorkshire blue-tan silky-coated terrier, was produced (fig. 5.1).[8]

Writing playfully about the breed's origins in the wool industry, Hugh Dalziel suggested that "his distinctive character is in his coat—well carded, soft, and long as it is, and beautifully tinted with 'cunning Huddersfield dyes,' and free from even a suspicion of 'shoddy.'" Yet, he felt that the dog's beauty was superficial, it being "a dandy" that "would look but a poor scarecrow in

"DUNDREARY," Blue Tan Terrier, the Property of Mrs. Foster, Bradford.

*Figure 5.1.* "Dundreary," Blue Tan Terrier, the Property of Mrs. Foster, Bradford. Stonehenge, *The Dog in Health and Disease,* 3rd ed. (London: Longman, Green and Co., 1879), 124

dishabille, and, one is apt to leave him with the scarcely concealed contempt for a scion of the 'Veneering family,' who in aping the aristocrat fails, as all parvenues do." Nonetheless, the new breed's amiable character, soft coat, and toy size made it a popular pet. The breed was a favorite of ladies who entered their dogs at shows, pointing to the increasing role that women, who typically favored toys, were playing in fancy culture.[9]

The acceptance of new breeds was, however, never straightforward. Dalziel was vexed about the differentiation of terriers and the confusion between the Scotch breeds and the new Yorkshire dog. His particular theory on the origin of the Yorkshire terrier was that "the dog was what gardeners call 'a sport' from some lucky combination of one of the Scotch terriers, either the genuine Skye or the Paisley toy, and one of the old soft and longish coated black and tan English terriers, at one time common enough, and probably one with a dash of Maltese blood in it."[10] Most variation between generations of animals was minor, and it took a number of breeding cycles to achieve a significant alteration in a feature. Occasionally, so-called sports occurred, where the young were very different from their parents, which gave breeders the opportunity for rapid change.

Improvement and novelty was not only pursued through breeding. With Yorkshire terriers there was agreement that selection and crossing had been supplemented by "warm water and brush . . . with the aid of elbow-grease, glycerine and a little dye" to improve condition and produce a beautiful animal. Dalziel explained, "Artificial means are used to encourage and stimulate the growth of the hair. The hind feet are kept encased in chamois leather boots, so that, even should they scratch, the claws being covered, the coat is neither broken nor pulled out, and the diet is carefully regulated so as to obviate heat of the blood and skin disease." The tail was docked in pups, as it was feared "some of them would reveal their Maltese origin by carrying the caudal appendage tightly over the hips." Breeders argued that in some breeds ears and tails had become adapted to cutting, either by selection or acquiring suitable tissues generation by generation. Breeders and many scientists continued to accept that acquired characteristics could then be inherited, mediated through "blood" that had circulated in the parents and thus was the agent of hereditary.[11]

## Recovery

The new fancy looked to the past as well as the present for novelty. They sought to return established breeds, like the bulldog, to their alleged original form, and they also hoped to recover extinct varieties, such as the Irish wolfhound. Once breeders had decided the weight issue with the bulldog, they turned their attention to its form in order to address claims that the modern dog looked quite different from that which had fought with bulls before 1835. Aficionados were divided on whether the modern dog actually needed to be improved and, if it did, whether to look back rather than forward. If they did try to return it to its old form, who would know what to breed for? In his chapter in Dalziel's *British Dogs* in 1879, Bulldog Club secretary F. G. W. Crafer claimed special insight could be gleaned from men who remembered bullbaiting: "There still remain true representatives of the original bulldog for the use of those breeders who wish to preserve the correct type of the pure, old-fashioned dog, and who are wise enough to decline to be misled by false pedigrees and specious arguments into breeding from novel-shaped parents under pretence of improving the breed and restoring it to what it is alleged to have been before bull-baiting became a separate

BULLDOG.

Capt. G. E. A. Holdworth's SIR ANTHONY.  Sire, Crib, by Duke II., out of Rush ; Dam, Meg, by Old King Dick, out of Old Nell.

*Figure 5.2.* Capt. G. E. A. Holdsworth's bulldog Sir Anthony. H. Dalziel, *British Dogs; Their Varieties, History and Characteristics* (London: "The Bazaar" Office, 1879), opp. p. 218

sport."[12] One such "old-fashioned" bulldog was said to be Capt. G. E. A. Holdsworth's Sir Anthony, but unfortunately he had a short show and stud career due to a fatal accident (fig. 5.2).

With bulldogs, the old London Fancy remained active, exhibiting at the new shows and providing stud dogs, and they had their own ideas on conformation and how to make improvements. They used selective breeding but were also alleged to be changing the head by mechanical and surgical remodeling. In *The Illustrated Book of the Dog* in 1881, Shaw did not spare his readers the gory details of the operations performed on puppies, which included compressing the nose and applying "Jacks," "causing the mutilated parts to remain in their new and abnormal position." The Bulldog Club was expected to step in and rescue the breed from "its more mischievous and inventive patrons." Shaw was optimistic: "our national dog [is] slowly, though surely, emerging from the hands of the residuum of the canine world, and taking its proper place in the kennels of a superior class of breeders and exhibitors."[13]

It was not just the bulldog's conformation that had changed; equally worrying was the alleged loss of courage and aggression. While breeding for conformation had become dominant, there continued to be a demand for sporting dogs with appropriate instincts and character, and bulldogs were still used in crosses to introduce aggression and courage. However, some aficionados felt that character had been lost as the breed became more fragile due to ill treatment and the misguided ideas of previous generations. George Roper, a leading figure in the Bulldog Club, wrote that "of all breeds none are more liable to deterioration than the Bull-dog. In a litter you seldom find more than one specimen up to the mark when arrived at maturity." His answer was "judicious in-breeding" to secure improvement, affirming that it was a "much wiser plan to breed from reliable and good blood than to admit questionable blood into your strain."[14]

The breed remained controversial to the end of the Victorian period. In January 1890, the *Kennel Gazette* introduced a feature, "The Past Year," in which well-known figures reflected on the quality of each breed in recent shows. The first report on bulldogs was unflattering, maintaining that many dogs at shows had frogginess, narrow under jaws, button ears, and houndlike sterns carried like a barber's pole. Opinions continued to differ on features such as the jaw, nostrils, length of back and legs, and "what degree of 'out at elbows' is the correct thing."[15] In one sense the desired conformation came from the imagined calisthenics of bullbaiting. The softest flesh on a bull for a dog to latch on to was said to be under the jaw, hence, the dog's protruding lower jaw. Wearing down the bull took considerable time and effort; a flat face and flush nostrils facilitated the dog's breathing.

There were a growing number of regional bulldog clubs, each with its own idea on conformation. London claimed to be the home of the breed, such that when the Manchester-based British Bulldog Club applied to affiliate with the Kennel Club in 1892, the Bulldog Club and South London Bulldog Society petitioned against it.[16] Another problem facing breeders was that in recent years many champion stud dogs had been lost, either dying or exported to America, where wealthy fanciers were building up kennels. With fewer good dogs to breed from, inbreeding increased.

In his review in January 1894, Edgar Farman, who was then vice-president of the Bulldog Club and had published a monograph on the breed, dwelt on character not conformation, stating that "many are soft-hearted and the listless way in which they stalk into the ring exhibits an utter absence of that quality for which the breed has so much credit." He blamed the prevalence of "milk-and-water animals" on inbreeding and the emphasis on external appearance since its vocation had been banned. However, by the end of the 1890s, there was greater optimism among aficionados, with progress on standards, numbers of dogs, and public acceptance. The only threat was the purchasing power of "the glittering dollar": "Are we breeding and spending . . . time and pains in rearing for love of the breed or for money?"[17]

The most controversial recovered breed was the Irish wolfhound. It was disputed whether the enterprise was legitimate, and some asked whether there was pure blood in any of the dogs produced. These debates all took place in a politically charged context of nationalism and Anglo-Irish enmity. For decades the Irish wolfhound had been written about as extinct, having disappeared when hunting had eliminated the wolf from Ireland in the eighteenth century. However, in the 1860s a group of enthusiasts set about its recovery, led by Capt. George Augustus Graham, a retired Indian Army Officer living in Gloucestershire. Graham, who kept greyhounds and Scottish deerhounds, was sold a dog, Faust, that his breeder in Cheshire alleged had Irish wolfhound blood. Graham was intrigued. With his friend Maj. John Garnier, and following research in books and journals, he became convinced that pure bloodlines remained in Ireland and that from these the breed could be recovered "in its pristine grandeur." He then set out to find hounds with the right blood, traveling around Ireland to buy several dogs whose owners claimed had the blood of true Irish wolfhounds.[18]

Graham's venture was speculative for several reasons. First, he had to trust the owners' claim that there was continuity of inheritance over nearly a century in dogs that often bore no resemblance to the types described in writings or represented in paintings and drawings. The size of the dogs meant that they had been mainly kept on the larger estates; hence, it was likely that a gentleman's word could be trusted. Second, there were great differences in the dog's representations in paintings and descriptions in literature. The main divergence of opinion was whether the Irish wolfhound had been greyhound-like, with the speed to catch a wolf, or mastiff size, with the strength to bring down a wolf. Working with the essentialist notion of breed, Graham was seeking the one true type and was unable to conceive that there had been variation among the dogs once styled Irish wolfhounds. The only agreement in past accounts was about height, but some claims seemed fantastical. The eighteenth-century Irish author Oliver Goldsmith, for instance, wrote that they were forty-eight inches tall at the shoulder, while the French naturalist Comte de Buffon had them as sixty inches tall when seated.[19]

Graham's initial attempts to breed from the dogs he acquired in Ireland had mixed results. Two dogs were "eunuchs" and one bitch a "non-breeder."[20]

The puppies of successful matings were prone to "death and disease," mostly due to distemper, and those that lived were not the desired article, having "plenty of bone, but ungainly and ill-shaped." Graham seemingly had more success when he crossbred his Irish dogs with his largest Scottish deerhounds, believing that this was legitimate, as the breeds were related. Indeed, there had been speculation that the Scottish deerhound was a descendant of the Irish wolfhound and that hence there was common blood.

By 1879, the same year that crop failures in Ireland led to the formation of the Irish National Land League, Graham was ready to exhibit at the Irish Kennel Club Show in Dublin. Controversy broke out when the Irish wolfhound class replaced that for Scottish deerhounds. Graham was behind the change, which he justified by arguing that the modern Scottish deerhound was "a representative of the old Irish Wolf-dog," though "of less stature, less robust, and of slimmer form" than its Irish ancestor. Many breeders were sensitive about the "nationality" of their dogs and nowhere more so than in Ireland at this time of growing nationalist fervor. Scottish deerhound enthusiasts were adamant that their breed was distinct and not "a degenerate descendant" of its Irish cousin. In the Irish wolfhound class at the Dublin Show, Graham entered his—perhaps provocatively named—dog Scot and was awarded third place (fig. 5.3). In his chapter on the breed in Dalziel's *British Dogs,* Graham grudgingly accepted the judgment. He wrote that while his dog had "more authentic Irish wolfhound blood in him than anything shown, and [was], in shape and style, correct," Scot was "wanting in coat, and, what is more important, size and substance, for he was small almost to weediness."[21]

The reporter in *Freeman's Journal,* Dublin's oldest nationalist newspaper, recognized only one dog as the true breed.

> These are not the dogs that Pliny tells us (if we can believe Pliny) were singly a match for a lion or an elephant. The only specimen of Irish wolf hound that Buffon ever saw, a snow white dog, sat five feet upon his haunches, and was almost as tall as two of the dogs of our degenerate days. The term "Irish Wolf Hound" applied to many specimens exhibited at the Exhibition seem only to be a euphemism for mongrel, but Mr Percy H. Cooper's Brian, to whom the first prize was deservedly accorded, very fairly embodies our conception of the race and might have his portrait painted as a "national emblem," with the harp, the "sunburst" and the "full length figure of Erin."[22]

There was then scepticism in Ireland about Graham's enterprise. He was, after all, an English gentleman of Scottish descent, living in Gloucestershire.

**Figure 5.3.** Capt. G. A. Graham's Irish wolfhound "Scot." H. Dalziel, *British Dogs; Their Varieties, History and* Characteristics (London: "The Bazaar" Office, 1879), opp. p. 76

Why look beyond Ireland for the country's iconic canine? The Irish wolfhound had already been adopted as a nationalist symbol, having been put, for example, on the gravestone of Stephen O'Donohoe, a nationalist who lost his life in an attack on a police barracks in Tallaght in 1867.[23]

In the same year as the Dublin show, Graham published a pamphlet on his research and breeding. His text was similar to that in Dalziel's volume, discussing the written and visual sources on the dog and the many variations in reports of the hound's size and type. In his final judgment on the preferred form, he relied most heavily on Richardson's *The Dog: Its Origin, Natural History and Varieties,* published in Dublin in 1845. Graham was persuaded that the defining characteristics of the Irish wolfhound had been size, power, and especially height: he made much of achieving a shoulder height of over thirty-five inches. However, he was clear that his aim was bolder—nothing less than "to recover the old Irish Wolf-dog rather than simply to breed a dog to give a good account of a wolf." He defended his work by comparing it to the recent improvement of other breeds.[24]

Why should not, then, such measures be taken to recover the more ancient, and certainly equally noble, race of Irish wolfhounds? It may be argued that, the services of such a dog no longer being required for sport, his existence is no longer to be desired; but such an argument is not worthy of consideration for a moment, for how many thousands of dogs are bred for which no work is provided, nor is any expected of them, added to which, the breed would be admirably suited to the requirements of our colonies. One after another the various breeds of dogs which had of late years more or less degenerated, as, for instance, mastiffs, fox terriers, pugs, St. Bernards, colleys, have become "the rage," and, in consequence, a vast improvement is observable in the numerous specimens shown from time to time. Let us, then, hope that steps maybe taken to restore to us such a magnificent animal as the Irish wolfhound.[25]

To guide his breeding, Graham had a model made, and, similarly, a German breeder commissioned a painting of his ideal, which was described as "a portrait of an Irish Wolfhound of 35 inches, life-size, of a grey colour, and it presents to the vision a most striking and remarkable animal of a very majestic and beautiful appearance, far, far beyond any dog the writer has ever seen in grandeur of looks."[26] Although Graham's model has not survived, there is an extant photograph of Graham with a scale drawing of his ideal stuck on (see plate 7).

While Graham was confident in his ideal and the possibility of recovery, there were many doubters. A particularly animated debate took place in the columns of the *Live Stock Journal and Fancier's Gazette* in 1879 and 1880. The mastiff aficionado Rev. Malcolm B. Wynn had his say, stipulating that the ancient type had been a heavier wolfdog, not a lithe wolfhound. G. W. Hickman, a Birmingham breeder of deerhounds, argued that "as such an animal is now extinct, any attempt to revive it will simply be a manufacture more or less conjectural." He was certain that the old Irish dog was a Great Dane type and that Graham's dogs were creatures of "inference, supposition and conjecture." He concluded, "That a gigantic rough-coated dog of the deerhound type may be produced by judicious breeding I do not doubt, but it must be by a still further large addition of foreign blood." Walsh had written in the third edition of *The Dog* in 1879 that what he termed "resuscitation" was "a most absurd idea" and Graham's dogs could only be "a modern breed, to which any other name may be given except the one chosen for it." Later he styled Graham's Irish wolfhound a "manufactured modern copy," doubting that there was any survival of "the old strain." In reply, Graham asserted that his dogs had "more or less true and authentic blood," adding, "I hardly think it any more manufactured than many other breeds that are now looked upon as 'pure.'" In this dispute, blood and conformation had equivalence in defining the authenticity of the breed.[27]

An Irish Wolfhound Club was founded in 1885 and flourished with Graham at the helm, demonstrating the power of enthusiasts to reinvent tradition and its material culture. However, improvement was slow and uneven. In the *Kennel Gazette*'s annual review of breeds in 1891, Graham reported that "comparatively little progress has been made in the fuller development of this noble breed" and that "the efforts of breeders, sad to say, have been futile this year," with "disasters and disheartening failures." However, at the population level, "the breed has 'come on' in a very decided manner." Graham assured his readers that "the type arrived at is much more uniform, and the size much more nearly approaching what it should be." He bemoaned the lack of progress again the following year, pointing to issues with both size and coat in individuals, and evenness across the breed. By 1897, he was more sanguine: "During the last three or four years great strides have taken place, both character and size becoming each year more fully established, as well as much greater unanimity of type. Indeed, the advance has been so decided and satisfactory that there is every reason to hope that we may see a race of Irish Wolfhounds, averaging near 33 inches (beyond which it is hardly likely we shall go), that shall combine all the desired characteristics, within the next few years, if breeders will exercise a careful discrimination in their proceedings." Graham's work exemplified how seeking improved breeds was also about producing a uniform population to the conformation standard, not just ideal individual dogs.[28]

The 1899 *Kennel Gazette* review of the breed acknowledged—and not before time in the view of Graham's critics—that an important factor in the recovery of the Irish wolfhound had been the introduction of "alien blood" from other breeds. Allegedly, the resuscitated Irish wolfhound had benefited from outcrosses with Great Danes to meet the standard's requirement on size, and with the Tibetan mastiff and the Russian borzoi to "improve," respectively, coat (long hair) and physique (greyhound type).[29] For most breeders this outcrossing was acceptable and to be expected as particular points were sought. They were after all only working toward a physical standard for show dogs; the dogs were never expected to hunt wolves, though Graham clung on to the claim that his dogs had authentic blood and might be able to work in the field. How the modern Irish wolfhound had been created in recent decades perhaps no longer mattered; in the future the influence of outcrosses was "being gradually absorbed, and few traces are now visible, but the good effects of its introduction remain in the increased size and stamina of the hounds of the present day."[30] It seems that purity of blood could be generated in relatively few generations, by inbreeding within a limited group of excellent dogs.

## Inbreeding and Crossbreeding

The foundation for improving breeds and the nation's canines overall was first to produce ideal individuals. To assist the new generation of breeders and fanciers, the principles of breeding were regularly set out in all forms of literature on the dog. Such advice distilled the accumulated wisdom of livestock, horse, poultry, pigeon, and other breeders going back to the late eighteenth century, though inflected with new ideas from natural history, evolutionary debates, and physiological science. The chapter on breeding in Walsh's *The Dog in Health and Disease* remained unaltered from the first edition in 1859 to the fourth in 1879.[31] In it he set out six axioms.

1. The male and female each furnish their quota towards the original germ of the offspring; but the female over and above this nourishes it till it is born, and, consequently, may be supposed to have more influence upon its formation than the male.
2. Natural conformation is transmitted by both parents as a general law, and likewise any acquired or accidental variation. It may therefore be said that, on both sides, "like produces like."
3. In proportion to the purity of the breed will it be transmitted unchanged to the offspring. Thus a greyhound bitch of pure blood put to a mongrel will produce puppies more nearly resembling her shape than that of the father.
4. Breeding in-and-in is not injurious to the dog, as may be proved both from theory and practice; indeed it appears, on the contrary, to be very advantageous in many well-marked instances of the greyhound, which have of late years appeared in public.
5. As every dog is a compound animal, made up of a sire and dam, and also their sires and dams, &c, so, unless there is much breeding in-and-in, it may be said that it is impossible to foretell with absolute certainty what particular result will be elicited.
6. The first impregnation appears to produce some effect upon the next and subsequent ones. It is therefore necessary to take care that the effect of the cross in question is not neutralised by a prior and bad impregnation. This fact has been so fully established by Sir John Sebright and others that it is needless to go into its proofs.[32]

Walsh was clear that the bitch was most important and valuable to the breeder, not only because she carried and suckled her progeny, but economically, as she "usually continues to be the property of the breeder, while the sire can be changed each time she breeds."[33] However, there was a sexual asymmetry that worked the other way and made males more valuable. The prepotency of a male could be assessed speedily—indeed, in a single year—while with a female it took many years. For example, a single male dog could be mated with ten females in a season, and the results seen and compared that very year. It would take a minimum of five years to assess a female with ten males, and comparisons made years apart would be difficult.[34]

It was not just parents that needed to be assessed, grandsires and granddams were also known to have "influence." Indeed, breeders had to look back many generations. It was commonly reported that features of the seventh generation back on the dam's side would often appear in puppies. Breeders called this phenomenon *throwing back,* while biologists referred to it as *atavism*. These terms were given new and decidedly pejorative meanings in late nineteenth-century debates on the social meanings of Darwinism. *Atavism* was widely used by social and cultural commentators, especially for individuals who were said to be "throw backs" or "reversions" to an earlier stage of human evolution. Cesare Lombroso, the Italian criminologist, gained a following after he speculated that criminals were people who had regressed to an earlier, less disciplined stage of human evolution. Moreover, as their "wild" character was of longer standing and hence more fixed, they were more likely to pass their traits on to their children, potentially creating a growing race of criminals.[35]

The value of breeding in and in lay in concentrating blood to give to purebred dogs, which breeders believed had the greatest prepotency. Such dogs and bitches earned for their owners higher stud fees and sales income, especially when backed by a trustworthy pedigree. On the concentration of blood by in-breeding, Walsh was equivocal: "Like many other practices essentially good, in-breeding has been grossly abused; owners of a good kennel having become bigoted to their own strain, and, from keeping to it exclusively, having at length reduced their dogs to a state of idiotcy [*sic*] and delicacy of constitution which has rendered them quite useless." Thus, he recommended crossbreeding when dogs became too "high bred," nervy, and weak. There had been notable successes, as in the recent use of greyhound crosses to lighten the "heavy form of the bulldog." Outcrossing was also invaluable for acquiring behavioral features in sporting dogs, such as drawing selectively on "speed . . . in the grey-

hound, courage in the bulldog, and nose or scenting power in the bloodhound."
Features could be transferred across breeds—for example, the finding and
questing powers of terriers might be given to pointers and setters, or the "sa-
gacity" of the poodle, Newfoundland, and terrier to a more stupid breed.[36]

In their respective books, Dalziel and Shaw also gave advice on the principles of breeding. Dalziel drew the familiar distinction between the man
"breeding for sale" and the "genuine breeder." The latter not only had the best
motives but also used "science": "[He] takes hold of a variety and says, this
dog would be improved for purposes of utility and beauty, by the breeding out
or modifying certain points he exhibits strongly, and the development of
others of which he is deficient and who with this as his *primary* object sets
about the work on certain intelligible and accepted lines, which, however,
does not necessarily preclude experiment which reason, stimulated by observation, may suggest and approve." The other "altogether less worthy, and
sometimes even unscrupulous, class" of men bred, "with no reference to their
fitness to mate, and with no other object than to sell their produce at the
highest possible price. To select the good and put down the useless is never
dreamt of. The weedy and the ricketty, if they can boast of prize winning relatives, will bring so many pounds from some foolish person or another, and
so the dealing breeder does his best to degenerate whatever breed he takes
in hand." In this context, pedigrees with "prize winning relatives" were
signs of superficial celebrity, not purity of blood. Dalziel concluded, "It is
hopeless to reform these mercenaries; but as I wish this book to be really
serviceable, I warn the tyro, and all who desire to possess good dogs, to be-
ware of a class that is so widespread."[37] Again, the new fancy's perennial
anxiety about the commercial "dog dealer" resurfaced, with concern now
focused on the effect of profit seekers undermining bloodlines.

Walsh followed most breeders in believing that inheritance involved both
the blending of characteristics and dominance. "There is a remarkable fact . . .
which is that there is a tendency in the produce to a separation between the
different strains of which it is produced, so that a puppy composed of four
equal proportions of breed represented by A, B, C, and D, will not represent
them all in equal proportions, but will resemble one much more than the
others." Complexity and capriciousness were also evident "in relation to the
next step backwards, when there are eight progenitors." All breeders
were aware that luck played a large part in achieving the desired form, color,
and size of dog. In all editions of *The Dog in Health and Disease*, Walsh told
his readers, "As every dog is a compound animal, made up of a sire and a

dam, and also their sires and dams, &c., so, unless there is much in-and-in breeding, it may be said that it is impossible to foretell with absolute certainty what particular result will be elicited."[38]

Dog breeders had advantages over the breeders of livestock and horses, however. First, bitches could produce two litters per year, each with many puppies, which increased the chances of producing the desired conformation. Second, some conformation features, such as color, could be observed at birth or when the pups were very young, whereas livestock breeders seeking good meat or Thoroughbred horse breeders wanting speed might have to wait years to assess results. Third, dog breeders could save time and money by disposing quickly of puppies with unwanted traits by selling them or putting them down.

Vero Shaw, seemingly impressed by Graham's work "recovering" the Irish wolfhound, quoted at length his principles on breeding detailed in his *Illustrated Book of the Dog* (1881). Graham set out the rules that he had used in his enterprise, which were taken from his friend Major Garnier. They are interesting because they reveal assumptions about the contributions of each parent and sex-specific linkages.

1. Quality is very much more dependent on the dam than on the sire.
2. Bone or size, on the contrary, is far more dependent on the sire.
3. Colour is almost wholly on the sire.
4. The coat is almost wholly independent of the sire.
5. Muscular development and general form is dependent on the dam.
6. All these are modified by the fact that the pure bred will (other things being the same) influence the progeny more than the other.[39]

The only elaboration he made of his schema was with respect to his first point on "quality," which he described as having to do with "'blood,' nervous development, vigour, energy, and character." In the book's chapter on breeding, Shaw discussed the possibility of other maternal influences or antecedent impression, that is, the physiological and nervous effects of the offspring's connection with the mother in the womb and through feeding.[40]

In the last quarter of the nineteenth century, breeders wanting insights into "blood" were referred to a pamphlet published in 1874 by William Tegetmeier, the doyen of fancy poultry breeders, and Dr. William Whytehead Boulton, a general practitioner from Beverley in Yorkshire, who bred cocker spaniels.[41] Boulton was famed for having produced a kennel of jet-black spaniels that bred true to color "generation after generation." [42] Experience with black spaniels seems to have been particularly instructive, as another widely quoted expert on breeding was Thomas Jacobs from Newton Abbot, also known for his black spaniels.[43] Black spaniels had become popular at dog shows, but their dark coat meant they were hard to spot from a distance and not much used in the field. As a creation of shows, their status was much debated: Were they sporting or nonsporting? Sometimes they were considered a strain, which by the 1880s meant a sub-breed, though at other times they were termed "a mongrel." Jacobs agreed: "Much has been written and said on the purity of the breed, deprecating the means I have adopted to produce them as calculated to alter a presumed type, and frequent missiles have been hurled at me and my dogs from behind the hedge. But where is the pure-bred Black Spaniel so much talked about? Proof of the existence of the pure-bred one (if ever there was one) has not been forthcoming; like most other sporting dogs, they are the result of different crosses."[44] He went on to argue that, despite the modern fetish for purity, crossbreeding had been and was still essential to improving breeds.

> We may keep to one strain for many years, and, in time, call them a distinct breed, but what is the result? To preserve that strain we must continually breed in and in to one family, until we get them difficult to rear, weedy, and devoid of sense, when they become useless for the purpose they are required. Therefore breeders have to resort to the crossing with another family, which may be of a different type or colour; by doing so you raise a great "hubbub" and cry that your dogs are not pure. In spite of these cries I followed my own dictation; my great aim was to improve the breed of Spaniels.[45]

His use of "resort" confirms that while inbreeding was the norm, outcrossing was accepted to save a strain. Jacobs was not modest—he claimed that his spaniels "eclipse everything I have yet seen"—and yet he was not satisfied. He had further improvement in mind toward an imagined ideal, "the standard I have marked out for my beacon."

Jacob's views epitomized the belief that experience and common sense were more important than theory. He claimed the tacit knowledge of genuine breeders with a good eye allowed them to see beyond external appearances to the interior and the blood of a dog. Indeed, they were more likely to trust their eye than a paper pedigree. Breeders had to anticipate how a dog or bitch would "nick," the term used for best matching parents to produce puppies with the desired features.[46] For example, mating a champion dog and champion bitch might produce offspring that were "high bred," whose pure blood was over-concentrated. Thus, it was often better to mate a champion with ideal "symmetry," with a less exalted bitch with more "substance" in their physique and character. Also, breeders' wisdom was that one generation was too soon to judge, as particular inherited features might be dormant and "throw back" in later generations.

In September 1881, "Caractacus" published his "Observations on Breeding" in the *Stock-Keeper and Fanciers' Chronicle.* His main topic was the "diversity of opinion" on what he called "consanguineous, or kin-breeding"—in other words, inbreeding. For some, there were degrees of inbreeding; in-and-in breeding was with very close relatives, and linebreeding was between more distant relatives in the same bloodline. Caractacus referred to historical figures: against in-and-in was the renowned "scientific breeder" Sir John Sebright, and in favor, Meynell and Bakewell. More recently, Charles Darwin was against; a chapter in his book *The Variation of Animals and Plants under Domestication* was titled "On the good effects of crossing, and on the evil effects of close interbreeding." In favor was Edward Laverack, the leading breeder of English setters, who argued that it was generally safe to breed between closely related dogs, if they were purebred. Caractacus himself found merit on both sides, but if forced to choose, he favored inbreeding. However, he qualified this by observing that breeding was an "art," not a science, as it was "surrounded by almost as many problems as there are changes in the kaleidoscope, and the field for the exercise of discretion is so very extensive that it would be futile to attempt to meet them with fixed rules."[47]

The following week, leading veterinarian John Woodroffe Hill attacked Caractacus's endorsement of inbreeding: "Only one answer is needed to the advocates of in-and-in breeding and, although comparative, it is true. Inspect our lunatic asylums, trace the marriage relations of parents of the scrofulous, deficiently functioned, or intellectualled [*sic*] human beings, and invariably it will be found in-and-in breeding has occurred." As an example of the benefits of outcrossing, he observed that "blood other than blue has strengthened our aristocracy." Hill wrote that in his London canine practice he had seen "diminutive black-and-tan toys, destitute of hair on the ears and skull, the latter unsightly large, the eyes painfully prominent and expressionless, the body deficient in symmetry, and the limbs distorted." Perversely, some owners considered these indications of "purity of strain"; he was clear that they had been "bred to death" and that such "insults to nature" would become barren and soon die out. A vigorous correspondence followed, which became increasingly personal, with Hill's professional competence challenged. The views expressed revealed tensions between the theories and principles of medical and veterinary expertise, and the experience and tacit knowledge of practical men.[48]

## Antecedent Impressions

Blood was not the only factor influencing the form and character of offspring; there were also the effects of gestation in the womb—so-called antecedent impressions—and milk suckled from their mothers. In gestation, the blood of mother and fetus were in contact across the placenta, making hereditary and physiological influences highly likely. Nervous impressions also might be transmitted from the mother to the offspring if, say, the pregnant mother had suffered a trauma or displayed hysterical symptoms. Such ideas were congruent with ideas about human heredity and had popular currency, although the medical profession was skeptical.[49] Another form of maternal-mediated impression consisted of features taken from "uterine brothers and sisters in the litter" and from the litters of previous pregnancies; a common metaphor was that the womb had been "stained."

Authors of dog books continued to cite breeding folklore with aphorisms and anecdotes. Stories were shared in part to entertain their readers and to show how far breeding had progressed, but many presented these as revealing truths about breeding rather than tall stories. Such was Hugh Dalziel's repetition of an old Delabere Blaine case, an instance in which "the mental impression made on the mind of a bitch by a dog she has been denied sexual intercourse with, affects most sensibly the progeny resulting from a sire of a totally different form and colour." His narrative was detailed.

> I had a pug bitch whose constant companion was a small and almost white spaniel dog of Lord Rivers's breed, of which she was very fond. When it became necessary to separate her on account of her heat from this dog, and to confine her with one of her own kind, she pined excessively; and, notwithstanding her situation, it was some time before she would admit of the attentions of the pug dog placed with her. At length, however, she was warded, impregnation followed, and at the usual period she brought forth five pug puppies, one of which was perfectly white, and although rather more slender than the others, was nevertheless a genuine pug. The spaniel was soon afterwards given away, but the impression remained; for at two subsequent litters (which were all she had afterwards) she again presented me with a white pug pup, which the fanciers know to be a very rare occurrence.[50]

There were other stories of bitches "impressing" upon their progeny the conformation features of past lovers.

Breeders, scientists, and doctors took one form of male influence more seriously, what would later be called "telegony." Dalziel explained the phenomenon in 1879: "One of the most strange and remarkable facts, as it is one of the least understood in connection with breeding, that the union of a bitch for the first time with a dog by which she conceives frequently exerts an influence on subsequent litters." Again, the metaphor was that the womb had been "stained," and crucially it was an influence that persisted, only gradually reducing with successive pregnancies. Scientists had widely discussed the phenomenon since the 1820s, when the Earl of Morton had reported that one of his mares, having previously borne a foal from an experimental cross with a quagga, a type of zebra that became extinct in the 1880s, produced foals with "a striking resemblance to the quagga," when subsequently mated with a black Arabian horse. Telegony was also assumed to occur in humans. Thus, a mother with a firstborn illegitimate child would not only suffer moral and social condemnation but would have to endure a "biological punishment," deterring any future husband because his children would "inherit" features of the (likely disreputable) man who fathered the first child. There has been speculation that Thomas Hardy had the phenomenon in mind when writing *Tess of the d'Urbervilles*. Stephen Kern has suggested that Tess thus bore a double stigma: "Tess inherited not only inherited her remote ancestor's destiny to kill ("a murder committed by one of the family, centuries ago"), but she presumably also absorbed some of the traces of her rapist Alec, who left her pregnant with a child who died in infancy. On her wedding night her husband Angel is shocked to learn of her former pregnancy and rejects her in disgust, charging she is not the woman he thought she was." In the dogs of fancy breeders, the taint of a promiscuous mating—say, by a bitch that had escaped for just a few short minutes—would pollute her bloodline and lessen her value.[51]

Breeders continued to report instances of telegony in dogs and other species through the final decades of the nineteenth century. Both scientists and veterinarians were increasingly skeptical of the phenomenon; indeed, the term itself was invented in the 1890s by August Weismann, a founder of modern genetics, only for him to dismiss it as "never [having] been known to occur." For biologists, the Penicuik Experiments, conducted by the Glasgow University biologist James Cossar Ewart in 1894–95, conclusively disproved any effects on progeny from previous maternal matings.[52]

## Conclusion

In the 1890s, a new name came to the fore in doggy literature: Rawdon B. Lee. He had been editor of the *Kendal Mercury* and "gave much time to cricket and field sports, especially fishing and otterhunting, and he became also an authority upon wrestling. In spite of defective eyesight he was one of the finest fly-fishers in England." He moved to London in 1883, succeeded Walsh as kennel editor of the *Field,* and began to produce books that displaced those of Dalziel and Shaw.[53]

In his 1890 book on the collie, Lee returned to Darwin's ideas in his reflections on breeding and recent changes in dog breeds. He noted that Darwin had stated that the possibility of creating new species by crossing had been greatly exaggerated. Lee disagreed, conflating species and breeds: "During the past quarter of a century extraordinary progress has been made in the matter of breeding to type in dogs and in other animals. The various toy spaniels, including the King Charles and the Blenheims, are very different in appearance, and especially in the shape of their heads and their faces, from what they were a generation past. The bulldogs are the same, and to a smaller degree the change is noted in other varieties." He wrote that these changes had been brought about in the first instance by chance variation or "sports," but then "the breeder has continued to, as it were, manufacture by judicious crossing other exaggerated properties, which have come to the neglect of other, possibly equally important, points." He observed, "With the exception of the shepherd's dog, deerhound, foxhound, and greyhound, there is not one strain of British dog at the present time which bears much more than a great resemblance to his variety as it was known a hundred years ago." Lee looked forward: "A hundred years hence more changes in our dogs will have been wrought, and may-be the so-called fancy varieties will come to be 'improved' until past recognition."[54] One way that this endeavor was being increasingly pursued was by the direct application of science to breeding, keeping, treating, and feeding dogs.

# CHAPTER SIX

# Improving Breed II: Science

In April 1880, the lead editorial in the first issue of the *Kennel Gazette,* the Kennel Club's journal, was titled "Scientific Breeding." The central argument was that while dog shows had certainly improved the nation's dogs, the improvement had largely been achieved through trial and error. Now, it was time to step up the endeavor by applying scientific approaches. The empirical data for such a venture had been accumulated in the pedigrees and results available in the Kennel Club stud book, and now was the moment for systemic investigations and scientific principles. This call was backed up in the third issue of the *Gazette,* in which another editorial asked veterinarians to take the dog and its ailments more seriously, and for canine pathology to be taught in the nation's veterinary schools. These aims linked canine improvement to major scientific developments of the age: evolution and heredity, germ theories of infectious disease, and experimental biology. Both the metaphorical blood of heredity and the actual blood circulating in dogs' bodies, it was thought, could be systematically improved.[1]

During this period, scientists and veterinarians increasingly interacted with breeders, exhibitors, dog-food suppliers, and other canine interests to better understand and in turn enhance the conformation and health of dogs. Charles Darwin and Thomas Henry Huxley (who was known as "Darwin's bulldog") set the context of evolutionary ideas and the new experimental biology. The most developed application of the new biology to dog breeding was by Everett Millais, set out in his book *Rational Breeding,* in which he utilized Francis Galton's ideas on ancestral heredity. Millais also pioneered artificial insemination and made laboratory investigations of dog distemper, the latter being seen as the bane of dog shows. We discuss changing ideas on the nature of this infection and responses to its threat, including Millais's campaigning for the sanitary reform of dog shows. Finally, we consider how the keeping and feeding of dogs was changed by the industrialization of dog-food manufacture and attempts to develop a mass market. This change was led by the largest supply company, Spratt's Patent, which claimed that its food suited the physiology of dogs and their transformation from savage carnivores to civilized omnivores.

## Artificial and Natural Selection

Darwin devoted the first main chapter in *On the Origin of Species by Means of Natural Selection, or the Preservation of Favoured Races in the Struggle for Life* to artificial selection. He had collected masses of evidence on the subject, and working out how to include it in his "big book" was one factor in its long gestation. In the end, he devoted just the opening chapter to artificial selection, later publishing his more extensive research on the topic in an expansive two-volume book in 1868. Darwin's "one long argument" in *On the Origin of Species* began by detailing the great changes made by humans in selecting particular features to domesticate animals (and plants too) over ten of thousands of years. He then asked readers to imagine what changes might be wrought over millions and millions of years by natural forces. One of the most controversial features of Darwin's theory was that evolution was not progressive: which animals and plants survived—*the fittest*—was contingent. Survival depended on how well organisms were adapted to their environment. Adaptation was dynamic and relative to conditions at a specific moment. When climate change, geological activity, and other forces altered the natural environment, selection pressures changed, and different organisms, produced by normal hereditary variation, survived in the new conditions, making them the *fittest*. Artificial selection, however, was progressive because there was an absolute goal: to change organisms to meet human needs.[2]

The different timescales of artificial and natural selection were crucial to Darwin's views on species. In the timescale of human history, species were essentially fixed; only over very long periods could they change or undergo, to use his term, "transmutation." He drew parallels between the limits of changes with species and with dog breeds: "Thus, a man who intends keeping pointers naturally tries to get as good dogs as he can, and afterwards breeds from his own best dogs, but he has no wish or expectation of permanently altering the breed."[3] From this he deduced that the wide variations in the size, form, and character of domesticated dogs meant the animal must have had more than a single origin. In fact, he suggested that dogs across the world had their roots in three earlier groups of canids, represented in modern times by wolves, jackals, and coyotes.

Dogs and cats were the subject of the first chapter of his 1868 book, *The Variation of Animals and Plants under Domestication*. In 1875, a slightly revised second edition followed, in which Darwin was agnostic on the origins of the different types of dog: "With respect to the precise causes and steps by which the several races of dogs have come to differ so greatly from each other, we are, as in most other cases, profoundly ignorant." By this time, Darwin had elaborated his theory of natural selection, adding new forces, many of which sped up evolutionary change. He was less insistent that evolution occurred through a myriad of gradual variations, and he admitted the possibility of more rapid changes; here again, artificial selection offered support. He wrote, "Those who have attended to the subject of selection will admit that, nature giving variability, man, if he so chose, could fix five toes to the hinder feet of certain breeds of dogs." Another example was how the underhanging jaw of the bulldog had been "increased and fixed by man's selection." Darwin cited Youatt's observations that greyhounds had become slimmer and faster, with greater symmetry, since they were used in racing and coursing rather than for hunting deer. The Newfoundland dog was said to have been "so much modified that, as several writers have observed, it does not now closely resemble any existing native dog in Newfoundland." Darwin had become more interested in heredity, as the emergence of new species required advantages to be fixed generation by generation in order to secure survival. Ensuring the survival of a new species seemed to require close or inbreeding among the "fittest," yet he was aware of its dangers from the experience of animal breeders, as the whole chapter dedicated to the problems of inbreeding in the book on domestication demonstrates. For example, the small populations of the old English hound and the old Irish greyhound were said to have accelerated their demise due to "the evil effects of . . . close interbreeding."[4]

Darwin had long speculated on questions of what is now termed *heredity* but what for him was "generation," which involved "mating and maturation." He had recognized from his extensive natural history studies that reproduction was essentially conservative and that "like begat like." However, one source of variation for nature "to select" was sexual reproduction and the mix of the two different parental forms. The results were unpredictable, but in some instances patterned, and Darwin explored one practical breeder's maxim (Yarrell's Law) that, in crossbreeding, the older established breed tended to show dominance. Darwin was also fascinated by "reversion," that is, when offspring showed the features from distant generations, or what breeders called *throwing back* and scientists termed *atavism*. Speculation on this phenomenon contributed to what might be termed Darwin's theory of heredity—pangenesis—which imagined that cells throughout the body threw off subcellular "gemmules" that in animals found their way into eggs and sperm and passed on parental features. Contemporary scientists were not persuaded by this hypothesis, and it was not pursued, but it is interesting how it is congruent with the notions of blood as hereditary matter and of the possibility of inheriting acquired characteristics.[5]

In 1879 and the spring of 1880, Huxley gave a number of lectures on dogs. These were given separately at the Royal Institution and in the Working Men's Lectures at the same venue.[6] He had been researching the fossils of doglike creatures, and his talks were part of his larger project to promote what he called "Darwinismus," today known as Darwinism.[7] When considering the genealogy of the dog, he asked rhetorically, "What is a dog?" His answer was to contrast a "common sense" viewpoint with a biologist's answer. The "common sense" viewpoint would say, "A dog is a hairy, four-footed, tailed animal with sharp teeth and fond of meat and bones; which brings forth its young alive as blind puppies and which the mother suckles; which barks and howls and is often singularly intelligent and affectionate. A dog fancier or a sportsman, or a shepherd will probably add to this a disquisition on the particular breed of which he favours." Huxley stated that the "scientific zoologist," with his deeper knowledge, would not only stress the unity across breeds but also point out the similarities between dogs and humans as mammals. From an evolutionary perspective, differences within the single species *Canis familiaris* were "rather of degree than of kind."[8]

Huxley wrote to Darwin in May 1880 on the likely origin of dog breeds in different canids, explaining, "I think I can prove that the small dogs are modified jackals, and the big dogs ditto wolves." With regard to more recent domestication, low- and high-class dogs were of common descent. "The zoologist knows that the menial cur that slinks along the street sniffing post and gutter is no accidental blot on the face of nature, but an integral part of that vast and varied assemblage of living things which constitutes the animal world [and] is connected by no very remote ties with even the highest and noblest of its members." Other writers drew direct parallels between the evolutionary history of dogs and humans, with references to the human race and the canine race, and—within these—races of humans and races of dogs. The conflation of human and canine groupings had been evident in George Jesse's popular *Researches into the History of British Dogs* (1866). He wrote that some saw the greyhound as "the inferior species of the canine race" and the origins of different types of dog with the "nations, tribes and ranks of the human race." *Race* and *breed* were used interchangeably for types of dog at this time, but *breed* was rarely, if ever, applied to human grouping. One assumed difference in the late Victorian era was that races of human were natural and stable, whereas breeds of domesticated animals were unnatural, mutable, and manufactured.[9]

## "Rational Breeding"

The plea for "scientific breeding" that launched the *Kennel Gazette* was not taken up by any dog breeder until 1889, when Everett Millais published *The Theory and Practice of Rational Breeding*. As we have shown, until then advice relied upon the "experience" and tacit knowledge of breeders, not theory. Millais was a Kennel Club member and leading breeder of basset hounds (fig. 6.1). The eldest son of the Pre-Raphaelite painter John Everett Millais and his wife Effie Gray, he had become well known in the late 1870s for breeding dachshunds and introducing the basset hound to British dog shows. He built up a kennel but had to abandon it in 1880 when he left the country for Australia. The reason given publically for this journey was that it was to improve his physical health; privately, however, his parents were concerned about his drinking and mental health, and the insinuations that there was degeneracy in the family. During his exile, Millais spent time exploring and hunting, and developed an interest in natural history and science. On his return, he took up dog breeding again, enrolled as a medical student, and worked as a sculptor, mainly of dogs. He became what can best be described as a "gentleman scientist," associating with leading biologists; undertaking over fifty breeding experiments for George Romanes, a friend of Darwin; working in the new laboratory of physiologist Charles Scott Sherrington at St. Thomas's Hospital, London; and corresponding with Francis Galton, who used Millais's basset hound pedigrees in his work on ancestral heredity. Millais's connections saw him elected to the prestigious Physiological Society in 1890.[10]

Millais's first scientific work, published in the *Veterinary Journal* in 1884, was on what he called "artificial impregnation," now known as artificial insemination. A persistent problem for breeders of all domesticated animals

*Figure 6.1.* Everett Millais. *Kennel Review,* May 1884, 8

was ensuring that mating was followed by conception. A guaranteed result was important economically, to avoid wasted stud fees, but also to ensure the veracity of pedigrees so that there was no doubt which individuals were the parents. Owners of large kennels worried that "the carelessness of servants" should not allow bitches "the slightest chance [to] steal away in search of a mate of her own selection."[11]

Millais had problems securing fertile mating with his basset hounds and tried introducing, with a syringe, semen he had collected from a dog into a bitch in season. His first attempt in 1884 resulted in successful conception, but all the puppies were stillborn. He tried again the following year, inseminating three bitches, and this effort produced live young. He tried seventeen "experiments" over twelve years, with the results eventually included in a paper by Walter Heape, the leading embryologist, who has been celebrated as a "pioneer of reproductive biology." Francis Galton read the paper to the

Royal Society. Since then, this paper has secured Millais a place among such pioneers of artificial insemination as the eighteenth-century Italian priest-naturalist Lazzarro Spallanzani, the London surgeon John Hunter, and Heape himself.[12]

In *The Theory and Practice of Rational Breeding,* Millais nailed his colors firmly to the mast of Darwinism: "Darwin's theory is the stupendous work of a stupendous mind, and, in making use of it as the base of this book, the reader will agree with me that it is the only theory put before the world that will satisfactorily account for the multiplicity of life-forms at present extant, and the variation which is constantly going on in nature, whether the reader be a student of nature or not." His overall aim was to show breeders "how pure stock of all kinds and of correct type may be bred," and how this was achievable through a scientific approach that monitored and balanced in-breeding and outcrossing. More widely, he wanted to show that "breeding is not a lottery, it is a certainty, and if conducted on a proper basis with a defi-nite object in view, and on such principles as I have endeavoured to show in these chapters, it will repay a rational breeder, and do more to prove that life and its varied forms are not the outcome of laws which man's understanding cannot fathom, but built up upon a definite system of rules and progression such as I have ventured to portray." The book had begun life as a guide for members of the Basset Hound Club but grew in size and ambition with Mil-lais's engagement with scientists.[13]

Millais had three principles from "experience" for "science": first, that in-breeding was essential for pure blood and quality; second, that it must not be taken to extremes; and third, that occasional outcrossing was essential and nothing to be worried about, because the original type could be reestab-lished in a few generations. In fact, Millais did not need his new experi-ments to show how it was possible to return swiftly to the original type. Back in the late 1870s, he said that he had crossbred his first basset hound with a beagle bitch because at the time there were few other bassets to breed with. He claimed that he bred the crossbred dog back with a basset, to produce a dog that was so basset-like in conformation that at the Islington Show in 1877 it had been "impossible to distinguish from pure bassets." Subsequently, he bred his basset-beagle crosses only with pure bassets, increasing the "basset factor until the beagle type was destroyed."[14]

On his return to England in 1884, Millais had helped found the Basset Hound Club. As with other specialist clubs, it was a fraternity of enthusiasts who exchanged stud dogs and breeding experiences, as well as setting standards. Many members reported the same problems in achieving conception, along with small litters and many stillborn puppies. Millais supposed that these problems were due to the degree of inbreeding, which came from the limited number of dogs in the country, and the practice of going to the small number of class-winning dogs to improve an owner's kennel.[15] The problems were, he contended, relatively easily solved by applying what he termed "Equal and Un-equal Factor" principles.

Millais began *Rational Breeding* with a natural history of types within a species, so that his readers might understand the place of breeds in taxonomic and hereditary structures. He was never economical with words and complicated matters further by using the term *varieties* for types of dog, not *breed*. Millais elaborated a hierarchy useful for all domesticated *species*. The scheme was illustrated by one of his hounds, Fino IV, whose *species* was the dog and *variety* (breed) the basset hound; he was from Millais's *strain*, his *family* was that of his father Fino de Paris, and he was, of course, the *individual*. All types of breeding (in, cross, inter, and line) had to be considered against this hierarchy that defined degrees of "in-ness."

Like all fancy breeders, those working with basset hounds had two aims: individually, to get their dogs to the highest conformation standard, and collectively, to reproduce the standard across the breed, which meant a more uniform-looking population that matched the breed standard. Inbreeding was the accepted method of best reproducing type. However, Millais argued that when taken to extremes either in closeness or frequency between dogs of the same *strain* or *family,* the practice led to "deterioration, degeneration, difficulty in rearing, and, finally, non-reproduction." His answer was scientifically monitored, measured, and controlled crossing. He developed a formula, which he termed the "Equal-Factor" and "Unequal-Factor" systems, drawing on Galton's ancestral law. The basis of the law was the assumption that "the two parents contribute between them on the average one-half, or (0.5) of the total heritage of the offspring; the four grandparents, one-quarter, or $(0.5) \times (0.5)$; the eight great-grandparents, one-eighth, or $(0.5) \times (0.5) \times (0.5)$; and so on."[16]

Galton first articulated this scheme in 1865, and it went through a number of iterations before the publication of its fully developed form in a paper, read at the Royal Society in 1897, in which he used Millais's basset hound pedigrees and their records of color as one evidence base. Galton is best known for his work on and advocacy of eugenics, which he defined as "the science of improving stock." In its earliest versions, he advocated positive eugenics, which he described as "judicious mating . . . to give the more suitable races or strains of blood a better chance of prevailing speedily over the less suitable than they otherwise would have had." Back in 1865, he had speculated on what would later be termed the eugenic possibilities of dog breeding.[17]

> We breed dogs that point, that retrieve, that fondle, or that bite; but no one has ever yet attempted to breed for high general intellect, irrespective of all other qualities. It would be a most interesting subject for an attempt. We hear constantly of prodigies of dogs, whose very intelligence makes them of little value as slaves. When they are wanted, they are apt to be absent on their own errands. They are too critical of their master's conduct. For instance, an intelligent dog shows marked contempt for an unsuccessful sportsman. He will follow nobody along a road that leads on a well-known tedious errand. He does not readily forgive a man who wounds his self-esteem. He is often a dexterous thief and a sad hypocrite. For these reasons an over-intelligent dog is not an object of particular desire, and therefore, I suppose, no one has ever thought of encouraging a breed of wise dogs. But it would be a most interesting occupation for a country philosopher to pick up the cleverest dogs he could hear of, and mate them together, generation after generation—breeding purely for intellectual power, and disregarding shape, size, and every other quality.[18]

Neither Galton, nor any other scientist or breeders pursued this idea, but it had a tangential relation to later work on domestication and selection for the enhancement or loss of certain instincts, as explored by Millais's collaborator Romanes. There was also interest in whether breed-specific behaviors, such as herding and pointing, were instincts that had been *artificially* selected, or whether dogs had acquired true "intelligence" and the ability to learn and understand during domestication. Conwy Lloyd Morgan, the scientist credited with founding animal psychology in Britain, was skeptical of animal intelligence. He cited some experiments that pointed to this conclusion, but seemingly just as telling were his personal experiences with his fox terrier Dan. The dog "learnt" to open a gate, which Morgan argued was not learning in the sense of understanding how the catch worked; rather, it came from first stumbling on the action by "trial and error" and then the action becoming a "habit."[19]

Millais illustrated his factor system with arithmetic fractions. The simplest equal factor was when two dogs of the same breed mated:

$$\left.\begin{array}{l} \text{Basset} \\ \text{Basset} \end{array}\right\} \frac{\text{Basset}+\text{Basset}}{2} = \text{Basset}$$

A cross between breeds was the simplest unequal factor:

$$\left.\begin{array}{l} \text{Basset} \\ \text{Spaniel} \end{array}\right\} \frac{\text{Basset}+\text{Spaniel}}{2}$$

However, if this cross was then mated with a basset hound, the resulting "composite" (Millais's term) would be predominantly or three-quarters basset:

$$\left.\begin{array}{l} \dfrac{\text{Basset}+\text{Spaniel}}{2} \\[2ex] \text{Basset} \end{array}\right\} = \frac{3\times\text{Basset}+1\times\text{Spaniel}}{4}$$

Furthermore, if each time the unequal-factor dog was put to a basset hound its spaniel factor was diluted, then "in a short time the spaniel prototype would be reduced, attenuated, and, in consequence, lose its hereditary power entirely," and "return to the original type."[20] Thus, after four further matings with a basset hound, the puppies would be 63/64 basset hound and 1/64 spaniel.

After the publication of *The Theory and Practice of Rational Breeding,* Millais wrote articles for the dog press, enjoying the position of an expert whose views were science based. He entered into vigorous and lengthy correspondence on many issues, particularly inbreeding and prepotency. One outlet was the *Dog Owners' Annual,* published in Manchester by Theo Marples, who had established *Our Dogs,* a weekly paper, in 1890.[21] Millais's profile grew in May 1890, when he published an article in the *British Medical Journal,* claiming to have discovered the microbial cause of dog distemper and to have developed a vaccine. Had his results been confirmed, he would have almost certainly become the "British Pasteur." They were, in fact, largely ignored. However, they were the beginning of a one-man campaign for the sanitary reform of dog shows.[22]

Millais returned to the subject of scientific breeding in February 1895 in a lecture delivered at St. Thomas's Hospital; that he spoke in that venue testifies to his status among metropolitan scientists. His title was "Two Problems of Reproduction," and the lecture discussed inbreeding and the

**Figure 6.2.** Photo 11, Bloodhound and Basset Hound. Millais used this illustration in his lecture and reproduced it in his pamphlet. The published photograph is poor quality. These are not the dogs used in Experiment 12; rather, they illustrate why artificial insemination was used. E. Millais, *Two Problems of Reproduction* (Manchester: "Our Dogs," 1895), 9

*influence* of the sire, including telegony and greater prepotency of wild over domesticated animals—a version of Yarrell. However, his audience was most interested in what he had to say on inbreeding, particularly on one of his outcrossing experiments. The experiment was the mating of a male basset hound to a bloodhound female—appropriately named Inoculation—using "artificial impregnation" to overcome the size difference (fig. 6.2). Details were subsequently reported as "Experiment 12" in Heape's 1897 paper on artificial insemination.[23]

Millais's specific goal was to counter the "deterioration" in his kennel, and in British basset hounds more widely, by introducing new blood. He chose not to go to French bassets, as they were "unimproved," and instead chose a British bloodhound. Mating dogs that were so different in size was justified on the grounds that, following Millais's schema, both were *varieties* of hound, meaning there was "a natural alliance." The sire was Nicholas, whose pedigree exposed a high degree of inbreeding. Millais claimed, no doubt for effect, that the fractional method, counting back five generations, revealed that Nicholas had just eight rather than the maximum 128 possible ancestors.[24]

This was not literally the case; rather, the same eight dogs appeared time and again in different bloodlines. Breeders referred to varying degrees of "in-ness," from close (e.g., parent-offspring matings) to more distant intergenerational (uncle-niece matings) relatives, sometimes termed line breeding.

"Experiment 12" produced twelve puppies, "all anatomically nearer the basset than the bloodhound, but in colour they took after their mother 'Inoculation.'" He then bred one bitch from the litter (Rickey) with Nicholas's father (Champion Forrester). This choice reveals that following just one input of new blood, he was unconcerned about returning to close inbreeding. This mating produced seven puppies that were "3/4-bred basset with 1/4 bloodhound"; all had basset anatomy, and six had basset coloring. After two further crosses, Millais had a litter that was "15/16 of Basset blood to 1/16 of Bloodhound." He summed up thus: "The result of this set of experiments has brought about animals which cannot be distinguished from Bassets, and they can be used throughout the breed to bring in the trifling quotum of fresh blood necessary without damaging or altering the existent type in the slightest degree."[25]

Millais compared his efforts to "an old adage, that it takes four generations to make a gentleman, and if this is so, I have indeed, bred Bassets from Bloodhounds." Perhaps ahead of his time and anticipating the twenty-first-century designer crossbreeds, he called for a new class for Basset Bloodhounds at dog shows. Millais, like other writers on dogs, occasionally drew direct parallels between humans and animals, and reflected on these with regard to races. Discussing why the wild parent showed prepotency more strongly than the domesticated one, he wrote, "There is but one answer to this, whether it has to do with man or the lower animals, and that is, that the domestic animals and the white man—that is, the higher types of man,—are of newer creation of evolution than the wild or dark; consequently their type is less fixed than the older ones, and when they by any circumstances are bred together the newer type goes down before the old." He went on: "If a man wishes to succeed as a breeder, or, rather, as a cross-breeder with such as I entertained when crossing the basset and the bloodhound, he must, to a certain extent, be conversant with such questions." Like that of many other writers of his period, Millais's thinking was framed in relation to prevailing ideas of "savagery," civilization, and the colonial project. Both the desire to control outcrossing by bringing breed back to the "superior" type within four or five generations, and the perceived potency of the "wild" over the domestic, speaks to the insecurities of Empire.[26]

Millais died in September 1897, and with him went his version of "rational breeding." George Krehl later wrote that "the bloodhound experiments of Sir Everett filled his most earnest friends with regret and despair. France is full of basset outcrosses, so there was no need to create canine nightmares." Nonetheless, in his lifetime, Millais's experiments represented a pioneering attempt to place dog breeding on a scientific footing, collaborating with biologists and using new technologies. "Experiment 12" is remembered in the British Basset Hound Club in the twenty-first century, with claims that its legacy remains in bloodlines back to Nicholas: "It is thought that it was this out-crossing that led to the present day Basset Hound being significantly larger than its French counterpart. *Interestingly, the Basset Hound has the greatest bone to weight ratio than any other breed of dog*" (italics in original.) This is fanciful thinking, given the generations of basset hounds bred since "Experiment 12." However, it is typical of the uses of the Victorian past in the present that we take up in our conclusion.[27]

## Health and Disease

Breeding was by far the most important way that Victorian doggy people endeavored to improve dogs, but they were also mindful of other ways that gave quicker results. Preparing a dog for showing involved ensuring they were in top condition, which meant management of their health and diet. From William Youatt's *The Dog* in 1845 and continuing with all editions of Stonehenge's *The Dog in Health and Disease,* most dog books included sections on canine diseases and advice on their treatment. These sections' length and detail indicate that most owners managed the health of their own dogs, and support the claims that the employment of veterinary surgeons was limited, in part due to cost and in part due to veterinarians' unwillingness to take on canine patients.[28] However, toward the end of the century, in London in particular, a number of veterinarians established canine (and to a lesser extent feline) practices to treat show dogs and the pets of the well-to-do. For the working and lower middle classes, there were pet shops whose proprietors gave advice on treatments, as well as selling foods and other supplies.

The medical pages of dog books were a mix of pathological science to aid recognizing signs and symptoms, and remedies that could be homemade or bought from chemists. Apart from Youatt, none of the authors of the principal dog books were veterinarians, though Walsh and Stables were medically qualified, and Dalziel may have had some medical training, as he had begun his career as a pharmaceutical salesman. The most common cure advocated for all manner of ailments was a change of diet, more specifically, a regimen of either stimulating or calming foods. One of the first popular tracts on dog health, published in 1884 by Spratt's Patent Company, was *The Common Sense of Dog Doctoring,* by Gordon Stables. It aimed to fill the gap between consulting a veterinarian and relying on popular nostrums. Needless to say, Spratt's products were recommended for most ailments, which included cures for distemper, mange, worms, rheumatism, lumbago, and jaundice; purging pills; and "Tonic Condition Pills—For debility and preparing Dogs for Exhibition." Spratt's also catered to wealthy owners, opening a sanatorium for the hospital care of dogs in 1877 by Clapham Common, London; it moved to the edge of "airy" Mitcham Common in Surrey in 1893. Its chief veterinarian was Alfred J. Sewell, who had a large private canine practice in London and whose grandfather William had been head of the Royal Veterinary College.[29]

For exhibitors, medical care focused on external, conformation-affecting conditions, such as skin and eye infections. However, an equally important function of veterinary measures, amateur or professional, was to improve the "condition" of dogs so that they showed their best at exhibitions. There were tangible aims—such as giving foods that would improve the luster and the condition of a dog's coat, and "manipulations" to ensure dogs walked elegantly in the ring—and less obvious aims, such as getting bright eyes.

The great bane of dog shows was distemper, an infection spread by close contact. It was second in prevalence only to the skin disease of mange. Shows were ideal sites of contagion, bringing together hundreds of dogs in a confined, ill-ventilated space for days. But the streets of towns and cities were dangerous too. Any dog out with its owner might be accosted and infected by street dogs. And if a show dog escaped from its kennel, it could be sniffed at or fought with, or might mate with its disreputable peers, and enjoy temporarily a feral life of scavenging and promiscuity. The greatest plague on the streets was rabies, against which local authorities took specific measures, such as rounding up street dogs and requiring all dogs to be muzzled and led.[30]

While rabies was a sporadic threat, distemper was a permanent danger. Veterinarians believed distemper was relatively new to Britain, having been imported from Spain, via France, in the second half of the eighteenth century. Initially, it was only a problem in hunt kennels, where it spread rapidly, particularly through poorly fed packs housed in ill-ventilated, damp buildings.

In 1876, the leading veterinarian William Hunting reflected the dominant view of the profession that distemper was "a contagious catarrhal fever." He set out its course: "An ordinary attack of distemper is ushered in by dullness, shivering, and sometimes loss of appetite. There is a disinclination to move, and a desire for warmth. The nose becomes hot and dry, and the eyes bloodshot, and there is either sneezing or a dry, husky cough. Next notice a thin watery discharge from the eyes and nose, and increased sensitivity to light. The respiration and pulse are quickened, and the bowels are usually constipated. Accompanying these symptoms there is always a rapid and marked loss of strength and condition." Complications included "pneumonia, jaundice, enteric disease, epilepsy, chorea, or paralysis." Walsh, obviously drawing on his medical experience, regarded distemper as a form of typhus fever. He listed five variations: "1st, Mild Distemper; 2nd, Head Distemper; 3rd, Chest Distemper; 4th, Belly Distemper; and 5th, Malignant Distemper." The prognosis varied; some dogs recovered, with or without treatment, though often in a debilitated condition. If the disease spread to the lungs and brain, it was almost always fatal. In years when it was virulent, distemper could destroy a whole kennel, either directly or through secondary infections. Dogs that recovered had no immunity to further infection; indeed, being in a weakened state, they might take the disease again more readily.[31]

The best advice from veterinarians and doggy people on how to combat distemper was prevention—avoiding exposure to infected animals and practicing strict hygiene when risk could not be avoided. Veterinarians had no proven cure; rather, their remedies offered the amelioration of symptoms or constitutional measures to strengthen a dog's system to fight the infection. They excoriated the claims of cures advertised by the makers of patent medicines and specifics. The best known of these were "Heald's Distemper Powders," "Rackham's Distemper Balls," and "Spratt's Patent Distemper Powders." One of the most widely advertised remedies was "Gillard's Compound," available by mail order from Frank Gillard, huntsman to the Duke of Rutland, testifying to the continued authority of the landed estates in all canine matters.[32]

Veterinarians and breeders accepted that susceptibility to distemper varied across breeds and by age, with puppies being most susceptible. Less-domesticated curs, mongrels, and shepherd's dogs were least affected, while highbred animals were most likely to succumb. Youatt had given a hierarchy of immunity: curs and shepherd's dogs were the most resistant, then terriers and hounds, with pugs and Newfoundlands most vulnerable. He gave no explanation for the sequence, though the implication was that an unnatural, indoor life or poor acclimatization to a new country made dogs less resistant. Generally, inbreeding was understood to make dogs "nervous," and given the persistence of humoral notions of disease in veterinary and popular animal medicine, a nervous body would be more likely to catch the disease. Also, in a single bloodline there was less chance of acquired or inherited immunity. Mongrels and curs had the advantage of mixed blood and immunity from their diverse forebears that had been inured to filth and infection.[33]

Distemper was particularly prevalent in the 1880s, leading to calls for greater veterinary vigilance at shows. Inspection of dogs had been a feature of shows from the start, and the Kennel Club's rules stated that any dog suffering from "mange or any other contagious disease" could be excluded on the decision of the show's veterinary inspector, though he only made recommendations to the show committee, who had the final say.[34] There were many complaints about dogs returning from shows with distemper and other diseases, raising questions about veterinary expertise and the laxity of show committees in applying rules. Distemper was so feared that many breeders refused to send their best dogs to shows or were selective about those entered.

Show organizers tried to reduce the spread of distemper, mange, and other infections by liberally spreading disinfectant on the benches, floors, and even on the dogs themselves. Fancy dog shows had always been an assault on the senses. The normal scent of the dogs' bodies was augmented by the stench from urine, feces, and fetid skin lesions, as well as the vapors from the pungent disinfectants applied to deodorize the event. In their marketing, disinfectant manufactures appealed to old and new ideas: their products were both antidotes to chemical miasmas and antiseptic killers of living germs. The products of the Sanitas Company were used at most shows in the 1880s, and they consolidated their position by offering prizes. By then, all disinfectants contained carbolic acid or similar antiseptic chemicals, and Sanitas faced competition from companies like Jeyes, Izal, and the makers of Condy's Fluid, which used violet-colored potassium permanganate. In June 1890, an advertisement for Jeyes stated that at the recent Liverpool show, which had not been disinfected with its fluid, there had been twenty-six cases of distemper among sixty-two puppies, whereas at the earlier Manchester show, where Jeyes was used, there had been just four deaths among a similar number of entrants.[35]

Undaunted by the skepticism of scientists toward his distemper "discoveries," Everett Millais continued to advance his views on prevention in the press and by lobbying the Kennel Club. For a while he enjoyed the support of Sidney Turner, a leading figure in the club, who published scientific and medical articles in the *Kennel Gazette* under the pseudonym of "Este." They soon parted company, as Millais wanted radical rather than gradual reform. In his view, no one took the disease seriously enough, from the amateur breeder right up to the leadership of the Kennel Club (despite Turner's insider pressure). Outraged by its members' continued indifference, Millais very publicly resigned from the Kennel Club in 1893. In his resignation letter, he wrote of distemper causing "massacres" at and after shows and said that the "continual destruction of young dogs" was "a scandal" and "a disgrace." His departure was even reported in the *British Medical Journal,* perhaps because Millais argued that the club's antivivisection stance also showed its disregard for science.[36]

After his resignation, Millais announced that he would use his parliamentary connections to promote a Distemper Protection Act. With an implied threat of legal actions against the Kennel Club, he also promised to found an exhibitors' protection association. The Royal Society for the Prevention of Cruelty to Animals (RSPCA) announced its support for the act as part of its larger campaigning to promote dog welfare, which had previously focused on urban strays, cropping, and the muzzling orders enacted to control rabies. However, Millais was no friend of the RSPCA, being a licensed vivisector and supporter of muzzling to control rabies, but his rhetoric on welfare at dog shows was similar. He cut a lonely figure, becoming alienated from the Kennel Club and missing no opportunity to condemn all aspects of its work. To Millais, the club's inaction on distemper was symptomatic of its overall failings as an organization for "not only the breeding of high class dogs, but the improvement of dog shows."[37]

## Feeding

All forms of advice literature on dogs offered recommendations on feeding, from how to wean puppies, to the feedstuffs needed for packs of hounds. Such advice was typically linked to wider care, giving, for instance, recommendations on the number of meals each day in a domestic routine or the design of provisioning facilities in the kennels of hunting packs. Whether a dog slept in a box under the table in an overcrowded house or in commodious buildings on a large estate, it needed a holistic regimen. Healthy and fit dogs were assumed more likely to show prepotency if their blood had vitality. Food alone would not be enough to ensure good health; dogs had to be kept in clean, well-ventilated surroundings, with comfortable bedding, and given plenty of exercise. What *not to do* was epitomized by the stereotype of the lady owner of a tiny toy dog, who kept her cosseted pet indoors, indulged it with sweets and luxury foods, and only took it out in her handbag. This caricature was still going strong in the 1930s, according to the reminiscences of country vet Alf Wight, who wrote as James Herriot, when he fictionalized the woes of a wealthy widow's overweight and underexercised Pekinese, Tricki Woo. What *to do* was demonstrated by leading exhibitors, who kept their dogs in specially designed kennels and employed dedicated staff to ensure their dogs were exercised and had a mixed diet of meat, cereal, and vegetables, with specific supplements to improve coat and vitality ahead of shows.[38]

The consensus among canine experts was that the dog, while once "naturally carnivorous," had become "by long domestication and habit, to all intents and purposes as omnivorous as man." For evolutionists such as Herbert Spencer and Samuel Butler, "habits" were one way in which acquired characteristics could be inherited, and they showed evolution to be progressive. Thus, the ideal diet for the modern dog was a mixed one that needed to be varied between individuals and breeds. In *The Dog in Health and Disease*, Walsh recommended that setters and pointers be fed mostly cereals and, very occasionally, weak broth, so that their scenting ability was not compromised by exposure to flesh. For the greyhound, who ran well under "a regimen of gelatine and farina [cereal grains]," meat was desirable because it was "nutritious and stimulating" and gave "fire to his temperament." However, this needed to be given carefully to avoid the blood becoming "too rich."[39]

From the founding of the Kennel Club in 1873, "Spratt's Patent" had a near monopoly providing food at its shows, and the company expanded its operation to provide the benches and other services to exhibitors. James Spratt was an electrician from Ohio who traveled to England in 1860 to sell his "Spratt's Patent Lightning Rods." After docking at Liverpool, he is alleged to have observed the sailors on his boat throwing their unwanted hardtack biscuits to dogs on the quayside, which inspired him to add dog food to his patent products. Spratt did not invent the dog biscuit. Cereal biscuits for dogs had been made by flour and livestock feedstuffs merchants for many years, their products being variations on the cereal feedstuffs made for horses, poultry, and cattle. Dog biscuits were initially sold to estates to feed packs of hounds, then firms diversified into the domestic market. In 1859, dog biscuits could be bought in London from R. T. Smith's, and John Martin also sold biscuits, along with "the best town made graves, 14 shillings per cwt." Graves (or greaves) was the residue of skin and connective tissue left after animal fat was rendered down, and it was formed into cakes as food for pigs as well as dogs.[40]

Spratt's business was successful because of three innovations: (1) combining cereal and meat in one biscuit, (2) extensive advertising of the company's products and creating a brand, and (3) using science to sell and differentiate its products from competitors'. An early 1864 advertisement for bulk purchases captures the pitch.

> KEEP YOUR DOGS for 1d. a DAY—SPRATT'S PATENT FIBRINE DOG and POULTRY CAKES, contain 20 per cent. of Fibrine, the dried unsalted gelatinous parts of the finest prairie beef cattle. Ground wheat, oat and malt meal. Poultry powder. The above cakes require either cooking, scalding or soaking. Either 20s. per cwt. (for cash) carriage free or 28s. free by London Parcel Delivery.[41]

The beef was sourced in dry bales from meatpackers in the Chicago stockyards, and cereals were purchased locally. The addition of poultry powder is curious. It was a dietary supplement given to birds to shorten the molting season, invigorate the growth of plumage, and renew the "fullest laying powers." Spratt's did produce poultry food, for which there would have been an urban market with fancy breeders and domestic hens. With both products the company had a consistent branding: a square biscuit, clearly marked "SPRATT'S—X—PATENT." The term "patent" signaled that the formula and appearance were protected by "letters patent," and it may also have suggested health-giving properties, as "patent medicines" were popular remedies bought directly from pharmacists, though disapproved by the medical profession. Interestingly, Spratt's advertised their dog, poultry, and game foods in the *Yearbook of Pharmacy*, with the claim to pharmacists and chemists that they "will be found to have a ready sale." The promotional activities of the business were led by Charles Cruft, who joined the company as office boy in 1865 and rose to general manager before becoming what he is known for to this day, a leading promoter of dog shows.[42]

Cruft's initial sales strategy was to obtain bulk orders from hunts and shooting estates, but the company soon added saturation advertising, aimed at breeders and pet owners, in books, specialist weeklies, and daily newspapers. Full-page advertisements appeared in 1872 in the second edition of

**Figure 6.3.** Spratt's Patent, Meat Fibrine Dog Cakes. Stonehenge, *The Dogs of the British Islands* (London: Horace Cox, 1872), 287 and end pages

Walsh's *The Dogs of the British Islands* and Idstone's *The Dog*. The advertisement in the former also carried details of "Spratt's Patent Great Challenge Poultry Food—the greatest egg-producing food and fattener," and "Spratt's Patent Date Farina Horse Biscuits," which enabled a horse to "sustain his mettle on a long journey" (fig. 6.3). Endorsements were an important feature of the marketing. Well-known personalities reported that their dogs had benefited from the biscuits, including the leading dog-show organizer John Douglas; the Queen's canine veterinarian, Charles Rotherham; and soon-to-be Kennel Club leaders Frank Adcock, J. H. Murchison, and Sewallis Shirley. The first Kennel Club stud book in 1874 carried a full-page promotion, which mentioned that Spratt's was enjoyed by the dogs at the Royal Kennels at Sandringham. In 1892, Annie Oakley, a markswoman and star of Buffalo Bill's Wild West Show, endorsed Spratt's biscuits as ideal for her St. Bernard, and the following year the Norwegian explorer Dr. Fridjtof Nansen reported that he fed them to the huskies during his Arctic adventures.[43]

The constitution of dog biscuits reflected domestic feeding regimens such as the meal Charles Dickens prepared for his dogs in the mid-1860s, a mixture of "2 pints oatmeal, 1 pint barley meal, and 1 pound Mangel Wurzel boiled together, and then mixed with pot-liquor which is poured over it. If there be no pot liquor in the house a sheep's head will make it very well. Any bones that happened to be about, may be put into the mixture to exercise the dog's teeth. Its effect upon the body and spirits of the creature is quite surprising."[44] In the 1880s, by which time biscuits had become a staple, Gordon Stables was recommending soaking them in meat or fish liquor. Overall, he wrote that dogs needed variety in their diets and confirmed their position as omnivores. Meat with insufficient exercise could be harmful, and it was essential that they ate their greens. Another feature of his dietary advice was discipline; dogs should have one main meal a day, both to avoid the Victorian dread of constipation and to teach discipline.

Spratt's asserted that it had "revolutionised the mode of Dog Feeding" by adding vegetable matter to the meat biscuit. As omnivores and in parallel with human evolution, dogs had become less savage and more civilized. Vegetables were said to be essential in keeping "the Dog's blood . . . in a cool state or their bowels in perfect order."[45] Spratt's had experimented with different vegetables and fruits, including imported dates, but had settled on beetroot, as its nutrients were said to best survive the production process. Spratt's used science to legitimate its products, making much of the results of tests undertaken on behalf of the *Field* in 1871. Different dog breeds had been given four formulations of Spratt's Patent containing 7, 10, 20, and 40 percent meat, respectively, and named Nos. 1, 2, 3, and 4. The results were positive, though qualified.

> We find every variety of dog can be kept in good health on the first of the above proportions, which may be shortly increased to the second; but none can take 20, and still less 40 per cent of meat, without at first showing by their heated noses the over-stimulating nature of their food. The most delicate lady's pet may soon be induced to eat No. 1, and will require nothing else except a little green vegetable occasionally; and either this or No. 2 is what we would recommend for all dogs that are not required to display an extraordinary amount of endurance, such as foxhounds or greyhounds. No. 1 suits the most delicate-nosed setter or pointer, and is quite sufficiently furnished with meat for any dog used with the gun, in which nose is the *sine qua non*.[46]

The conclusion was that meat was the most sensitive component in a dog's diet, and too much, rather than too little, was often the problem. Over the 1870s and 1880s, Spratt's moved to greater product differentiation, formulating special foods with varying amounts of meat for greyhounds, puppies, and dainty and delicate toy dogs.

Spratt's faced competition from companies old and new. The second edition of Walsh's *The Dogs of the British Islands* in 1872, as well as the full-page Spratt advertisement, carried promotions for "Slater's Meat Biscuit for Dogs," endorsed by the Rt. Hon. Lord George Cavendish and the gamekeeper of the Earl of Durham; "Petherbridge's Meat and Oatmeal Biscuits"; and "W. G. Clarke's Buffalo Meat Biscuits." Another brand widely advertised was Chamberlin and Smith of Norwich, which boasted of being "Purveyors to the Royal Kennels at Sandringham." Their most common advertisement suggested that they were endorsed by Parliament too, though the MPs were unflatteringly depicted as bulldogs (fig. 6.4).[47]

Spratt's routinely warned that "unprincipled dealers may serve you a spurious and dangerous imitation"; one veterinarian reported finding "pipe

*Figure 6.4.* A Canine Parliament, Purveyors to the Royal Kennels at Sandringham. *Stock-Keeper and Fancier's Chronicle,* 2 Dec. 1893, 608

clay and sawdust" in some brands, a problem similar to the adulteration of human foods. In a context where brand protection was still relatively new, Spratt's was very litigious, taking out injunctions and usually winning, against any firm that infringed its patent, including, in 1875, W. G. Clarke's. Competitors were undeterred, however, and both Clarke's and Norfolk-based Chamberlin's became national brands. They never rivaled Spratt's in size or profile, and in the 1890s Spratt's expanded its operations to the United States and Russia.[48]

## Conclusion

Charles Darwin put artificial selection at the heart of his theory of evolution by natural selection, corresponding and writing extensively on the subject. Inheritance was a key issue in all aspects of his work on evolution, because as species changed (speciation), features had to be passed on from generation to generation to effect transmutation. However, Darwin's writing on inheritance, particularly his theory of pangenesis, was speculative and seems not to have drawn much upon the ideas of practical breeders or found its way into their thinking. Francis Galton was one of the few late nineteenth-century biologists to engage with breeders, and it was his ideas that Everett Millais sought to bring to dog breeders. Although Millais had limited success in convincing other doggy people of his ideas, his efforts to fuse dog breeding and health with science were pioneering, opening up a place for science and technology in the dog fancy. However, his efforts were not helped by his personality or his reputation in the doggy world as a maverick, a factor that also worked against his campaigning to rid dog shows of distemper. The main effect of science on the doggy world occurred not directly through the ideas of scientists, veterinarians, and other experts but indirectly through the knowledge embodied in commercial products, particularly Spratt's Patent Dog Biscuits for physiology and nutrition, and Sanitas Disinfectant from hygiene and bacteriology. Efforts to market such hybrid products drew upon and combined recent experimental science and the experience of the fancy. In the show world, Spratt's also represented modern technology and the place of commercial interests. An article, "The Rise of the Dog Show," in the *Stock Keeper and Fanciers' Chronicle* in July 1895 was illustrated with a drawing taken from a Spratt's advertisement, which showed metal, easily disinfected benches with an separate enclosure for each dog, with the company's logo painted on the sides.[49]

There were no sections on health and disease or on breeding, keeping, and feeding in Rawdon Lee's three-volume *Modern Dogs of Great Britain,* published in 1894. For the first time, a major Victorian publication on dogs was just about the history and description of breeds. The book was novel in another way; the illustrations were no longer of show prizewinners; rather, they depicted ideal, "typical specimens of the various breeds." There was no shift to modern photography but instead a reiteration of the value of a sharp drawing with clear lines. Lee stated that that "such pictures must be more useful than any portraits of individual dogs could be—dogs whose prominence before the public is more or less ephemeral." This was coded criticism of modern breed standards, and he made further references to "exaggerated properties" and "the neglect of . . . important points" in many modern breed champions. Lee was reflecting a new questioning of whether the dog shows and the changes they had wrought had, in fact, brought "improvement" to the nation's dogs. As we show in the next chapter, fundamental questioning of the whole enterprise of conformation breeding and showing, not just of individual standards and "improvements," became a major political issue within and beyond the dog fancy from the late 1880s onward.[50]

# CHAPTER SEVEN

# *Whither Breed*

In January 1889 the magazine *Punch* published a drawing by George Du Maurier of a well-to-do lady leading a group of cleverly imagined new breeds: the Dorgupine, Crocodachshund, Pomme-de-Terrier (Black-and-Tan), Ventre-à-Terrier (Scotch), Hippopotamian Bulldog, German Sausage Dog, and the Hedge-Dog (fig. 7.1).[1] Although this image satirizing the "Dog Fashions for 1889" has been reproduced many times in books and articles on pedigree breeds, Harriet Ritvo is the only historian to link the satire to late nineteenth-century complaints about the wider fin-de-siècle concern with degeneration.[2] The cartoon reflects the power of breeders to produce new forms and the growing influence of women as owners, breeders, and exhibitors of fancy dogs, and hints at masculine dog fanciers' disdain for this trend. The leaders of the Kennel Club, particularly its sportsmen, worried about the feminization of the fancy. They complained that it was turning some breeds into mere fashion accessories and that the dogs were losing their character and working abilities. The Kennel Club also viewed as villains some specialist clubs for encouraging the proliferation of "unnatural," "unhealthy," and "infantile" breeds, and the perennial enemies of canine improvement, exhibitors and "dog dealers" who were interested only in profit. However, another view of the cause of the deterioration of the nation's dogs was that the whole dog-show enterprise, which the Kennel Club had fostered, had been misguided.

In this chapter we consider reactions inside and beyond the dog fancy to criticisms of its aims and methods, beginning with the tensions in the Kennel Club over the fate of sporting breeds. The club's founders had been sportsmen who favored gundogs, but there were accusations that their new focus on shows had led to the loss of working abilities in pointers, setters, and

DOG FASHIONS FOR 1889.

Dorcupine, Crocodachshund, Pomme-de-Terrier (Black-and-Tan), Ventre-à-Terrier (Scotch), Hippopotamian Bulldog, German Sausage Dog Hedge-dog.     (*By Our Special Dog-fancier.*)

**Figure 7.1.** George Du Maurier, "Dog Fashions for 1889." *Punch,* 25 Jan. 1889, 42

retrievers. More widely, many thought that the Kennel Club's unwillingness to set standards allowed conformations to be needlessly embroidered by specialist clubs and that this practice was contributing to the deterioration of the nations' canines. Cropping ears and docking tails were two ways of altering points that had been continued from pre-show days, but many viewed these as "mutilations" that were no longer justified. In the late 1880s a campaign to end these practices originated from within the fancy but failed; a decade later another effort from outside the fancy was successful.

These controversies highlight the shifting sensibilities around the fancy and expose the different interests at play in regulating its norms and values. In the rest of the chapter, we discuss the challenge of women to the authority of the Kennel Club, which ran from the issues about the operation of shows, through attempts to create an alternative vision of the fancy, right up to wholesale condemnation of the dog-show fancy. For an example of how breed development engendered debate, we can look at controversies about how Collies had changed in form and character from the 1860s to the 1890s, and what to do about it. The breed was among the most popular, in both number and esteem, and it was hotly contested whether in its various roles—in work, at shows, and as a pet—the breed had seen "improvement" or "deterioration." As such, the debates around the "modern" Collie at the end of the

Victorian era exemplify all of the issues that had animated the fancy over its invention of the modern dog.

In this chapter, we use capital letters for breed names because by the end of the Victorian era, they had become proper nouns, indicating that breeds were individual and individualized entities—for example, a shift from pointers to the Pointer.

## Shows and Sporting Dogs

In what was probably John Henry Walsh's last editorial in the *Field* on canine matters before his sudden death in February 1888, he reflected on changes in dogs and shows since the 1860s. He had started his career in rural sports, but his interests, publications, and expertise had widened with the growth of leisure and competitive sports. The verdict of this authoritative observer was that there had been progress, but he remained critical of the Kennel Club, which he felt was failing in its aim of improving the nation's dogs. At shows, Walsh felt that there were fewer "bad dogs" and mongrels appearing on the benches and that most improvements had been in nonsporting breeds; for example, the St. Bernard had been developed from the dogs first imported in the 1860s. However, with sporting dogs, while their "outward appearance" had been improved, their "working quality [was] quite another matter." Setters, for instance, "in certain circles" were now "looked upon with suspicion, especially by those admirers and breeders" who were "as much in favour of him as a working animal as one for the show bench."[3] Like other sporting breeds, Setters did not have clubs to promote their position at shows or in field trials. Walsh argued that the Kennel Club ought to step in and support "clubless" breeds by offering better prize money to encourage the exhibition of sporting breeds.[4] One worry was that imbalance in the populations of show and field dogs would mean that the latter would suffer from inbreeding or that their abilities would degenerate or be diluted by crossing with the larger numbers of fancy animals. One solution to the conflicts between sport, exhibitions, and work would be to have different standards for each role.[5]

Types of Spaniel had their origin in different styles of shooting in different places, which had led to the designation of several breeds, but some fanciers worried about needless proliferation. The first Kennel Club stud book in 1874 had listed four breeds (or six, depending on how they were counted): Spaniels (Field, Cocker, and Sussex), Clumber, Irish Water, and Water Spaniels, other than Irish. A decade later, the Cocker had been dropped, while

Sussex and Field were listed in their own right alongside Clumber Spaniels.[6] Hugh Dalziel and Vero Shaw, in their volumes on British dogs in 1879 and 1881, respectively, added the Black Spaniel and the Norfolk Spaniel. The 1882 edition of Walsh's *The Dogs of the British Islands* also included "The Modern Field Spaniel" and "The Modern Cocker," the latter with regional varieties from Norfolk, Wales, and Devon. Walsh was ambivalent about the new varieties. He pointed to the arbitrary character of many breeds, observing, "Throughout the country there are numberless breeds of cockers of all colours, varying from white, black, or liver to red and white, lemon and white, liver and white, and black and white."[7] At shows, the number of Spaniel varieties meant low numbers in each class; to confuse matters further, ad hoc groupings often were put together to make a class. Walsh concluded that Spaniels were in need of a club, though that might be a mixed blessing.

A Spaniel Club confined to sporting breeds was founded in 1885, as was a Toy Spaniel Club for King Charles and Blenheim Spaniels. Spaniel Club members took a while to agree on standard points and in the spring of 1888 approached the Kennel Club to alter the classification in its stud book.[8] Typically, the Spaniel Club settled differences by splitting breeds, and overall it proposed nine. Dalziel described this approach as creating "an unmeaning and absurdly arbitrary jumble" and was typically scathing when the Kennel Club accepted the recommendation, showing it had "a narrowness of viewing the whole subject, and indefinite weakness of purpose." No doubt having a dig at the Kennel's Club continued unwillingness to adopt standards, he stated that no one could learn from the stud book "what particular breed of dog any one of those registered may be."[9]

The Spaniel Club wanted to see the Cocker accepted into the stud book, and the first step was for members to agree among themselves what the breed was. They decided on a split between "Black" and "Any other Variety of Cocker Spaniel." Critics asked whether the Cocker was now just a show dog, with its working role taken by the Field Spaniel, regarded as a smaller dog but with the same "lines." The name did not help, as "cocking" and "cockering" referred both to finding game, especially the woodcock, and to indulging and pampering. In 1888 the *Stock-Keeper and Fanciers' Chronicle* published an article on the Cocker Spaniel with the title "Show v. Working Spaniels," in which the author wondered whether the working Field Spaniel was to become extinct as Cockers took over. A follow-up letter from "One who does not believe in Clubs" questioned "the 'bastard' class" for Cockers at the upcoming Warwick Dog Show, which had adopted the new standards of the

Spaniel Club. The writer thought their standard was a mistake, as it went back to a "light-boned, short-bodied, short-headed, short-eared dog," abandoning recent improvements. Another correspondent referred to the Spaniel Club as "a body of out-and-out lunatics for framing such an abominable standard," because they were tending to make "*all* Spaniels Cockers, and exterminate the rest."[10]

The reply from "A Member of the Spaniel Club" claimed that their standard was not only about conformation; rather, it defined a dog that was made for "activity" and avoided the trend toward twenty-five-pound "Dachshund monstrosities." Critics, like Thomas Jacobs, countered with the observation that "long, low, heavily feathered dogs" could work all day, and the Spaniel Club worked with the "*theorist's* ideas of what a dog with certain physical characteristics ought . . . be able to do or not do, why, of course, 'so much the worse for the facts.'" The short-term solution was to divide classes at shows by weight, but by the 1890s complexities multiplied further as the lighter Field Spaniel had become a type of Cocker. At this time, black dogs were prized in both breeds, but by the 1900s lighter tricolors were in vogue for Cockers, while the Field Spaniel remained black, becoming in time the Black Field Spaniel. The Spaniel case indicates the scope for bitter conflict between sporting and conformation interests and the complicating role of breed clubs in arguments about canine improvement.[11]

### "Dogdom's Deterioration"

An 1894 article in the monthly *National Observer* mused on "Dogdom's Deterioration."[12] The anonymous author—almost certainly Vero Shaw, who wrote regularly on horses, poultry, and dogs—argued that while it had taken centuries for "the local dog to build himself up to local requirement" in "shape and make," this development had been undone in recent decades. The Mastiff, which had once been able to tackle bears and lions, had become "a leggy, loose jointed, flat flanked, slack boned creature," with the head of a Bulldog, the coat of an Italian Greyhound, and "the disposition of a congenital imbecile." With the Bulldog, exaggeration of form had produced "a helpless toad-like monstrosity," and the Dandie Dinmont Terrier was being converted into "a fake-coated, silky-eared, wooden-jointed lap dog." The causes were the "sacrificial show-bench and interbreeding," which had produced dogs of "pseudo-symmetrical imbecility," deficient in body and mind. Indeed, the author wondered if the "best hope lies in the mongrel." If high-bred dogs were nervy, disease prone, and weak, then perhaps low-bred dogs were resilent, healthy, and strong? Significantly, Shaw's reference was to mongrels not curs, that is, to features imparted by inheritance and blood, not geography and behavior. As we have seen, *mongrel* had changed from being descriptive of intentionally crossbred dogs to suggesting dogs that were the product of multiple, promiscuous matings and showed no recognizable breed conformation. Accordingly, within the dog fancy, *mongrel* increasingly became a pejorative term, and it was this meaning that spread into popular usage for "nonbreed" dogs.

In a signed article the following year, Shaw continued to worry about "the gradual effacement of old-fashioned breeds" and the degeneracy of others. Citing many reasons for the decline, he was most concerned about "alien immigration from without," seen in the effect of the popularity of "new varieties from abroad" on "English Dogs." This echoed wider fin-de-siècle fears that linked immigration and degeneracy, and saw political expression in the 1905 Aliens Act. The number of dogs in many breeds was falling and with it their quality, and breeders with "an experimental turn of mind" were "intermingling" English and foreign blood. Shaw argued that British dog lovers, whom he termed "philokuons," were also ruining foreign breeds. He suggested the "correct type" of Dachshund at British shows would "excite the surprise and ridicule of German breeders" and that English admirers of Schipperkes had adopted "an inferior type," while "guileless Belgians" kept "the correct article" at home. He expected the same to happen with the latest fashion, the Dogue de Bordeaux: "In fact a Dogue de Bordeaux 'club' will assuredly be formed, and then gentlemen who a couple of months ago had never set their eyes upon a specimen will draw up and issue a 'standard of points,' etc., etc., for the guidance of their countrymen and edification of the French originators of the variety." Shaw observed that whatever the variety, the position of foreign dogs in Britain was assured because fashion favored the exotic.[13]

Implicit in Shaw's parody was the view that the leaders of the Kennel Club, by concentrating on governance, had ceded too much power to breed clubs, which pursued their own specialist interests. These clubs could apply for affiliation with the Kennel Club, but their leaders only had a say in policy if appointed to committees as individuals. The club still operated as an executive council body and court of appeal, not a representative parliament. Functioning as "the Jockey Club of the kennel world," the club's committees still focused on administration, the organization of shows, and revisions to its rules. Complaints that came before the club's committees were rarely upheld;

most lacked the evidence needed to adjudicate on competing claims. Many complaints raised matters that seemed petty or based on private grudges, and these were usually politely dismissed. Some within the fancy felt the club, "the oligarchy of Cleveland-row," had become too powerful, with too much influence on the press. The *British Fancier,* which had relocated from Newcastle upon Tyne to London in 1890, claimed that it was the only independent voice, as Kennel Club leaders had pecuniary interests in the influential *Stock-Keeper and Fanciers' Chronicle,* and that *Our Dogs* was an ally of the club—a more doubtful claim.[14]

The dog press, especially in gossip columns such as "Pet Dog Patter" in the *British Fancier* and the "Whispers of the Fancy" in the *Stock-Keeper and Fanciers' Chronicle,* published protests about the behavior of specialist clubs. Within the Kennel Club, there remained mixed feeling about "specialist" clubs and "specialist" judges. An 1890 editorial in the *Kennel Gazette* argued that they "have been and are of very great use in the advancement of scientific breeding"; however, they were "apt to get into grooves."[15] In other words, specialists tended to exaggerate particular points and lose sight of the all-around dog, ideally one "as good for his work as for the show bench." To ensure the prestige of their event, show promoters wanted the top experts deciding the prizes, which often led to the same men making awards to the same dogs. Complaints about this practice echoed the accusations of favoritism that Kennel Club founders had faced in the 1870s before they mostly absented themselves from officiating at their own shows.

One indicator of the shift in show culture toward nonsporting breeds was the increase in toy dogs exhibited at the major events. To Walsh and the older fraternity, the Kennel Club's recognition of more fancy shows in London and around the country signaled, if not a takeover, then too many concessions to exhibitors, dealers, and their like. Most promoters had added a third toy dogs division to those for sporting and nonsporting breeds, and there were exclusive toy breed shows. There had long been single-breed events for aficionados of the Bulldog, and the new single-breed and group shows were also for enthusiasts. The Toy Dog Club had been established in 1884, the year that also saw the first Pug Dog Show. From the outset, the Toy Dog Club shows were grand affairs, held over several days at the Royal Aquarium at Westminster. The First Great Terrier Show, organized by Charles Cruft, was put on at the same venue in February 1886. This event is often cited as the first Cruft's show and the start of the series that continues to this day. However, Cruft had been active in the organization and promotion of dog shows from the 1860s, and the 1886 event was not promoted with his name. That brand-ing was first employed for the 1891 Terrier Show, which was retrospectively styled as Cruft's seventh.[16]

The mid-1880s saw other evidence of the rise of the toys. In most of the fancy and popular press, the Pug and Toy Dog shows were covered quite fac-tually in the same way as all-breed shows. However, the Second Toy Dog Show in 1886 received a critical, principally misogynistic, review. The *Daily News* referred to it as "a drawing room dog show," with ladies "everywhere." The report mocked the fact that allegedly "professional coiffeurs were engaged making ready the Maltese Terriers," while many dogs "repose[d] upon elegant cushions or handsomely-whorled antimacassars," and others had lace curtain in front of their pens (fig. 7.2). The correspondent of the *North Eastern Daily Gazette* described the exhibits as "little monstrosities" that looked "like embodied imps of weak mind and are in all respects excel-lently adapted for drawing room cultivation." The *British Fancier* carried an article that warned that producing toys was "a worse evil than inbreeding in as much as it buys its penalty sooner." The penalties were "delicacy . . . bulging heads, indicating weaker brains, protruding eyes and shaking limbs."[17]

It was alleged that Pugs had been reduced in weight from thirty to less than twenty pounds, and they still seemed to be shrinking. One typically misogynistic correspondent on the question, "T. A. T.," speculated that ladies' "kindlier feeling" might lead to toy classes for St. Bernards, Great Danes, Borzois, Setters, and Pointers. There were warnings that the trend toward "Tom Thumb" dogs was producing new problems: small dogs did not seem to transmit their "best points," and small bitches had difficulty giving birth. Possibly taking inspiration from the taxidermist Walter Potter, whose popu-lar museum of anthropomorphic taxidermy tableaux opened in 1880, the Third Toy Dog Show in 1887 pioneered a new class: the best arrangement of taxidermied former pets, arranged in domestic settings. Winners included a group of Pugs playing cards and another enjoying a tea party; all dogs were clothed, and many were posed with cigarettes. It might have been expected that newspapers with a sporting origin like the *Field* would have considered this a nadir for the dog show, but the report surprisingly welcomed it as an "innovation." In the fancy, dog shows exclusively for toys were initially seen mainly as entertainment, but concern about their influence grew with the proliferation of classes and the number of entries at the grand all-breed shows in London and Birmingham.[18]

NOT FOR COMPETITION

MOTHER AND CHILDREN

1st PRIZE

MELANCHOLY

THE LAP OF LUXURY

TOILETTE

FATED TO BE FREE

THE TOY DOG SHOW AT THE ROYAL AQUARIUM, WESTMINSTER

*Figure 7.2.* The Toy Dog Show. *Graphic,* 30 Oct. 1886, 1. © British Library Board

Fashion was one motor of these developments, but so too were the long-standing issues of vulgar commercialism and social climbing. The conflict continued to be framed as between the "dog lover" who wanted canine improvement, trusted his eye, and looked for intelligence, and the "dog dealer" or exhibitor who valued paper pedigrees and admired "peculiarities" and profit. In 1886, Hugh Dalziel wrote about "exhibitors" in an article for New York's *Harper's New Monthly Magazine* titled "Dogs and Their Management." His main aim was to instruct Americans on every aspect of breeding and care, from selecting mates to training. The difference between "dog lovers" and "exhibitors" was, he explained, in their social class, values, and character. Dalziel observed that domestication had taken place over thousands of years and had been of equal benefit to humans and dogs but that shows had ended the mutuality. Shows had "brought to the front, a class of men who had no merit in themselves and kept dogs, often spending much money on them, but with no love and but little understanding of the animals, the sole object being to secure to themselves a fame or a notoriety through their dogs." Dalziel was severe: "I sincerely wish they [exhibitors] could be eliminated, and business left to those who are capable of appreciating the dog, and whose desire is to improve each breed."[19] Needless to say, exhibitors did not accept this description and maintained that their dogs were the best, having being tested in open competition at shows and in the market.

Against the exhibitors stood true dog lovers. They styled themselves as amateurs, not professionals, a division that was of major importance in British sports at this time and long after.[20] Dog lovers were not interested in exaggerated points, knew the dangers of excessive inbreeding, and believed that winning prizes in competitions was a means to improving dogs, not an end in itself. But competition generated its own tensions and divisions, with continuing disagreements about breed standards and their interpretation. The proliferation of breed clubs, mostly dominated by exhibitors, made matters worse. Some were formed when members could not agree on a standard and malcontents broke away, while others were provincial, but they all had different interests that shaped inter- and intrabreed politics.

An editorial in the *Kennel Gazette* in July 1890 cautioned that exhibitors' preoccupation with paper pedigrees and their fixing of exaggerated points was threatening the purity of bloodlines. Too often they made an acquisition to their kennel "simply because he is a prize dog, and without considering at all whether such a sire will suit his strain." The author drew comparisons with breeding in other domesticated animals. The aim with Thoroughbreds was only to breed for "qualities possessed and not for mere outward appearance." Horses were bred for stamina not skin and courage not contour," while breeders of show dogs looked "at the hide not the heart, at the shape and not the spirit."[21]

There were, of course, positive assessments of the effect of fancy shows. Socially, the supporters of the Kennel Club claimed that its leadership had improved the culture of shows. Participants no longer conformed to the stereotypes of the upper-class swell and the working-class rough; rather, there was a greater mingling of classes and leveling up, with "all sorts" coming along. There had been "Canine Progression" too, as "Ædipus" argued in the *Kennel Gazette* in September 1894. He insisted that the registration of pedigrees and results had produced "a greater regularity of type" and the elimination of "objectionable features," and that "each race has become more defined." Acknowledging that breeders had produced more handsome dogs adapted for showing but not for working, the author suggested that their endeavors were unfinished. One breed at issue was the Fox Terrier, which had become too large for going to ground to chase foxes out of their dens. However, Ædipus argued, "Is it not better to accept the Fox Terrier as it is seen today as an example of the advance of scientific breeding, and trust to its supporters to, in the future, add the finishing touch which will connect the elegance of the present with the varmint of the past?" Other breeders argued that "an improved type" of Fox Terrier was not needed, as near perfection had been reached.[22]

One breeder's progress was another's regression. Harding Cox, a stalwart of the Kennel Club since the late 1870s and owner of "thousands of dogs," had changed his mind about the benefits of shows. In 1897 he wrote that "breeders have, so to speak, become fanatics or 'faddists,'" and judges had encouraged particular points in conformation to encourage the production of "monstrosities and cripples." Fox terriers were his favorite; the best he had ever seen, Mister Chippy, had a conformation that at show would "cause any self-respecting judge to be carried out insensible," but in the field he was "magic," as in his "veins ran the bluest of blood."[23]

By the end of the century there was, tellingly, little written on the improvement of the nation's dogs overall. Instead the commentaries were breed by breed, as in the *Kennel Gazette's* annual series "Year in Retrospect." In 1899, Walter F. Jefferies controversially claimed that Bulldogs had reached a state of perfection and that if the "old celebrities" were alive today, they would not be in the money. Others saw the breed as still in crisis and the modern type as something between "a dog and duck." Sidney Turner wrote that commenting on Mastiffs was "a melancholy task"; he asked that "a few good men should take up this magnificent old breed in a patriotic spirit" and save this once-noble dog.[24] There was most ambivalence about the Great Dane, a breed that had divided opinion for many years. Walsh had refused, despite pressure from aficionados, to include it in any of the four editions of *The Dogs of the British Islands*.[25] The breed's size and perceived aggression made it hard to control, which caused it to go out of favor in the rabies scares of the late 1880s. However, its numbers had revived in the 1890s, with the alleged advantage that rabies quarantines had prevented importation of dogs from the Continent and allowed the establishment of a smaller, neater type. The Mastiff was remodeled again in the "revolution made by the non-cropping of ears edict" in 1895.[26]

While many expressed concern about the deterioration of dogs within the fancy, others were optimistic about the wider effect of their enterprise. There had been, according to F. A. Manning, an improvement in street dogs: "Those of the past were, by comparison, little better than mongrels; whilst at the present time we constantly see dogs of all kinds of the purest type. The mongrel, except as the homeless and outcast, has almost disappeared, and its place taken, by any rate, decently bred dogs." He claimed that shows had educated the general public about dogs' "real nature" and that well-bred dogs were better behaved and more adoptable as human companions. Indeed, the public could now acquire any dog they wanted, and "we owe this to dog shows, and the interest taken in all matters pertaining to the animal world."[27]

In "Dogdom's Deterioration," Shaw had asked his readers to imagine dog-breeding standards applied to the English people. "The races of men are as varied as the breeds of dogs, yet happily there is as yet no set endeavour to bring each to one immutable type. Fancy every Englishman brought to the dead level of a Dumaurier [*sic*] drawing or still worse improved into a young Greek god of the lady novelist!" He was puzzled why so much reliance was now placed upon pedigrees and purity of blood, which he described as the "inherited contact of generations." Many historians have pointed to parallels between the aims of fancy dog breeding and eugenics, and to Francis Galton's brief interest in fancy dog shows. There were similarities in terms of the aim of producing "good stock" by selective breeding, and it is interesting that these linkages have not been explored by historians of eugenics, who largely ignore animal breeding. Dog breeders can be seen as hardnosed positive and negative eugenists: keeping the puppies they wanted, discarding those they did not, and choosing which dogs and bitches to breed. In fictional versions of eugenics, as in Aldous Huxley's novel *Brave New World,* breeding policies resulted in segmented and standardized social strata, with individuality forfeited for utilitarian goals. In the social and political sphere in Britain (other countries were different), eugenists mainly advanced positive rather than negative eugenics policies. They were largely silent on the inevitable class and gender inequalities that their policies would produce, which were hidden in the overall aim of race improvement.[28]

## Cropping I

The attempt to ban the cropping of dogs' ears was the "great controversy" in canine affairs in the late 1880s and early 1890s, dividing opinion in the Kennel Club, and within and between breed clubs. The controversy began as an issue about standards in individual breeds but escalated to challenge the aims and reputation of the modern fancy and put these on trial in the courts of law and public opinion. Was the modern fancy a humane, progressive enterprise, or was it backward looking, perpetuating cruel and barbarous practices?

In June 1886 the Irish Terrier Club proposed to ban from shows all crop-eared dogs of its breed—that is, those with surgically reshaped ears. Previously, the ears of many breeds of show Terriers had been trimmed to give them a pointed shape and make them stand up. The Kennel Club was asked to endorse the ban and add it to their rules. In turn, this request prompted leading lights in the fancy to ask for a wider ban. Sidney Turner, who at that time was deputy chair of the Kennel Club, was joined by leading figures in dog publishing: Hugh Dalziel, Gordon Stables, and George Krehl, the editor of the dog section of the *Stock-Keeper.* Turner had already proposed a total ban in an article in December 1887.[29]

Some supporters of the ban insisted that the practice had actually been illegal since the passage of the 1835 Animal Cruelty Act, under which the Royal Society for the Prevention of Cruelty to Animals (RSPCA) had brought a number of cases to trial. Although cropping was practiced widely, those prosecuted were typically associated with the working-class urban fancy, confirming the point made by many historians that the RSPCA was as much, if not more, concerned with stopping the "brutal customs" of the poor, as with protecting animals. However, from the 1860s the RSPCA had taken a critical stance on other aspects of dog shows, including travel arrangements, feeding, chaining on benches, and overnight conditions. They had supported

Edwin Landseer when he had objected to the "cutting" of ears when judging at the National Dog Show in 1862. The defenders of the practice, including a group who had formed the Dog Cropping Society, contended that it was not cruel, especially when carried out under anesthetic, and that dogs had adapted to the procedure.[30]

The original reason for cropping and docking (shortening or removing tails) was to prevent injury in fights or when tackling game. The practices continued after dog fighting was outlawed because shortened ears and tails had become the accepted look of certain breeds. The main objection of the RSPCA was that cutting was mostly carried out on sensitive young dogs and was painful; veterinarians only rarely used anesthetics, and working-class croppers never did. In the 1880s the breeds regularly shown with cropped ears were Great Danes, Bull Terriers, Irish Terriers, Old English Terriers, Black-and-tan Terriers (sometimes called the Manchester Terrier), and Toy Terriers.[31] The fashion had seemingly passed for removing the whole ear of Pugs and clipping back those of Dalmatians.[32] In 1888 members of the Black-and-tan Terrier Club disagreed over whether to follow the lead of the Irish Terrier Club in banning cropping. Hoping to outflank their opponents, the reformers asked the Kennel Club to initiate a ban across all breeds.[33] They were encouraged by a view, certainly mistaken, that gentlemen shunned cropping, and not much was at stake, given that shows under Kennel Club rules had small classes for breeds on which cropping was practiced.

In the months before Turner's motion was discussed in February 1889, the dog press was full of correspondence on the matter, with most asking for reform and supporters of cropping largely silent.[34] "Humanity," a vocal proponent for change, called for the RSPCA to bring more prosecutions, and asked who the croppers were. Were they the amateur owners who did not know better and were simply following fashion, or were they the professional, "exhibiting dealers" who were able to make "bad ears good" and profit from "faking." A third group was implicated, too: the gentlemen of the Kennel Club (including the president, Sewallis Shirley) who kept Bull Terriers and were alleged to be the "greatest offenders" because they cropped their own dogs and yet were now hypocritically contemplating a ban. All critics contended that the croppers were out of line with modern times, public sentiment, and good sense.[35]

In the debate on Turner's motion, Shirley moved that it was too controversial and important for the committee. So at a general meeting on 5 March 1889, the following motion was proposed: "That any dog born after July 1

1889, which has been cropped shall be ineligible to compete for any prize at any show held by the Kennel Club." The birth date meant that dogs with cropped ears born before then could still be entered and could be for many years. Members asked what would happen during this transition, when cropped and uncropped would be together on the benches. Would shows introduce different classes for the two types? If not, how would judges choose between the two? Prospectively, what shape and size of ear should be adopted for uncropped dogs? Breeders who accepted that it would be better to "improve rather than mutilate" wondered how long it would take to agree upon, select, and crossbreed to produce a new ear.[36]

At the general meeting, Turner claimed that the tide of opinion was with him and that the initiatives of the Irish Terrier and Black-and-tan Terrier Clubs meant that, in future, cropping might only be required for Great Danes and Bull Terriers. Drawing upon his medicoscientific credentials, Turner argued "that Nature in her work never erred" and that they should accept dogs' ears as they were. There was strong opposition to the proposal from the Bull-terrier Club, especially its president, who suggested that the ears of dogs bred to the current standard had thickened to stand up after cutting. Moreover, no one knew what its "natural ears" might look like, if indeed, they had ever been anything other than made for cropping. Shirley was again able to engineer a delay, proposing that it was such an important decision that the Kennel Club should consult with breed clubs, breeders, and exhibitors. Tellingly, setting vested interests against emotion, he commented that "the Kennel Club should be the guardian of the rights of specialist clubs and exhibitors, and while not wishing to ignore the sentimental part of the matter, still [cropping] should be dealt with cautiously, as greatly affecting exhibitors."[37]

Many in the Kennel Club and outside claimed that it should be leading, not following, opinion; after all, its first rule was to "endeavour in every way to promote the general improvement of dogs." The result of the consultation was considered at summer meetings. A majority wanted to continue cropping in every breed, except, unsurprisingly, Irish Terriers.[38] Silent supporters of the practice had found their voice and won the day. The Kennel Club abandoned reform and continued to endorse what their critics considered a "barbarous" practice.[39]

Faking was the other burning question of the hour and, while nowhere near as controversial as cropping, also raised questions about the aims of dog shows. As one correspondent to the *Stock-Keeper and Fanciers' Chronicle* in

1888 asked, "Why are dog shows held? Are they intended to encourage the skill of the hairdresser, or are they meant to improve the various breeds of dog?" The Kennel Club once again declined to alter its rules and left judges at local shows to decide what were acceptable "alterations." While chalking, dyeing, and altering features other than ears and tails were condemned, fanciers mostly tolerated the "minor crimes" of "removing soft, long dead hair," shaving ears, and making "little finishing touches to Collies, Great Danes, Mastiffs, Spaniels, curly Retrievers, etc., etc." Some fanciers wanted a rule that only allowed "natural" dogs at shows; however, the dominant view was that certain faking practices were near universal and often impossible for judges to spot.[40]

Controversies over cropping, trimming, and cosmetics caused problems for the acceptance of certain breeds, for example, the Old English or White English Terrier.[41] In the first edition of *The Dogs of the British Islands* in 1867, Walsh distinguished three types of the "vermin killing tribe" of Terriers that were not Skyes—Dandies, Fox, or Toys—but there were more types recognized at shows, including the English White. This dog was smooth coated and, because of low numbers, was often shown in the class for "Terriers except black-and-tan." By the fourth edition of Walsh's book in 1882, the White English Terrier was one of thirteen Terrier breeds, now distinguished from the Black-and-tan (or Manchester) Terrier by color. Despite being bred by well-known figures, such as Vero Shaw and Gordon Stables, the breed never enjoyed popularity. In 1894, Rawdon Lee explained why, quoting Dr. Andrew Lees Bell, a Terrier breeder from Dunfermline. "The great amount of trouble requisite to keep white English in form and to prepare them for exhibition naturally exercises an influence inimical to the popularity of the breed. The cropping of the ears, the trimming of the tail, shaving the ears, the washing and general anxiety to keep the dog spotless till after the show, all combine to make the hobby too tiresome to allow the breed to be popular with those at any rate who have little leisure for the indulgence of their pet hobby." The breed's popularity declined further with the eventual banning of cropping. Lee wrote of the "decadence" of "the most fragile and delicate of our terriers," due to small numbers leading to inbreeding. Bell saw the "decay" and disappearance of the breed as inevitable, as their place had been taken by dogs "better suited to the wants and conveniences of the present day than they unfortunately are."[42]

## Cropping II

Cropping hit the headlines again in November 1894, when three people appeared at Worship Street police court in London, charged with cruel treatment of a Bull Terrier in a prosecution brought by the RSPCA. Typically, the working class was in the dock. Harry Cooper, bar steward at the United Radical Club, was a "dog dealer" who had allegedly commissioned Carling and his wife, Mary, to do the cropping. At the hearing, the procedure and the dog's suffering were described in graphic detail. In their defence, Cooper and the Carlings maintained that they were responding to the demands of owners and, being experts, did a better job, with less suffering, than owners themselves would have done. Carling had confirmed the extent of the practice when he went to the Kennel Club's December Show at the Crystal Palace with an RSPCA officer and identified sixteen dogs he had cropped.[43]

In the early months of 1895, the dog press piled pressure on the Kennel Club to halt cropping, using the issue to demonstrate that its leadership was a closed group, out of touch with the modern fancy and popular sensibilities.[44] The Manchester-based newspaper *Our Dogs,* edited by Theo Marples, led the opposition. Marples was brought up in Derbyshire and Yorkshire before working his way up in textiles to have his own business in Blackburn, Lancashire. He followed a familiar route in the doggy world, from exhibiting and judging at shows, to journalism with the *Field* and the *Stock-Keeper and Fanciers' Chronicle,* to contributing to the books of Vero Shaw and Hugh Dalziel. In the 1890s he turned to publishing, beginning with *The Dog Owner's Annual* in 1890, which was followed a year later by the *British Fancier,* devoted to all fancy animals. He began the weekly *Our Dogs* in 1895 and later wrote books.[45] The *Annual* and *Our Dogs* became main outlets for critics of the Kennel Club, not least Everett Millais.

On cropping, Marples claimed to speak for two constituencies who had not been heard: dogs and British society. He pointed out that dogs "of course, alas! have no voice in the matter, except while the barbarous operation is being performed," while society was outraged by the deeds of a "sordid section of interested dog owners," which were a "stain upon the national character." *Our Dogs* assumed the role of canine "protector," defending dogs against those for whom animals represented nothing more than their pecuniary value—"exhibitors" again. In the *Field,* most correspondents wanted to keep cropping, but its editorials were for banning, arguing that the practice was "as useless as it is barbarous."[46]

The full hearing of the Cooper and Carlings case took place on 18 January 1895 and attracted the attention of the national press. John Colam, secretary of the RSPCA, prosecuted, calling leading veterinarians to testify that the operation was cruel and especially so when performed by unqualified men like Carling. The defense also relied on veterinarians and the views of Shaw, who argued that cropping was for a dog's well-being, protecting a "fighting dog" from future and repeated injury, not in the ring, but in their everyday lives. Witnesses from the fancy testified that an uncropped Bull Terrier was not recognized at shows and had never won a prize—cropped ears were required by the breed standard. A week later the court found all the defendants guilty. They were unable to pay the fine and went to prison. The Bull-terrier Club had financed their defense but did not offer to pay the fine. Some critics mischievously suggested the Kennel Club should have paid, as many of its leaders had previously had their dogs cropped. A week later the RSPCA brought a second case to the Worship Street court. Following the conviction of the Carlings, the defendant, William Storer, agreed to plead guilty. His solicitor asked for mercy on the grounds that "there was no doubt that the [Carlings] judgment was the death sentence of the dog-cropping practice" and that it was now probable that the Kennel Club would act.[47]

The Carling case had reanimated opinion within the dog fancy and outside, such that the Kennel Club came under immediate pressure to change its stance and revisit the reforms abandoned in 1889. Well-known dog owners, sensitive to public opinion, began to distance themselves from the club. One such was the popular author and Bulldog aficionado George Sims, who wanted to make it clear that he was "not a dog fancier, but a dog lover." The Ladies Kennel Association (LKA), established in 1894, lobbied the club for the introduction of a ban to stop any cropped dog born after 31 January 1895 being exhibited under club rules. The Bull-terrier Club finally changed its position and asked the Kennel Club to stop the cropping of their breed at an early date. On the other side, the White English Terrier Club asked for no change and maintained that breeders should not give in to a "few faddists." When the secretary of the Kennel Club announced that it had resolved not to oppose any ban on ear cropping, Rawdon Lee proposed a counterpetition, suggesting that his "private zeal" had "outrun his discretion." However, on realizing that more than half the club backed the ban, Lee had to withdraw. As *Our Dogs* commented, Lee "might just as well command the sea to stand still as attempt to stem the ocean of indignation that has arisen against the inhuman practice."[48]

The turning point came when Edgar Farman, an aficionado of the Bulldog, a breed with which cropping was supposed to have ceased long ago with the banning of dog fighting, sought support for a ban from the Prince of Wales, still the Kennel Club's patron. He ensured that the prince's sympathetic reply was widely reprinted in the national and dog press. The letter stated that the prince had kept dogs for many years and exhibited many "but that he had never allowed any to be 'mutilated.'" It concluded, "His Royal Highness has always been opposed to this practice, which he considers was causing unnecessary suffering, and it would give him great pleasure to hear that owners of dogs had agreed to abandon such an objectionable fashion." An editorial in the *Stock-Keeper and Fanciers' Chronicle* stated that "cropping was all over bar the shouting." A week later it was said to be "not merely moribund, but dead." As well as celebrating its success, the RSPCA welcomed the support it had enjoyed from the *Field,* having "drawn over to its side the sporting forces that have hitherto allied themselves against humanitarianism."[49]

However, there were still formalities to complete. On 26 February, Shirley proposed a resolution, seconded by Farman, that "No dog born after 31st March 1895, nor Irish Terrier born after 31st December 1889, can, if cropped, win a prize at any Show held under Kennel Club Rules." In the debate that followed, cropping was painted as a hangover from the "bad old times" of dogfights and bullbaiting. Speakers claimed that opinion among genuine fanciers had shifted, due not to cases in the metropolitan police courts but to the "dictates of a humane nature." Farman claimed that the Kennel Club had, all along, planned to table this motion at its annual general meeting and said that cropping was a "blot" on the "fair fame of the fancy." This humane measure was symbolic as well as practical, exemplifying how the club had made dog fancying "respectable and responsible." The motion was carried unanimously. During the changeover period, when cropped and uncropped dogs were allowed, with the former still likely to be favored, Charles Cruft, ever the opportunist, introduced a special medal at his shows for the best Bull Terrier, Black-and-tan Terrier or White English Terrier "with unmutilated ears." The members of the Great Dane Club could not reconcile themselves to the change, particularly as the practice continued on the Continent and the United States, and the club dissolved itself at the end of 1895. A new club was formed a year later, whose promoters decided to the make the best of a rule with which they still disagreed.[50]

Debates continued about cruelty in the modern fancy and who was responsible for its practice. The RSPCA had brought cases to court against docking tails, but these were rare and met with little success. Critics of docking were ridiculed by stockkeepers, farmers, and landowners, for whom the docking of horses, sheep, and pigs was commonplace, and seen to cause little suffering. Some veterinarians concurred with this view and argued that if dogs were to be docked, there was less pain if the operation was performed on puppies when the tail was less developed. However, one consequence of the Carling ruling was that it made some people uncertain as to the legality of docking. Indeed, an editorial in *Our Dogs,* published in February 1895, had fueled the close association between the two practices, claiming there was no more justification for docking than there was for cropping.[51]

In the discussions that led to the ban on cropping, Kennel Club member J. Thorpe Hincks, a devotee of Spaniels, proposed an amendment to end the "whole system of mutilation," including docking and the removal of dew-claws. Hincks maintained that Spaniels suffered greatly from docking and called for the Kennel Club's leadership to act, rather than prevaricate as they had over cropping. His amendment was voted down, but he tried again in 1896, arguing that the new rules adopted after the Carling case applied to all surgical alterations. He suggested that breeders should apply their skills in the "scientific method" of selective breeding instead of trying to acquire features through surgery and argued that, "there was nothing scientific in cutting a dog's tail off; there was nothing achieved towards improving the breed simply by cutting off a part it was not wanted to see." The points system exposed the absurdity of docking in some breeds. In the case of Spaniels, judges had to score for the tail and its carriage, the latter being difficult to say the least, when it was little more than a stump. There was also concern about the manner in which Schipperkes were being docked, with the tail "cruelly removed by being literally torn out by the roots."[52] Hincks's motion was again lost on the grounds that docking was a trivial operation. However, the club passed an amendment targeted at Schipperkes and their native Belgian breeders; it was proposed by George Krehl and seconded by Turner, who said that "it was cruel and inhuman to take out the whole tail."[53] A subcommittee was formed to consult with specialist clubs on further action. It decided in October 1896 that docking was only to be allowed with specified breeds—Spaniels, Poodles, and Fox, Irish, Sealyham, and Airedale Terriers.[54] The leaders of the Kennel Club could congratulate themselves that with these bans and other changes, they had made the modern fancy more humane and given it greater respectability. Some critics thought that reforms should go further, though. In April 1897, Harold Warnes called for the end to all forms of "trimming"—surgical and cosmetic—and for dogs to be "shown as Nature intended."[55] Particularly offensive to him were Poodles, which were "clipped, singed and shaped in the most grotesque fashion" by their owners. Needless to say, Poodle owners were predominantly women, and Warnes's remark was another dig at their growing number in the fancy. But women, too, were unhappy about the show fancy and were campaigning for its reform.

### Ladies, Shows, and Breeds

In January 1886, the *Morning Post* published a letter from the well-known author "Ouida" (Maria Louise de la Ramée), complaining that "a dog show is a painful and brutal thing, exciting the passions, desires and nerves of a dog in a most cruel and most injurious manner." In the 1860s, Ouida had begun publishing a series of melodramatic novels that divided literary opinion but sold very well. Thought by some to be "snobbish, intolerant, and rude," she was reportedly "a difficult hostess and a demanding, insulting guest, yet she still attracted enough important people to hold a salon." Financial difficulties led her to move in the 1870s to Florence, where she continued to write and took up various causes, including antivivisection and antimuzzling, about which she regularly inveighed the national press. Her letter in January 1886 was prompted by rabies-control measures in London, which she described as the "muzzle-and-slaughter craze." Ouida was disappointed but not surprised that fanciers supported such measures, which in her eyes proved that they were not real dog lovers. With characteristic invective, she complained that show dogs led a life of "irritation, fatigue, alarm and weariness," being subject to ill treatment and neglect when transported on the railways, and were only kept for the vanity owners and as "a machine for grinding guineas." Three years later, in the context of a 1889 rabies scare, she drew a parallel between the cruelty of the muzzle and the confinement of show dogs in railway vans, chained to benches and left unattended overnight. Ouida aligned the damage to dogs with that to national character, declaring that animal cruelty was "doing so much to destroy the naturally sturdy, independent, and generous temper of the country." Despite her self-proclaimed antifeminism, Ouida's critique of shows and their effects on the nation's dogs was implicitly gendered and similar to female critiques of vivisection in science and medicine, which were portrayed as calculated male cruelty.[56]

An explicitly female alternative to the Kennel Club, with different values, saw institutional expression in 1894 with the creation of the Ladies Kennel Association. One of its aims was to counter the misogyny at all levels of the fancy, routinely demonstrated by "slighting remarks" and sneering at "doggy" women. For members of the nascent association, a civil and humane fancy was a precondition for women's presence and defined their role as reformers. In its campaign against cropping, the LKA argued that women had a particular sympathy "because they are liable to look on the dog as something better than a medium for betting and a money making piece of property." By presenting themselves as moral interpreters and authorities on animal suffering, members provided a rationale for their inclusion in shows as agents of greater respectability. Their arguments were not entirely dissimilar to those of social purity or "equal but different" suffrage campaigners, and it cannot be coincidental that the LKA was formed as "the woman question" gathered momentum, even though some of the women involved distanced themselves from feminism. Initially, the Kennel Club was open to giving women a recognized place in the fancy, accepting the registration of the association in August 1894.[57]

The idea for a ladies organization came from a meeting at the Toy Dog Show in May 1894 and was orchestrated by Mrs. E. J. Thomas, Miss Darbyshire, and, the effective leader, Mrs. Alice Stennard-Robinson (née Cornwell and often styled Stannard-Robinson) (fig. 7.3). A colorful and controversial figure who had shown Pugs in Australia and London, Mrs. Stennard-Robinson would meet many challenges as honorary secretary of the LKA. She was born in 1852 in West Ham, Essex, to a working-class family, which emigrated to Australia when she was a year old. They moved around, including to New Zealand for a while, experiencing mixed fortunes until, quite literally, her father's company struck gold.[58] Alice had returned to London and made a career in music, but when her father died she returned to Australia to take over his company. She had great business acumen and, with several large gold finds, became exceedingly rich. Returning to London at the end of 1887, she floated the newly named Midas Company and, seemingly on a whim, bought the *Sunday Times*. This acquisition was suggested by her fiancé, Philip Robinson, an author turned adventurer-journalist. In succeeding years, the pair traveled between England and Australia, before settling in London in the early 1890s. Alice's great wealth brought her a high social position, which was seemingly unhindered by her

***Figure 7.3.*** Alice Cornwell, later Mrs. Alice Stennard-Robinson (1889). National Portrait Gallery, London

working-class, colonial origins or by her partner's notoriety; he was a bankrupt divorcee and had been sacked as editor of the *Sunday Times* for publishing an article on the Prince of Wales's alleged debts. They were referred to disparagingly as "The Swish Family Robinson."[59]

The name Ladies Kennel Association was adopted after the Kennel Club objected to it being called the Ladies Kennel Club. The LKA's initial goal was to have better facilities for women at shows and special classes for ladies' dogs, and in its early months the Kennel Club offered support. So too did the *Stock-Keeper and Fanciers' Chronicle,* which made available its offices and rooms for the LKA's officials to use. At the Kennel Club Show in October 1894, the LKA offered forty-eight silver medals for women exhibitors. However, the honeymoon was short-lived, with sentiment turning against it ahead of its first show in June 1895, which was to be held in Ranelagh Gardens and limited to lady exhibitors. The change of sentiment was led by the Duchess of Newcastle, described in Lane's *Dog Shows and Doggy People* in 1902 as "the most popular of her sex of the ranks of Doggy People" after Queen Alexandra. Famed for her Borzois and Fox Terriers, the duchess lived at Clumber Park, though she seems not to have bred or exhibited its Spaniels. Only a few days before the first LKA show, she announced that she would not be entering her dogs. Stating her disapproval of the principles of the show, she wrote that she could "see no good in a Show held entirely by ladies, for as a rule their dogs are the worst in the class, and seldom get more than vhc [very highly commended]."[60] She was also against the award of special ladies' prizes at other shows.

The LKA show went ahead and was, according to one report, "a brilliant social function" (fig. 7.4). It attracted large crowds and a visit from the Prince and Princess of Wales, and had novel features, including an art show and a whippet race, no doubt to the surprise of many, as it was a working-class sport. If anything, the show was too successful, and overcrowding led to charges of chaotic organization. The *Sunday Times*—naturally, given its ownership and editor—praised Mrs. Stennard-Robinson as "a brilliant organiser and untiring worker." However, the Duchess of Newcastle was unforgiving about the LKA and its show. Writing to the *Stock-Keeper and Fanciers' Chronicle,* she opined that "dog shows, I imagine, were started to improve the breed of dogs—this association seems to me started for the sole object of giving prizes and specials to dogs mostly not worth their entry fees. I consider a show held only for ladies' dogs is very liable to set the wrong value on dogs shown." Mrs. Stennard-Robinson replied politely, yet firmly, refuting her insinuations. The reaction in the dog press was divided, but those supporting the duchess typically fused criticism of the show with prejudices about "ladies." One suggested that the prizes had been bestowed "not on dogs" but "on their owners," which was a bold claim, as among the judges were Kennel Club stalwarts like J. H. Salter and Sidney Turner. F. B. Lord from Chudleigh in Devon wrote that an LKA show was "an utter

Mrs. Willie Temple's Pomeranian

Miss Mackenzie's Italian Greyhound "Hero."

Lady Deerhurst.

Miss Mackenzie's Italian Greyhound "Mario."

Mrs. Andrew's Pug "Samson of Swanland."

Miss Mary Moore (*Photo by Alfred Ellis*).

Mrs. Andrews and "Champion Nancy."

Mrs. Carstairs.

Mrs. Hannay.

Mrs. Borman and her Borzois "Statesman" and "Statlight."

Miss May Palfrey (*Photo by Alfred Ellis*).

Mrs. Preston Whyte.

Mrs. Claude Cane.

Mrs. Willie Temple.

The Countess of Aylesford.

Mrs. Panmure Gordon.

Mrs. Peel Hewitt.

THE LADIES' KENNEL ASSOCIATION SHOW: WINNERS AND EXHIBITS.

*Figure 7.4.* The Ladies Kennel Association Show: Winners and Exhibits. *Hearth & Home: An Illustrated Weekly Journal for Gentlewomen,* 13 July 1899, 387. © British Library Board

absurdity" and could not "be seriously supposed to advance the cause of scientific breeding; it can only be regarded as a foolish fad very likely to bring dog-showing and prize-winning into contempt." Despite (or, perhaps, because of) the controversies, the Princess of Wales agreed to become the LKA's patroness.[61]

The initially cordial relations between the LKA, the Kennel Club, and the dog press did not survive the Duchess of Newcastle's intervention, which seemed to start open season on women's role in breeding, showing, and keeping. Only the *British Fancier,* which had a weekly "Ladies Page," came to the LKA's defense. A common complaint was that ladies' favorites were tiny, fragile dogs with small brains that made them nervous and stupid, and long coats that hid pests and produced mange. This complaint's parallels with arguments advanced against women's access to higher education and the professions on account of "nerves" and fragile bodies are striking. For critics, toy dogs were less companions than decorative and indulgent possessions. A comic piece in the *Strand Magazine* in 1897 observed, "The people who are showing all these dogs are not their masters—they are simply their lackeys, valets, cooks, hair-dressers, shampooers and bottle washers." On the other side, "Sad Dog" maintained that women could and should compete equally with men, and had "never heard of a dog deteriorating through being owned by a woman." There were also claims that men were abusing LKA events by entering dogs under their wives' name for easy wins.[62]

Mrs. Stennard-Robinson bore the brunt of the criticism of the LKA, as well as many personal insults. A pivotal event was the association's show in Holland Park in June 1896. The arrangements came in for wide criticism, though these seem largely due to the event's success in attracting large attendance. The report in the *Ladies Kennel Journal* annoyed all sides; on the one hand, it glossed over the problems, and on the other, it provided completely misjudged flattery of Mrs. Stennard-Robinson, observing that she went around the show "like Ariel sailing majestically across the meadow . . . laden with a whole silversmith's shop in the way of prizes."[63] Although she saw herself as a strong woman, leading from the front and not afraid of confrontation, too often she was ahead of the LKA membership, prompting comments that she was autocratically operating a "reign of terror." On several occasions she resigned from her position as honorable secretary, but she was always invited back, gaining enough support from loyal members. The LKA fell out with the Kennel Club, largely over technical issues, such as objecting to the publication of ladies' private addresses in exhibition catalogues, entry conditions, and the status of "Champion Classes" at LKA shows. But underlying these differences was hostility on the grounds of sex and social class, as was made evident in abusive personal comments. As "a new woman" and a rough "colonial" with "new money," Mrs. Stennard-Robinson was the target of explicitly misogynistic and class-based comments. But she gave as good as she got, allegedly calling the Kennel Club's committee "baboons." When *Our Dogs* complained of her "use of the lowest slang expressions," she threatened to sue, and the magazine published an apology.[64]

The LKA continued to hold annual shows in London and around the country. Relations with the Kennel Club did not improve, and in 1897 the club announced its intention to establish its own Ladies Branch (KCLB). There was no hurry; the branch was not formally instituted until two years later. In the meantime, LKA shows continued to be social successes, a regular part of the London summer season, and to be innovative about what constituted a dog show. Although the LKA had its origins at a Toy Dog Show and Mrs. Stennard-Robinson kept Pugs, its shows included many breeds, and large dogs were always included. Celebrity and glamour were important. There were special classes for the dogs of actors and actresses, with the "Sarah Siddons Classes" judged by the socialite Lady Colin Campbell. These aspects were reported on sarcastically, as in December 1897 when it was suggested that a cloud hung over the LKA's show due to the recent death of Fussy, Sir Henry Irving's Fox Terrier, given to him by his legendary acting partner, Ellen Terry. At this event, the *Daily Mail* also reported seeing classes for low comedy dogs, professional beauty dogs, general utility dogs, and paying amateur dogs.[65]

An increasingly prominent feature at LKA shows and one said to distinguish them from self-interested Kennel Club events was that profits would go to good causes, such as the RSPCA and the Great Ormond Street Children's Hospital. There were also special prizes for lifesaving and charity-collecting dogs, which were common at railway stations, wearing identifying coats and boxes strapped to their backs. During the South African War (1899–1902) prizes went to dogs that collected the most money for injured soldiers' hospitals. This charitable orientation reaffirmed the moral status of lady breeders and exhibitors who considered themselves separate and above the business of mainstream shows. In the end, such shows lost money, and debts, blamed ultimately on Mrs. Stennard-Robinson's mismanagement, closed down the LKA in 1903. It was quickly relaunched under new leadership and within the Kennel Club's ambit, thus no longer offering an alternative doggy world.[66]

Ouida did not accept that the LKA had created an alternative. In July 1899 she returned to the columns of the London press bemoaning "the encouragement given [to dog shows] by women of rank." She asked, "How much longer must these foolish and cruel exhibitions go on? What is their motive? A paltry vanity, or a love of lucre. It can be no other." To her former criticism of these "most harassing, injurious, unnatural, and sexually irritating gatherings," she added objections to the commodification of dogs. Ouida satirized upper-class women exhibitors, positing that the shows were put on "in order that Lady Gay Spanker, or Mrs. Teddy Cheeke, may add to the cups on the buffet, and sell the pups from her kennel at a higher figure. What do we hear constantly? 'Lady Moneyspinner has sold her miniature Pomeranian Crumb for three thousand guineas to Baroness Goldleaf.'" She maintained that in making claims for a greater role for their sex in canine affairs, the women in the LKA had not brought "sweetness and light" and the "mercy and tenderness of women" but had persisted in their efforts with the same values as dogmen.[67]

The LKA answered Ouida's claims about cruelty at their shows, arguing that they had set more humane standards, efforts that ladies' critics had mocked as indulgence of their dogs. A reply by "Leiothrix," also in the *Standard,* made larger claims about the recent influence of women on the modern fancy.

> Beyond a doubt, the intrusion of the softer feminine into dog-exhibiting has been entirely for the good of the dog, which used to be, for the most part, a mongrel more or less; which used to be, if pure bred, kept for sport, under sporting conditions and a gamekeeper's rule; which used to be exhibited by men of the lower classes, if at all; which if it were petted in any way, generally lived a life of fat and pampered inaction ending in painful disease; and which has benefited by the spread of knowledge brought about by its increased popularity.[68]

Leiothrix went on to argue the virtues of "breeds kept carefully pure," welcoming the takeover of shows by "men and women of refinement."

In these exchanges, no one raised the matter of what shows had done and were doing to breeds. Yet back in 1891, Ouida had written that dog shows had become "a pest" and that the whole enterprise was a sham:

"The dog-judges . . . are all of them dog-sellers, and set up an wholly arbitrary and often most foolish standard of excellence, setting down in true Justice Shallow fashion mere fantastical rules to what is or is not form in their sapient sight."[69]

In 1895, Frances Power Cobbe had made an equally fundamental assault from a contemporary feminist perspective. Cobbe, widely known for her campaigning against vivisection, had written to the LKA a letter in which she qualified her support of the organization: "I may boast of being a true Dog-lover: I am not a Dog-fancier. I take no interest in the various points of conventional beauty, which are the glory of Dog-shows. On the contrary, I am so deeply interested in the character of dogs and their intellectual possibilities that I rather *resent* the importance attached to the length of their tails or the precise angles of their noses." She echoed Galton's remarks on breeding "intelligent dogs": "I want to see breeds carefully formed of specially clever dogs, whether mongrels or otherwise: a clever male mated with a particularly intelligent female, and do so for some generations, no foolish beauties being admitted into the family. At the end we should, I think, see some remarkable results." An eccentric figure, Cobbe was nonetheless admired for her principled stance on many issues. By this time, however, she was a fading influence, having retired to the family estate of her companion Mary Lloyd, and it was ironic that her uncle Richard Lloyd Price had been a founding member of the Kennel Club.[70]

In the 1890s there was a vogue among ladies for larger dogs, especially those with long silky coats. The wealthy favored, for instance, the recently imported Borzoi, also known as the Russian or Siberian Wolfhound. The dogs of Princess Alexandra and the Duchess of Newcastle were regular prizewinners. The breed was initially reputed to be highly strung, ill tempered, and vicious, characteristics that some xenophobic commentators thought it shared with Russian peasants. However, in the Borzoi's few years in Britain, according to William Drury, it had acquired the civilized character of its new owners and become "a thorough aristocrat, quiet and dignified in his manner, never rushing about to the detriment of the household goods, and seldom given to unnecessary barking."[71]

Another ladies' favorite was the Collie, said to be second only to the Fox Terrier in popularity as a show entry and as a pet.[72] It was described as "beautiful," with a fine head, elegant body, and a long soft coat, with many colors acceptable. It was also intelligent and had exemplary character, features that Gordon Stables celebrated in the much-reprinted children's novel *Sable and*

*White,* an "autobiography" narrated by Luath, a show Collie.[73] The breed had its own weekly column in the *British Fancier,* an honor only enjoyed by two other breeds, the Bulldog and Fox Terrier. Collies were ideal pets, "companionable" and "petted by all classes . . . from the commoner to the Queen on the throne." Among fashionable ladies, the best known was Lady Burdett-Coutts, "the richest heiress in all England" and a long-time supporter of animal charities. She was always accompanied by a dog given to her by the actor Sir Henry Irving; it was no show dog, having been bought from a shepherd at Braemar, and presumably had the purest Highland blood.[74]

## Collies

As Collies became more popular among men and women of different classes, Hugh Dalziel and Rawdon Lee responded by publishing books on Collies (based on the Collie chapters in their previous books), and many features on the breed and individual dogs appeared in the canine press. Although most accounts acknowledged variations within the breed, their overall framing was essentialist, as they were about *the* Collie. In the early 1890s an alleged "Collie boom" drove the sales of prizewinners and stud dogs to reach new heights. Prices for Collies were high in Britain, and British breeders also found a ready and lucrative market for their dogs in North America as the breed became fashionable among the super rich on the East Coast.[75] In 1893 the American banker J. P. Morgan bought three dogs in London for £1,000. (Morgan's New York estate, Cragston, had heated kennels, with electric light and porcelain baths, that could house up to one hundred dogs.) In 1897, Ormskirk Herald was reported to be the most expensive dog ever sold when Mr. A. H. Megson, "The Collie King," paid £1,600 for him.[76]

Collies were in a unique position among breeds. They were working dogs, but not nonsporting. They came from the country and were associated with rough shepherds, not the landed classes. The Kennel Club and most books categorized them as nonsporting, reflecting their growing presence in homes and on show benches, which posed the old question of the extent of a conflict between breeding for fancy points and breeding for ability and character. While Dalziel claimed that shows had led to the deterioration of working breeds, Lee disagreed: "Dog shows have added to the beauty of the collie; the [sheepdog] trials must add to his intelligence." A decade later he held the same view, writing that "there is no doubt that our collie or sheep dog is one of the most useful of the canine race, and within the last quarter of a century he has likewise been made ornamental."[77]

How had Collies been altered over the Victorian period? In 1845, Youatt had stated that "the sheep-dog possesses much the same form and character in every country" across the civilized world. Mainly concerned in his description with herding ability, he acknowledged variations between sheep-dogs from different countries, with the dogs' size and form developing relative to the terrain and type of sheep they herded. Drawings of the English and Scotch sheepdog were included, the latter having the expected coloring of black or black mixed with grey or brown. Walsh continued with the same division in both *The Dog* (1859) and *The Dogs of the British Islands* (1867). Charles Pearce (Idstone), writing in 1872, denigrated the English sheepdog as "inferior." It was rough in body and spirit, "slower and heavier," and had taken on the character of his teacher, the English shepherd, who was "surly, silent and for the most part ignorant, and he has a special dislike to strangers." In contrast, the Scotch shepherd's dog, now called the Scotch Colley, was "one of the most perfect animals extant." The English sheepdog disappeared for a while, seemingly becoming the Bob-tailed Sheepdog or Drover's Dog, before reemerging, paradoxically, as the Old English Sheepdog, shaggy tailed and bearded. In his book *The Collie,* Lee wrote that there was "no record" of "how this old-fashioned variety came about," implying that it was a modern manufacture.[78]

In the 1870s, Walsh attacked the manner in which breeders were shortening and thinning the coat of Scotch Collies to impart beauty, because with this change the dogs were losing warmth and water protection. He believed that this result had been mainly achieved by crossing with Gordon Setters, to add "beauty," impart "brilliancy and rich colour to the coat," and give "feather" (fine hair) to the legs. Walsh warned of the growth of what he called "the 'toy dog' point of view" and suggested that "if a pet is wanted solely as such, the Gordon setter in his purity is a handsomer dog than the colley with a more pettable disposition, and it would be better to select him accordingly."[79]

In 1873 the breed arrived as a show dog and was given an aristocratic imprimatur when Sewallis Shirley's Shamrock won first prize in the open Sheepdog group at the first Kennel Club Show at the Crystal Palace. The report on the show was clear that conformation had trumped intelligence, remarking that Shamrock was "about as well-feathered a dog as we've seen at the show, but he has a soft 'daft' sort of look about him which is very different from the intelligent-looking 'Brush' of Miss Chappell, which was second." The following year there were two classes, rough and smooth coated, and soon a third was added: the English rough-coated short-tail variety. The prizewinner in 1875 was another of Shirley's dogs, Trefoil, which is now said to be foundational for the modern show Collie.[80] Throughout the 1870s and 1880s, the sporting press was more interested in the development of Sheepdog trials than in the progress of Collies at shows. The first recorded trial in Britain was at Bala in October 1873, organized by John Lloyd Price and Shirley.[81]

Dalziel was typically outspoken in his account of the Scotch Colley: "There has been almost as much nonsense written about this dog as on the subject of teetotalism." Claiming that "a usurper rules where the true colley should reign," he blamed this state of affairs on "three delinquents": "the incompetent judge," "the advertising and ignorant dealer," and the "too often superficially informed dog show reporter." Dalziel had strong views on color, he wanted "black on the body, with tan points, and white collar and chest and forearms, and at times a blaze up the face and white tip to tail." He bemoaned the change in fashion from "the high domed skull and more or less full forehead" to "triangular heads, with the foreheads planed down to a perfect level and tapering jaws as long as those of a pike." This was said to be typical fancy breeding, favoring the exaggeration of particular points and ignoring the overall balance of features. All that said, the Scotch Colley was one of his favorites: "There is no dog that excels the colley in good looks, high intelligence, and unswervable loyalty to his master," qualities that also assured its "high position as a general favourite with the public."[82]

The Collie Club, adopting a new spelling, was founded in 1881 and set a standard for shows, in which color was "immaterial," and the coat earned most marks. Its first show was held in June 1885 at the Westminster Aquarium, offering a social as well as canine contrast to the Bulldog Show the previous week. The *Daily News* claimed that few "real workers" were in the hall and that, for most of the one hundred dogs on show, "Their main business in life seems to be to fawn on their mistresses and masters, to be petted and to act as house guardians." Collie shows became an annual event and suffered the usual troubles of complaints about judges applying the breed standard erratically and unfairly (fig. 7.5).[83]

The past, present, and future of the Collie was the subject of an August 1888 article in the *Kennel Gazette* by William Arkwright (a descendant of Richard Arkwright of spinning jenny fame), coal mine owner and breeder of Pointers and other sporting dogs.[84] "The Fancier versus the Collie" targeted the effect of shows on working and sporting dog breeds in general.

THE COLLIE DOG SHOW AT THE ROYAL AQUARIUM, WESTMINSTER

*Figure 7.5.* Success—Failure—Spoilt—Neighbours—Finishing Touches: The Collie Dog Show at the Royal Aquarium, Westminster. *Graphic,* 14 Apr. 1888, 6. © British Library Board

Arkwright saw dog breeding going the way of fancying with pigeons, poultry, and rabbits, where "monstrosity manufacture is carried on." Concerning conformation, "it would be a reasonable and fair bargain to say that while the 'Toys' . . . might be handed over to the vagaries of the fancy and its priesthood," "useful breeds" should be protected. He characterized the fancier thus: "All his perverted ingenuity is summoned to the task of creating distortion, and he glories in a deformity which embodies his particular craze to excess: in fact, he is almost heroic in his resolve to sacrifice all else to gain the coveted exaggeration."[85]

Arkwright felt that sporting dogs, like Pointers and Retrievers, had largely escaped the attentions of fanciers but that the Collie had not and was being ruined. He went on in an imaginative vein that might have inspired the "Dog Fashions of 1889" that appeared in *Punch* a few months later: "Fanciers have recently determined that a Collie should have an enormous coat and enormous limbs, . . . so they have commenced to graft on to the breed the jaw of an alligator, the coat of an Angora goat, and the clumsy bone of a St. Bernard."[86] He warned that specialist clubs were becoming "hotbeds for the propagation of 'fancy.'"

Arkwright then turned his invective to loss of ability and intelligence. He contended that the Collie had been detrimentally altered into "a mere show bench ornament" and "little toy," useless for any work. Judges were allegedly concentrating on the head, ears, and coat and neglecting the body, which was losing bone and strength, such that Collies were "soon to be as small as a rabbit." William Rust wrote to the *Stock-Keeper and Fanciers' Chronicle* suggesting all was not lost and arguing that the Collie's body and character had been acquired through "continual activity and watchfulness," an achievement that, regardless of squareness of body or density of coat, should be recognized. A retrospective on "Dogdom in 1892" was also positive about the breed as a whole, maintaining that it had become "more uniform," with "fixity and assimilation of its type." One short-lived enthusiasm had passed: "Borzoitical" heads, which were "long and narrow . . . with prick ears" and low brows implying a loss of intelligence.[87]

The actions of Collie breeders, abetted by the Collie Club, were still being questioned a decade later. Gregory Hill, a Birmingham breeder, complained that the Collie fancy had followed the common pattern of plumping for "one special feature," in this case "the 'pretty face' craze" and a preference for an "abnormal length of head."[88] He accepted that in its natural, pre-show state the head usually lacked refinement and needed improvement, but the changes had now gone too far. More broadly, show Collies had become victims of arbitrary aesthetics, fashion, and commercial interests. Breeders were prone to "split hairs and finick about trifling facial differences," which might lead to "good bodied and headed dogs beaten by weeds with pretty blazes, or a sweetly pencilled or shaded eyebrow." He worried that special features were often "boomed for pecuniary reasons," with breeders—he called them "smashers"—whispering at shows to influence judges and convince buyers of the worth of the "special feature." Hill argued that improvement would have to "stand still" at some point but was pessimistic in the short term, as "alluring prizes in hard cash" would be directing efforts. The Collie exemplified the contradictions of fancy culture. Was there to be just one Collie, or two, or three? That is, was there to be a fancy dog for showing, a Sheepdog for working on the farm, and a companion dog for the home?

Many of the issues in Collie breeding and showing, and indeed of breeds as a whole, from faking to the role of the Kennel Club, were epitomized in a

THE CORRECT EAR–OLD STYLE.                THE WRONG EAR–NEW STYLE.

*Figure 7.6.* The Great Collie Ear Trial. *SK&FC,* 20 Jan. 1899, 41

controversy over a single breed point—Collies' ears. "The Great Collie Ear Trial" did not take place in the field with sheep, nor on show benches, but in the Edinburgh Court of Sessions. On 15 December 1898, the judge began to hear a case brought by Harry Panmure Gordon, a wealthy stockbroker, president of the Scottish Kennel Club, and leading figure in the London doggy world. He accused Andrew Watson, a baker, and James Montgomery, a powerloom tuner, for selling under false pretenses. He had bought a dog unseen on the strength of its record at shows and its pedigree, but when his purchase arrived, he was dismayed to find it had "prick-ears," not the correct "drop ears" (fig. 7.6).[89]

The implication was that Watson and Montgomery, typical for their class of fanciers, had misrepresented their dog on the basis of having manipulated—and thereby faked—its ears. Leading figures from the fancy and veterinarians gave evidence. The crux of the case was whether "training dogs' ears" was legitimate, it being custom and practice with the breed. Perhaps Panmure Gordon was behind the times, and small ears were in vogue, which meant using weights to train them to fold over was necessary to meet the breed standard. More controversially, there were suggestions that the standard be changed and prick ears accepted.[90] In the event, Panmure Gordon lost the case. The judge ruled that the shape and size of Collie's ears was not for the court to decide and recommended that Panmure Gordon take the matter to the Kennel Club. The judge was surprised that the Kennel Club had failed to show leadership on the question; its critics were not surprised.

## Conclusion

Despite all the questioning, dog shows continued to increase in number and popularity. In 1899 a record number of events were held under Kennel Club rules, and there were 2,457 entries in the club's forty-fourth show at the Crystal Palace that October. However, the number of classes was reduced, and one report stated that it was "painful to see so many old breeds giving place to foreign importations." The report went on to comment on the "wretched spectacle" of Pointers and Setters, revealing that "the 'leading society' has lost touch with its sporting men and county gentlemen." In London, Cruft's in February each year was the largest event, regularly attracting over three thousand entries, with its success attributed to "its great popularity among the ladies" and their favored breeds. However, in *Country Life Illustrated* in 1899, the report of Cruft's show was quite negative about the state of many breeds: "The fact is that, in their pursuit of something new and foreign, English breeders have neglected to support the old national varieties, which used to be the pride of their fathers, the results being that whilst the Mastiff has become moribund, the Bulldog has been metamorphosed into a caricature of what it used to be." Exhibition and entertainment had seemingly replaced competition as the principal raison d'être of shows. Furthermore, the improvement of native breeds had been forsaken for the importation of foreign dogs, the latest of which were the Bouledogues français, Elk-hounds, Eskimos, Lhasa Terriers, and Samoyeds.[91]

If the nation's, and now the world's, dogs were to be continually improved, and certain breeds rescued from deterioration, then breeders would have to respond. There were renewed calls for scientific breeding to avoid the pitfalls of inbreeding, linebreeding, and close breeding, and for breeders to choose healthy dogs and bitches, with pure blood evidenced by long pedigrees. However, in 1899 a debate at the Kennel Club revealed the limitations of its stud book as the guide to the improvement of breeds. A resolution put to the general meeting in October objected to the registration of a crossbreed dog, which, if carried, would have excluded all such dogs from competitions. As with the Newcastle upon Tyne Dog Show forty years before, the question of crossbreeds was key to the improvement of the nation's dogs and the meaning of breed. The case for allowing the registration was that crossbreeding was essential to the breeder's art and essential for improvement to change points and counter the deleterious effects of excessive inbreeding.[92]

There were other issues at stake in the debate. If crossbreeds were banned from the stud book, the implication was that its register would be seen as "an absolute guarantee of breed," which it was not. In fact, the Kennel Club stud book was just about the validation of breeders and owners, not their dogs. The club's leadership made it clear that "the primary object of registration . . . was to prevent a certain class of fraud, and . . . to enable dogs to be identified and to give the Kennel Club a clear idea and knowledge of, and therefore authoritative control over the dogs shown under its rules."[93]

The attempt to ban the registration of crossbreeds was roundly defeated. The status of entries in the stud book was clarified by a second resolution, which sought "to correct the misapprehension that registration is a guarantee of the purity of blood," when it was only meant "to furnish evidence of identity and a record of its pedigree." This tacitly confirmed that the Kennel Club's role had been social, concerned with governance and setting the stage for breeders, breed clubs, show judges, and, above all, competition to establish and maintain conformation standards as the best way "to promote the general improvement of dogs."[94]

# Conclusion

## The Past in the Present

Victorian dog shows and the culture they promoted fundamentally changed the dog, leading to what we have termed the invention of the modern dog as one whose form and identity are defined by its breed or lack thereof. What had previously been called *varieties* and *types* of dog were remodeled physically and culturally into breeds, which were defined by ideal conformations. The improvement of breeds, achieved by show competitions and scientific breeding, secured the best conformation points and concentrated pure blood. The new breeds were in time homogenized: externally, as breed populations became more uniform in look, and internally, as inbreeding reduced genetic diversity. The type of dog invented in the Victorian era spread across the world, and British dog show culture was emulated in North America, Continental Europe, and many British colonies. Many of top dogs made celebrity appearances, enjoyed stud tours, or were exported in a growing international trade.[1] The invention of the modern dog was a remarkable transformation, which we believe this book is the first to detail by focusing on the co-creation of new canine forms, and the people, institutions, and culture that produced the changes.

Our social and biological history placing the invention of the modern dog in the Victorian era is supported by recent work on the dog genome. A study led by Greger Larson and including leading experts on genetics, archaeology, and biogeography reviewed new data on the domestication of the dog and confirmed the modernity of breeds. Larson and his colleagues note that the term *breed* was "problematic" and that "claims for the antiquity (and long-term continuity) of modern breeds are based upon little or no historical or empirical evidence." Like us, they "only use the term *breed* when referring to modern dog breeds recognized by kennel clubs," and they express in biological terms conclusions similar to the argument advanced in this book: "The vast majority of modern breeds were only created in the past 150 years emerging from what was a relatively homogeneous gene pool formed as a result of millennia of human migration and the subsequent merging of multiple, previously independently evolving dog lineages. This history, along with the closed gene pools and small effective population sizes associated with recent breed formation, also explains the strongly supported genetic monophyly of individual breeds and the lack of resolved relationships between them." "Genetic monophyly" refers to shared descent from common ancestors and relates to the study's larger conclusion about domestication, which is that "there is a major disconnect between truly ancient dogs and modern breeds."[2]

Dog breeds were cultural and material creations that resulted from social negotiations, struggles, and accommodations, the terms of which were shaped by the wider Victorian contexts of industrialization, urbanization, and social class. The narrative set out in this book shows the contingencies of the co-production of breed standards and dogs that conformed to those standards. Breed standards were never static. They were not meant to be, as they were products of the age of improvement. Standards were contested at every level, from the judging of the competition classes at the many hundreds of large open shows held across the country, down to the negotiations among breed club leaders to set conformation standards. While there was agreement that all breeds should be improved, there was never agreement on what this meant, and complaints were especially strong about the "recovery" and "manufacture" of certain breeds. We have shown the plurality of relationships between Victorians and their dog breeds, and readers may recognize in these nineteenth-century issues and social tensions many continuities with recent debates about pedigree dog breeds.[3]

The leading authorities on dogs in the Victorian era had a sense of history. They mostly celebrated their dogs' modernity, in the sense that their dogs had been recently remade materially with morphologies that conformed to breed standards. As such, it was assumed that all breeds were "improvements" and were still being improved by the "art" and "science" of breeding, the elaboration of pedigrees, and the benefits of competition. As such, past breeding practices were cast as having been undisciplined and ill informed, with their mixing of different varieties and strains, and resorting to multiple crossings. The men and, later, women of the new dog fancy also saw themselves as an improvement on their doggy forebears, especially those in the

Fancy, being of higher social class and economic status, and better educated. They were, or aspired to be, gentlemen and ladies, people of integrity who could be trusted. These attributes were important because the accuracy of pedigrees was only as good as the trustworthiness of the breeder who supplied the information, and show judges had to present themselves as—and act as—honest, disinterested, and fair arbiters.

Twenty-first-century dog shows are shaped not only by the Victorian-born sensibilities, social fabric, and practices that we have inherited but by our own versions and interpretations of that past, an imagined past that continues to hold many of us and our dogs captive. The recent controversy on breed and breeds has often been framed as about "pedigree dogs," as on the blog "Pedigree Dogs Exposed" and in Michael Brandow's "biting history of pedigree dogs."[4] The implication is that inbreeding, from a limited genetic pool of dogs with the "right" pedigrees, is at the root of the health and welfare problems of certain breeds. Expert opinion is in broad agreement, though more nuanced.[5] We make no judgment on such claims but note that many experts, while acknowledging that there are gene-linked disorders in dogs, point to problems that arise from conformation standards set by breed clubs and applied by show judges.[6] Victorian dog breeders faced the same issues and had their own answers, based on their historically specific understanding of inheritance, which was quite different from that developed in genetics from the early twentieth century and genomics in the twenty-first century. From their own experience and that of livestock breeders, as well as assumptions about inheritance in human families, Victorians were aware of the consequences of inbreeding, evident in physical deformities or high-bred nervousness. Breeders had to balance the likelihood of such problems with the benefits of concentrating pure blood to ensure the expression of type. Crossbreeding was widely used to compensate for inbred weaknesses and to introduce a favored physical feature to improve—or ruin, as critics would claim—a breed.

In 2012, People for the Ethical Treatment of Animals (PETA) launched a campaign lobbying the public to boycott the Kennel Club's annual dog show, Crufts. Reminiscent of a mid-nineteenth-century advertisement for a freak show, the campaign's poster declared Crufts to be the "Annual Parade of Genetic Freaks Prone to Disease and Disability!" (see plate 8). The "headline" acts in the "Genetic Freak Show," the Cavalier King Charles Spaniel and the Pug, with the German Shepherd as support, referenced breeds that enjoy considerable public affection, as well as some of the most serious physiological conditions caused by inbreeding. The cleverness of this poster rested on the analogy between the pedigree dog now and the Victorian human "freaks of nature" displayed for entertainment and commercial gain, a phenomenon that should elicit distaste in a more inclusive post-Victorian society.[7] Indeed, Victorian critics of dog shows claimed that certain breeds were "monstrosities" and "freaks."

PETA's poster referenced a much broader Victorian legacy haunting the twenty-first-century dog fancy, too: commerce, consumption, and mass culture. Despite widespread criticism from animal welfare agencies and prominent social commentators, Crufts remains a huge popular phenomenon, with annual visitor numbers over 150,000. Sponsorship by major pet-product companies, television coverage, and over 400 trade stands—combined with the 22,000 dogs representing 200 breeds competing for "Best in Breed" in show and assorted agility and talent contests—make Crufts a jamboree of consumer-driven popular entertainment.[8] It surely was not accidental that PETA's poster used a font and format similar to those promoting P. T. Barnum's traveling-circus spectacular, which included dog shows as well as General Tom Thumb, in the latter half of the nineteenth century. If Crufts is the epitome of the dog fancy in Britain today, then it is also a window into a world where animals represent commodities, status symbols, and vehicles for entertainment. The foundations of this dog culture were laid in the Victorian period, and the legacy of that culture persists in current show and breed practices.

In many respects, the nineteenth-century dog fancy represented a microcosm of the broader issues at stake in Victorian society. The development of breed and the clashes between "doggy" interest groups both reflected and perpetuated the tensions between old blood (the aristocracy) and new money in the nineteenth century. Dogs were remade culturally with imagined pasts and assumed correlations between form and function, and the best dogs were thought to be those with high-class, purebred ancestries. However, they were also commodified in new and profitable ways as the dog market expanded to incorporate shifting notions, and differential values, of pedigree. The new dog fancy was riven with antagonism between those who proclaimed an interest only in improving breeds and those who sought to make money from doing so. Much of the fancy's history, and especially that of the Kennel Club, is one of competing groups seeking variously to regulate the commercial value of, and cultural capital invested in, dogs. Situating the dog fancy in a context of trade, industry, and finance shows how dog breeders debated the value of competition and contested the ethics of speculation against regulation.

Charles Dickens and Anthony Trollope wrote novels exploring the murky social and financial politics of speculation; their respective characters Merdle and Melmotte, enmeshed in webs of profit and being "in society," might well have kept pedigree kennels.[9] That both characters are corrupt, for all their "gentlemanly" veneer, indicates the lasting antipathy toward speculators and their social pretensions. If trade and wealth were volatile, however, *pedigree* enabled some preservation of privilege in the public domain. Dogs would never be a substitute for marriage into the landed gentry, but the language of pure blood and good breeding enabled, as Harriet Ritvo has also argued, entrepreneurs and middle-class professionals to participate in an imagined past.[10] This book's long chronology enables us to chart the fluctuating fortunes of "gentlemanly capitalists" in the context of dog breeding and their changing relations with landed society and popular culture.[11] The establishment and monopoly of the Kennel Club, with its emphasis on good governance, eventually resolved the tensions between competing groups of dog breeders. The club's leadership drew authority from the patronage of aristocratic exhibitors and breeders; however, its members also accepted that they could not determine the culture of shows that were open and meritocratic. This situation reveals the uneasy compromise at the end of the century between an old world of rural sports and the new one of interest in canine "improvement" and entertainment.

The long chronology of the dog fancy also highlights the shifting dynamic across the nineteenth century of the relationship between "the people" and the establishment. This period was one of rapid and far-reaching change. From the dark days of cruel sports, Bill Sikes, and murky "dog dealers," plebeian doggy folk were ostensibly refashioned by the end of the period into respectable consumers. Rising living standards over the period and the expansion of mass consumption meant that, on an informal and localized level at least, some aspirational working people could participate in the new fancy as small-scale breeders. On another level, the growth of a national and regional show culture, which combined sliding-scale admission to spectacular entertainments and exhibitions of "improved" dogs and people, reflected and contributed to a mass culture of leisure. The dog show's trajectory from agricultural, and largely economic, event, to an organized, regulated entertainment venture (which was not without its economic aspects) mirrored other facets of popular culture that commercial leisure providers sought to incorporate. Yet, as in other areas of popular life, stability was fragile.

The popularization of the dog fancy as a respectable endeavor guaranteed wide coverage of dog shows in the tabloid press, and by the latter decades of the Victorian era, the dog had been sufficiently anthropomorphized to bear the weight of British moral sensibility. But the press had a nasty habit of barking with bite. When the dog fancy overstepped the mark on issues of welfare, such as cropping dogs' ears, the mouthpiece of the people was quick to voice its disapproval, claiming affinity with oppressed creatures and declaiming the upper classes for their preoccupation with fashion over feeling. Although most breeders and exhibitors across social classes clearly held their show dogs in affection, cases of cruelty in the fancy, which often involved the collusion of working-class individuals, muddied a class-based antipathy to breeders and hinted at the continuity of an older, albeit smaller, plebeian dog culture.

Late-Victorian attacks on dog shows and the ways breeds had been "improved" mixed humanitarian concern with morality, gender, and class politics. Such critics were not preoccupied with canine health per se, except specific issues such as cropping and docking. What is so striking about the twenty-first-century assault on the dog show is the new focus on canine genetic health, which has led some to argue that conformation breeding and breed standards are incompatible with canine well-being. That this new criticism has destabilized the pedigree breeding industry testifies to the prominence and place of genetics in modern society and culture. As we have shown in this book, the original setting of breed standards was contested and the outcomes contingent. Standards came from contemporary social practices, not from nature or history. They were subject to change and revision throughout the Victorian era, with major disagreements over what was an "improvement." The main reason for the drive for uniformity across a breed population was competition at shows and the aspirations of breeders, exhibitors, and dealers to meet judges' expectations, even though these were capricious. Standards became ideals and drove convergence within breed populations, which in turn fostered reliance upon inbreeding. Had shows been differently constituted, without the proliferation of classes and, say, with a single class for all Spaniels, breed standards could have encouraged diversity and had greater tolerance for difference.

The changes wrought on dogs by doggy people in the Victorian era were revolutionary: first, with the very adoption of *breed* as way of thinking about and remaking varieties of dog; and second, with the material remodeling of dogs' bodies, and, wittingly or unwittingly, their character, behavior, and genetics. Kennel clubs across the world, responding to the recent critiques of "pedigree dogs," have begun to alter the conformation standards of some breeds and encourage genetic diversity. It remains to be seen how radical these changes will be. This book, by showing the historical contingencies that shaped the invention of the modern dog, can be read as giving license not only to the remaking of individual breeds but to reimagining the very category of *breed* itself.

## Introduction

1. J. Cooper, "Why Our Mongrels Are a Dying Breed," *Daily Telegraph*, 3 Mar. 2013, www.telegraph.co.uk; Statista, "Share of Dog Owners in the United Kingdom (UK) in 2011 and 2013, by Breed Type," www.statista.com; P. Mahaney, "What Is the Value in Knowing Which Mix of Breeds Make Up Your Mongrel?," PetMD, 27 Nov. 2012, www.petmd.com. The meaning of *mongrel* changed over the Victorian period. In the early years mongrels were mixed-variety dogs, often deliberately mated to combine features of different varieties to give a dog that suited a particular role. In a sense they were exactly like the designer dogs of today. Sometimes this was finely managed, and if a mongrel was bred with a different variety again, it was referred to as a triple mongrel. By the late nineteenth century and in large part due to the ascendancy of breed, *mongrel* had become more of a pejorative term. *Mixed mongrel* was coined to describe dogs in which there was little or no evidence in their conformation of any defined breed.

2. Cooper had previously published a book that was a humorous tribute to Britain's mongrels. J. Cooper, *Mongrel Magic* (London: Mandarin, 1992).

3. In the field of biology, *morphology* refers to external physical form, while *anatomy* refers to internal physical form.

4. *Doggy people* was a term that came into vogue at the end of the Victorian period, replacing the *dog fancy* of the collective of those who invested time, money, and interest in dog affairs. In the early nineteenth century, the Dog Fancy was a predominantly working-class phenomenon associated with dog fighting, ratting, dog stealing, and related activities. After 1850, such ventures went into slow decline, and the term *fancy* was applied to all those active in dog affairs. C. H. Lane, *All About Dogs: A Book for Doggy People* (London: J. Lane, 1900); C. H. Lane, *Dog Shows and Doggy People* (London: Hutchinson, 1902).

5. G.-L. Leclerc, Comte de Buffon, *Natural History, General and Peculiar,* vol. 4, trans. W. Smellie (London: T. Cadell and W. Davies, 1812), 344–45.

6. W. Secord, *Dog Painting, 1840–1940: A Social History of the Dog in Art* (Woodbridge, Suffolk, UK: Antique Collectors' Club, 1992); W. Secord, *Dog Painting: The European Breeds* (Woodbridge, Suffolk, UK: Antique Collectors' Club, 2000).

7. M. M. Taylor, "The Curious Case of 'Epitaph to a Dog': Byron and The Scourge," *Byron Journal* 43 (2015): 43–56.

8. We are grateful to Valerie Foss for sharing her knowledge of this painting.

9. W. Youatt, *The Dog* (London: Charles Knight, 1845), 53, 55. He is quoting from J. Macgregor's *Historical and Descriptive Sketches of British America* (London: Longman, 1828). Many authors have sought the origin of the Labrador breed in such speculations, which from our perspective is pointless as, at that time, breeds as we now know them did not exist in concept or conformation. See B. Fogle, *Labrador: The Story of the World's Favourite Dog* (London: William Collins, 2016).

10. Stonehenge, "The Newfoundland Dog," *Field,* 28 July 1877, 97; R. B. Lee, *A History and Description of the Modern Dogs of Great Britain and Ireland: Non-sporting Division* (London: Horace Cox, 1894), 99.

11. Lee, *History and Description,* 79; Secord, *Dog Painting,* 356; "Obituary: Arthur Wardle," *Times* (21 July 1949), 7.

12. R. B. Lee, *A History and Description of the Modern Dogs of Great Britain and Ireland: Non-sporting Division,* 3rd ed. (London: Horace Cox, 1899), 78, 89.

13. Ibid., 107, 108.

14. Ch. Merrybear D'Artagnan was bred by Gordon Cutts and Patrick Galvin, and is owned by Lyn and Terry Chapman. The Natural History Museum's branch at Tring in Hertfordshire has a display of taxidermied dogs from the late nineteenth and early twentieth centuries. This collection provides an excellent opportunity to see the conformations of yesteryear and compare them with those of today. For photographs and a commentary on the collection, see K. Dennis-Bryan and J. Clutton-Brock, *Dogs of the Last Hundred Years at the British Museum (Natural History)* (London: British Museum (Natural History), 1988).

15. Youatt, *The Dog,* 7, 73, 81, and 88–89; Lee, *History and Description* (1893). Every chapter in Lee's book details the long and recent history of each breed, ending with the arrival of the "modern" type and its points.

16. J. Clutton-Brock, *A Natural History of Domesticated Mammals* (London: Natural History Museum, 1999); M. Derr, *How the Dog Became the Dog* (London: Duckworth, 2013); K. Thomas, *Man and the Natural World: Changing Attitudes in England, 1500–1800* (London: Allen Lane, 1983); I. H. Tague, *Animal Companions: Pets and Social Change in Eighteenth Century Britain* (University Park: Pennsylvania State University Press, 2015).

17. N. Pemberton and J.-M. Strange, "Dogs and Modernity: Dogs in History and Culture," *European Review of History* 22 (2015): 705–8.

18. The best-known earlier use of the term is in Shakespeare's *Richard II,* in which John o'Gaunt speaks of "This happy breed of men, his little world." It is generally agreed that this remark refers to the English population and the country. The speech also mentions "this scepter's isle" and "This other Eden, demi-paradise, this fortress built by Nature herself." Later in the speech, breed refers to royalty: "Feared be their breed and famous by their birth."

19. N. Russell, *Like Engend'ring Like: Heredity and Animal Breeding in Early Modern England* (Cambridge: Cambridge University Press, 1986), 160–61. Russell's source, which uses an essentialist and ahistorical notion of breed, is M. L. Ryder, "A Survey of European Primitive Breeds of Sheep," *Annales De Genetique et De Selection*

*Animale* 13 (1981): 381–418. Also see M. L. Ryder, "The History of Sheep Breeds in Britain," *Agricultural History Review* 12 (1964): 1–12, in which Ryder tellingly concluded, "Not until the end of the eighteenth century did agricultural writers begin to give definite descriptions of different breeds. This was unfortunately just after many breeds had been changed" (7).

20. R. Trow Smith, *A History of British Livestock Husbandry to 1700* (London: Routledge and Paul, 1957).

21. D. L. Wykes, "Robert Bakewell (1725–95) of Dishley: Farmer and Livestock Improver," *Agricultural History Review* 52 (2004): 38–55; J. Humphreys, "Bakewell, Robert (1725–1795)," rev. G. E. Mingay, *Oxford Dictionary of National Biography* (Oxford: Oxford University Press, 2004), www.oxforddnb.com.

22. J. R. Walton, "Pedigree and the National Cattle Herd, circa 1750–1950," *Agricultural History Review* 34 (1986): 152.

23. H. Ritvo, *The Animal Estate: The English and Other Creatures in the Victorian Age* (Cambridge, MA: Harvard University Press, 1987), 106.

24. M. Wallen, "Foxhounds, Curs, and the Dawn of Breeding: The Discourse of Modern Human-Canine Relations," *Cultural Critique* 79 (2011): 125–51, quote on 127. Wallen's framing is developed from Jacques Derrida. See J. Derrida and D. Wills, "The Animal That Therefore I Am (More to Follow)," *Critical Inquiry* 28 (2002): 384.

25. M. Cobb, *The Egg and Sperm Race: The Seventeenth-Century Scientists Who Unravelled the Secrets of Sex, Life and Growth* (London: Simon and Schuster, 2007), 220–49.

26. Three key stallions were imported in the late seventeenth and early eighteenth centuries, and through their bloodline descendants have dominated Thoroughbred horse breeding. The horses were the Byerley Turk, acquired by Capt. Robert Byerley as his war-horse in the 1680s before becoming a stud stallion; the Godolphin Arabian, foaled in Yemen and imported to England in 1729 by Edward Coke; and the Darley Arabian, bought in Aleppo in 1704 by Thomas Darley. Recent studies of the genomes of Thoroughbred racehorses has revealed that 95% of the quarter of a million stallions worldwide can be traced back to the Darley Arabian and "ten founder females account for 72% of maternal lineages." E. P. Cunningham et al., "Microsatellite Diversity, Pedigree Relatedness, and the Contributions of Founder Lineages to Thoroughbred Horses," *Animal Genetics* 32 (2001): 360–64.

27. The recorded pedigrees of racehorses were published annually in an open stud book, the hope being that transparency would ensure honesty. The closure of the book was enacted by the so-called Jersey Act to prevent the entry of American-bred Thoroughbreds. The Jersey Act was abandoned in 1949. J. A. D. Wentworth Blunt-Lytton, *Thoroughbred Racing Stock and Its Ancestors: The Authentic Origin of Pure Blood* (London: G. Allen and Unwin, 1938).

28. J. Auerbach and P. Hoffenberg, *Britain, Empire, and the World at the Great Exhibition of 1851* (Aldershot: Ashgate, 2008).

29. G. Thomas, "Modernism, Diaghilev, and the Ballets Russes in London: 1911–29," in *British Music and Modernism, 1895–1960,* ed. M. Riley (Farnham: Ashgate, 2010), 67–92.

30. P. Howell, "The Dog Fancy at War: Breeds, Breeding, and Britishness, 1914–1918," *Society and Animals* 21 (2013): 546–67.

31. M. Brandow, *A Matter of Breeding: A Biting History of Pedigree Dogs and How the Quest for Status Has Harmed Man's Best Friend* (London: Duckworth, 2016). Also see K. M. Rogers, *First Friend: A History of Dogs and Humans* (New York: St. Martin's Press, 2005), 108–52; K. C. Grier, *Pets in America: A History* (Chapel Hill: University of North Carolina Press, 2006), 25–35 and 233–40. Good examples of popular histories of British dogs are A. Croxton Smith, *British Dogs* (London: Collins, 1945), and C. I. A. Ritchie, *The British Dog: Its History from Earliest Times* (London: Robert Hale, 1981).

32. The Kennel Club Library has a comprehensive collection of books on all aspects of dogs.

33. Ritvo, *Animal Estate*, 93. Also see H. Ritvo, *The Platypus and the Mermaid and Other Figments of the Classifying Imagination* (Cambridge, MA: Harvard University Press, 1997); "The Emergence of Modern Petkeeping," *Anthrozoös* 1 (1988): 158–65; and "Animals in Nineteenth-Century Britain: Complicated Attitudes and Competing Categories," in *Animals and Society: Changing Perspectives*, ed. J. Serpell and A. Manning (London: Routledge, 1994), 106–26.

34. Ritvo, *Animal Estate*, 82–121. Also see H. Ritvo, "Pride and Pedigree: The Evolution of the Victorian Dog Fancy," *Victorian Studies* 29 (1986): 227–53.

35. Ritvo, *Animal Estate*, 84. Also see R. Jann, "Animal Analogies in the Construction of Class," *Nineteenth Century Studies* 8 (1994): 89–101.

36. M. Rosenberg, "Golden Retrievers Are White, Pit Bulls Are Black, and Chihuahuas Are Hispanic: Representations of Breeds of Dog and Issues of Race in Popular Culture," in *Making Animal Meaning (The Animal Turn)*, ed. L. Kalof and G. Montgomery (East Lansing: Michigan State University Press, 2011), 113–26.

37. R. Wood, "Genetic Prehistory in Selective Breeding: A Prelude to Mendel," *Journal of the History of Biology* 35 (2002): 402–4; T. Ito, *London Zoo and the Victorians, 1828–1859* (Woodbridge, Suffolk, UK: Boydell, 2014); H. Cowie, *Exhibiting Animals in Nineteenth-Century Britain: Empathy, Education, Entertainment* (London: Palgrave Macmillan, 2014); N. Pemberton and M. Worboys, *Rabies in Britain: Dogs, Disease, and Culture, 1830–2000* (London: Palgrave, 2012); H. Kean, *Animal Rights: Political and Social Change in Britain Since 1800* (London: Reaktion, 1998).

38. Thomas, *Man and the Natural World*, 102–20. Anachronistically, from the argument in this book, Thomas writes about particular "breeds."

39. E. Fudge, *Brutal Reasoning: Animals, Rationality, and Humanity in Early Modern Thought* (Ithaca: Cornell University Press, 2006). Also see E. Fudge, *Renaissance Beasts: Of Animals, Humans, and Other Wonderful Creatures* (Urbana: University of Illinois Press, 2004); E. Fudge, *Perceiving Animals: Humans and Beasts in Early Modern English Culture* (Basingstoke, UK: Macmillan, 2000); L. Brown, *Homeless Dogs and Melancholy Apes: Humans and Other Animals in the Modern Literary Imagination* (Ithaca: Cornell University Press, 2010); G. Ridley, ed., "Animals in the Eighteenth Century: Special Issue," *Eighteenth Century Studies* 33 (2010): 431–683.

40. D. Haraway, *The Companion Species Manifesto: Dogs, People, and Significant Otherness* (Chicago: Prickly Paradigm, 2003); D. Haraway, *When Species Meet (Posthumanities)* (Minneapolis: University of Minnesota Press, 2008).

41. E. Russell, "Evolutionary History: Prospectus for a New Field," *Environmental History* 8 (2003): 204–28. In his earlier book, *Evolutionary History: Uniting History*

*and Biology to Understand Life on Earth* (Cambridge: Cambridge University Press, 2011), Russell wrote of dogs having undergone two radical transformations; the first was their descent from wolflike ancestors, while "the second transformed dogs into breeds that looked and behaved in different ways" (54).

42. E. Russell, *Greyhound Nation: A Coevolutionary History of England, 1200–1900* (Cambridge: Cambridge University Press, 2018).

43. P. Howell, *"At Home and Astray": The Domestic Dog in Victorian Britain* (Charlottesville: University of Virginia Press, 2015). Also see P. Howell, "June 1859 / December 1860: The Dog Show and the Dogs' Home," BRANCH: *Britain, Representation and Nineteenth-Century History,* ed. D. F. Felluga, *Extension of Romanticism and Victorianism on the Net,* online.

44. A. H. Skabelund, *Empire of Dogs: Canines, Japan, and the Making of the Modern Imperial World* (Ithaca: Cornell University Press, 2011); S. Swart, "Dogs and Dogma: A Discussion of the Socio-Political Construction of Southern African Dog Breeds as a Window on Social History," *South African Historical Journal* 48 (2003): 190–206; L. Van Sittert and S. Swart, "Canis Familiaris: A Dog History of South Africa," *South African Historical Journal* 48 (2003): 138–73.

45. M. E. Derry, *Horses in Society: A Story of Animal Breeding and Marketing Culture, 1800–1920* (Toronto: University of Toronto Press, 2006); *The Art and Science in Breeding: Creating Better Chickens* (Toronto: University of Toronto Press, 2014); and *Masterminding Nature: The Breeding of Animals, 1750–2010* (Toronto: University of Toronto Press, 2015).

46. M. E. Derry, *Bred for Perfection: Shorthorn Cattle, Collies, and Arabian Horses since 1800* (Baltimore: Johns Hopkins University Press, 2003).

47. Brandow, *A Matter of Breeding.*

48. Ibid., x. Bekoff was the cofounder, with Jane Goodall, of Ethologists for the Ethical Treatment of Animals; see www.ethologicalethics.org.

## Chapter 1  •  Before Breed, 1800–1860

1. J. W. Rogerson, "What Difference Did Darwin Make? The Interpretation of Genesis in the Nineteenth Century," in *Reading Genesis after Darwin,* ed. S. C. Barton and D. Wilkinson (Oxford: Oxford University Press, 2009), 75–91.

2. N. Pemberton, "The Rat-Catcher's Prank: Interspecies Cunningness and Scavenging in Henry Mayhew's London," *Journal of Victorian Culture* 19 (2014): 520–35.

3. K. Malik, *The Meaning of Race: Race, History and Culture in Western Society* (Basingstoke: Palgrave Macmillan, 1996); I. Hannaford, *Race: The History of an Idea in the West* (Baltimore: Johns Hopkins University Press, 1996); E. Beasley, *The Victorian Reinvention of Race: New Racisms and the Problem of Grouping in the Human Sciences* (London: Routledge, 2010).

4. J. Caius, *Of English Dogges, the diuersities, the names, the natures and the properties* (1576); G.-L. Leclerc, Comte de Buffon, *Natural History, General and Peculiar,* vol. 4, trans. W. Smellie (London: T. Cadell and W. Davies, 1812). Also see O. Goldsmith, *A History of the Earth and Animated Nature,* vol. 2, new ed. (London: W. C. Wright, 1824), 1–15.

5. Caius, *Of Englishe Dogges,* 21. The exemplar of this type is said to come from Malta, and many writers who have ahistorical and essentialist concepts of breed have claimed that Caius was referring to the modern Maltese terrier. See V. T. Leitch et al., *The Maltese Dog: A History of the Breed* (Bronxville, NY: International Institute of Veterinary Science, 1970), 39–46.

6. This combination of classifying and external form is an exemplar of the natural history "way of knowing" identified and discussed by John Pickstone in *Ways of Knowing: A New History of Science, Technology, and Medicine* (Manchester: Manchester University Press, 2000), 60–82.

7. H. Ritvo, *The Platypus and the Mermaid and Other Figments of the Classifying Imagination* (Cambridge: Harvard University Press, 1997), 85–130 and 131.

8. Comte de Buffon, *Natural History,* 344–45.

9. Ibid., 12, 358, 382.

10. R. B. Davies, "Edwards, Sydenham Teast (bap. 1768, d. 1819)," *Oxford Dictionary of National Biography* (Oxford: Oxford University Press, 2004), www.oxforddnb .com; S. Edwards, *Cynographia Britannica: Consisting of Coloured Engravings of the Various Breeds of Dogs Existing in Great Britain; Drawn from the Life* (London: C. Whittingham, 1800), 131–34.

11. Edward Jesse wrote in 1846 that turnspits were like automatons and lacked the character of other dogs, which made them vulnerable to ill treatment: "A pointer has pleasure in finding game, the terrier worries rats with considerable glee, the greyhound pursues hares with eagerness and delight, and the bull-dog even attacks bulls with the greatest energy, while the poor turnspit performs his task by compulsion, like a culprit on a tread-wheel, subject to scolding or beating if he stops a moment to rest his weary limbs, and is then kicked about the kitchen when the task is over." E. Jesse, *Anecdotes of Dogs* (London: Henry G. Bohn, 1858), 418–19.

12. Edwards, *Cynographia Britannica,* 5.

13. D. Blaine, *Canine Pathology, or a Full description of the diseases of dogs; with their causes, symptoms, and mode of cure . . . Preceded by an introductory chapter on the moral qualities of the dog* (London: T. Boosey, 1817), xii–l. On Blaine's career and work, see J. B. Simonds, *A Biographical Sketch of Two Distinguished Promoters of Veterinary Science, Delabere P. Blaine, William Youatt* (London: Adlard and Son, 1896).

14. W. Youatt, *The Dog* (London: Charles Knight, 1845). On Youatt's life and career, see E. Clarke, "Youatt, William (1776–1847)," rev. S. A. Hall, *Oxford Dictionary of National Biography.* Thirteen further editions of *The Dog* were published in Britain by Long, Brown, Green, and Longmans between 1851 and 1904. Five editions were published in the United States between 1847 and 1895.

15. Youatt, *The Dog* (1845), 100.

16. Ibid., 1, 2, 11, 45, 52–55, 11–12.

17. D. Gange, *Dialogues with the Dead: Egyptology in British Culture and Religion, 1822–1922* (Oxford: Oxford University Press, 2014); P. Howell, *At Home and Astray: The Domestic Dog in Victorian Britain* (Charlottesville: University of Virginia Press, 2015), 102–24; Youatt, *The Dog,* 38, 42, 72–86; P. Beckford, *Thoughts on hunting: In a series of familiar letters to a friend* (London: E. Easton, 1781); R. T. Vyner, *Notitia Venatica: a treatise on fox-hunting, to which is added, a compendious Kennel stud book* (London: Rudolph Ackermann, 1841); Nimrod [pseud.], *The Horse and the Hound:*

*Their various uses and treatment, including practical instructions in horsemanship and a treatise on horse- dealing* (Edinburgh: Adam and Charles Black, 1842). Nimrod was Charles James Apperley, who wrote for the press and authored many books on hunting and horseracing.

18.  R. W. Burkhardt Jr., "Constructing the Zoo: Science, Society, and Animal Nature at the Paris Menagerie, 1794–1838," in *Animals in Human Histories: The Mirror of Nature and Culture,* ed. M. J. Henninger-Voss (Rochester, NY: University of Rochester Press, 2002), 231–57.

19.  Phrenology was also a possible influence. Its leading assumption was that brain function was localized and its power reflected in the size of different parts of the organ, which was in turn imprinted on the shape of the skull. J. van Wyhe, *Phrenology and the Origins of Victorian Scientific Naturalism* (Aldershot: Ashgate, 2004).

20.  Youatt, *The Dog,* 11.

21.  Ibid., 43, 98; A. Feuerstein, "'I promise to protect dumb creatures': Pastoral Power and the Limits of Victorian Nonhuman Animal Protection," *Society and Animals* 23 (2015): 148–65; A. Moss, *Valiant Crusade: The History of the R.S.P.C.A.* (London: Cassell, 1961), 117–18.

22.  E. Mayhew, *Dogs: Their Management: Being a new plan of treating the animal, based upon a consideration of his natural temperament; illustrated by numerous woodcuts depicting the character and position of the dog when suffering disease* (London: George Routledge, 1854); D. Vlock, "Mayhew, Henry (1812–1887)," *Oxford Dictionary of National Biography*; G. Hodder, *Memories of My Time including personal reminiscences of eminent men* (London: Tinsley Brothers, 1870), 58–60; F. Forester, ed., *The Dog by Dinks, Mayhew and Hutchinson* (New York: Woodward, 1873).

23.  C. Hamilton Smith, *Dogs: The Naturalist's Library,* vol. 10 (Edinburgh: W. H. Lizars, 1840; H. D. Richardson, *Dogs: Their Origin and Varieties* (Dublin: James McGlashan, 1847); D. Low, *On the Domesticated Animals of the British Islands* (London: Longman, Brown, Green, and Longmans, 1853).

24.  Ibid., 12, 101; H. D. Richardson, "Dogs," in *A Cyclopedia of Agriculture, Practical and Scientific,* ed. J. C. Morton (Edinburgh: Blackie and Son, 1855), 666.

25.  Richardson, *Dogs,* 60. Richardson varies, with seemingly no significance, between the bloodhound and the Bloodhound.

26.  Ibid., 68–69.

27.  B. Moubray, *A Practical Treatise on Breeding, Rearing, and Fattening all kinds of Domestic Poultry,* 2nd ed. (London: Sherwood, Neely and Jones, 1816); T. M. Craddock, *Domestic Animals: Containing instructions for the . . . Management of Milch Cows, Pigs, Poultry, etc.* (London: T. M. Cradock, 1840); J. Rogers, *A Complete Directory for the Proper Treatment, Breeding, Feeding and Management of all kinds of Domestic Poultry, Pigeons, Rabbits, Dogs, Bees, and Instructions for the Private Brewery* (London: Thomas Dean, 1854), 5–6.

28.  Mrs. Loudon, *Domestic Pets: Their Habits and Management* (London: Grant and Griffin, 1851), 3, 4, 26–31.

29.  G. T. Bettany, "Jesse, Edward (1780–1868)," rev. A. Goldbloom, *Oxford Dictionary of National Biography*; E. Jesse, *Anecdotes of Dogs* (London: Richard Bentley, 1846), 467.

30. H. Ritvo, *Animal Estate: The English and Other Creatures in the Victorian Age* (Cambridge, MA: Harvard University Press, 1987), 95–96; M. Wallen, "Foxhounds, Curs and the Dawn of Breeding: The Discourse of Modern Human-Canine Relations," *Cultural Critique* 79 (2011): 125–51.

31. R. Longrigg, *The English Squire and His Sport* (London: Joseph, 1977), 665–68; R. Carr, *English Foxhunting: A History* (London: Weidenfeld and Nicolson, 1986), 38–40; J. Mills, *The Sportsman's Library, or Hints on the Hunter: Hunting, Hounds, Shooting, Game, Dogs, Fishing, &c.&c.* (Edinburgh, 1845), 102–3; Vyner, *Notitia Venetica*, 14.

32. A. N. Harvey, "Meynell, Hugo (1735–1808)," *Oxford Dictionary of National Biography*; J. L. Randall, *A History of the Meynell Hounds and Country, 1780–1901* (London: Sampson, Low, 1901). Meynell's position as the inventor of modern fox hunting has recently been challenged. I. M. Middleton, "The Origins of English Fox Hunting and the Myth of Hugo Meynell and the Quorn," *Sport in History* 1 (2005): 1–16.

33. E. Griffin, *Blood Sport: Hunting in Britain since 1066* (London: Yale University Press, 2007), 125–40; Richardson, *Dogs,* 64; D. C. Itzkowitz, *Peculiar Privilege: A Social History of English Foxhunting, 1753–1885* (Hassocks: Harvester, 1977), 24–29, 31–49, 58–62, and 176–77.

34. *A List of Mr. Smith's Hounds* (Lincoln: W. Brooke, 1817); Col. J. Cook, *Observations on Fox-hunting, and the Management of Hounds in the Kennel and the Field* (London: Edward Arnold, 1826), 142.

35. Nimrod [pseud.], *Hunting Reminiscences: Comprising Memoirs of Masters of Hounds* (London: Rudolph Ackerman, 1843), 39 and 130.

36. Middleton, "The Origins of English Fox Hunting," 1–6; Vyner, *Notitia Venetica,* 3, italics in original.

37. F. D. Radcliffe, *The Noble Science: A Few General Ideas on Fox-hunting for the Use of the Rising Generation of Sportsmen* (London: Rudolph Ackerman, 1839), 281, 282; Stonehenge, *Manual of British Rural Sports,* 4th ed. (London: Routledge, Warnes, and Routledge, 1859), 119, 168–75, and 337–42 (first published in 1856); Cook, *Observations,* 18. Fox hunting was expensive. In 1840, one estimate was that the annual outlay on a hunt that went out four days each week was £4,000. "Report," *Morning Chronicle,* 29 Aug. 1840, 4.

38. H. W. Herbert, *The Complete Manual for Young Sportsmen* (New York: George G. Woodward, 1856), 230–31. Variations in the height of dogs remained, as they were designed to suit local conditions, particularly the height of fences and hedges.

39. Stonehenge, *Manual,* 119; D. Birley, *Sport and the Making of Britain* (Manchester: Manchester University Press, 1993), 130–34; Griffin, *Blood Sport,* 165; "Floral Fetes and Agricultural Meetings," *Illustrated London News (ILN),* 17 Sept. 1859, 9.

40. Youatt, *The Dog,* 11. On the long history of greyhounds in Britain, see E. Russell, *Greyhound Nation: A Coevolutionary History of England, 1200–1900* (Cambridge: Cambridge University Press, 2018).

41. Stonehenge, *Manual,* 162–67, 187.

42. J. Mills, *Sportman's Library,* 275.

43. Stonehenge, *Manual,* 147–51; I. Roberts, "Starting a Hare: Exploring the History of Coursing since the Mid-nineteenth Century," in *Our Hunting Fathers: Field Sports in England after 1850,* ed. R. W. Hoyle (Lancaster: Carnegie, 2007), 212–40.

44. H. Cox and G. Lascelles, *Coursing and Falconry* (London: Longmans, Green, 1892).

45. Stonehenge, *The Greyhound, Being a Treatise on the Art of Breeding, Rearing, and Training Greyhounds for Public Running; Their Diseases and Treatment. Containing also, Rules for the Management of Coursing Meetings, and for the Decision of Courses* (London: Longman, Brown, Green, and Longman, 1853), 179.

46. "Coursing," *Bell's Life,* 20 Mar. 1836, 3; "Coursing," in *Encyclopedia of Traditional Rural Sports,* ed. T. Collins, J. Martin, and W. Vamplew (London: Routledge, 2005), 78–79; F. Lloyd, *The Whippet and Race Dog* (London: L. Upcott Gill, 1894); W. D. Drury, *British Dogs: Their Points, Selection, and Show Preparation* (London: L. Upcott Gill, 1903), 96–103.

47. C. Butler, *The Complete Dog-Fancier's Companion: Describing the nature, habits, properties, &c. of sporting, fancy, and other dogs* (London: William Darton, 1819), 85.

48. M. Derry, *Bred for Perfection: Shorthorn Cattle, Collies, and Arabian Horses since 1800* (Baltimore: Johns Hopkins University Press, 2003), 4–10; J. H. Walsh and I. J. Lupton, *The Horse in the Stable and Field: His Varieties, Management in Health and Disease, Anatomy, Physiology, etc.* (London: 1861), 150–54; T. Mason, "Dowling, Vincent George (1785–1852)," *Oxford Dictionary of National Biography*; Stonehenge, *The Greyhound,* v, 34. Walsh and Lupton, in *The Horse,* write, "When we say that a horse is 'in form' we intend to convey to our hearers that he is in high condition and fit to run" (84).

49. R. A. Welsh, *Thacker's Courser's Annual Remembrancer and Stud Book* (London: 1842–46); Stonehenge, *The Greyhound,* 372, 388, 197, 258, 16, 249. King Cob was a very successful stud dog, best known as the grandfather of Bedlamite— "undoubtedly the most successful stallion of modern times." Darwin mentions Bedlamite for his prepotency in later editions of *The Variation of Animals and Plants Under Domestication,* vol. 2 (London: John Murray, 1875): "A famous black greyhound, Bedlamite, as I hear from Mr. C. M. Brown "invariably got all his puppies black, no matter what was the colour of the bitch," but then Bedlamite "had a preponderance of black in his blood, both on the sire and dam side" (40).

50. Stonehenge, *The Greyhound,* 208–9.

51. J. L. Down, "Marriages of Consanguinity in Relation for Degeneration of Race," *Journal of Mental Science* 13 (1867): 1201; W. Adam, "Consanguinity in Marriage, Part I," *Fortnightly Review* 2 (1865): 710–30, and "Consanguinity in Marriage, Part II," *Fortnightly Review* 3 (1865): 74–88; "The Week," *British Medical Journal* ii (1863): 327. Séguin added, "This is, in fact, just what is observed in animals whose breed is improved by man." On Séguin, see J. W. Trent, *Inventing the Feeble Mind: A History of Intellectual Disability in the United States* (Oxford: Oxford University Press, 1995), 35–54. Such studies were contemporary with Francis Galton's first publications on heredity and intelligence (1865) and anticipated certain late nineteenth-century eugenic concerns. However, eugenists rarely discussed the dangers of inbreeding.

There is no mention of consanguineous marriages or inbreeding in A. Bashford and P. Levine, eds., *The Oxford Handbook of the History of Eugenics* (Oxford: Oxford University Press, 2012). We discuss these issues in chapters 6 and 7.

52. "Advertisement, "Stallion Greyhound," *Bell's Life*, 11 Mar. 1855, 1; "Coursing," *Bell's Life*, 2 Jan. 1853, 4; Stonehenge, *The Dog in Health and Disease* (London: Longman, Green, Longman and Roberts, 1859), 36, 35–37.

53. R. W. Hoyle, *Our Hunting Fathers: Field Sports in England after 1850* (Lancaster: Carnegie, 2007); M. Billett, *A History of English Country Sports* (London: Robert Hale, 1994).

54. Richardson, *Dogs*, 67–68; Youatt, *The Dog*, 93, 5; Howell, *At Home and Astray*, 50–72; Mills, *Sportsman's Library*, 244–46.

55. F. Butler, *The Breeding, Training, Management, Disease &c. of Dogs* (New York: Francis Butler, 1857), 116. Italics in the original.

56. Mills, *Sportsman's Library*, 242–43; Butler, *Breeding, Training*, 123; Col. W. N. Hutchinson, *Dog Breaking* (London: John Murray, 1848), 1–8.

57. J. Mason, *Civilized Creatures: Urban Animals, Sentimental Culture, and American Literature, 1850–1900* (Baltimore: Johns Hopkins University Press, 2005), 69–71; C. Lyell, *Principles of Geology*, vol. 2 (London: John Murray, 1835), 455.

58. Stonehenge, *The Dogs of the British Islands* (London: Horace Cox, 1867), 95; Youatt, *The Dog*, 92; Stonehenge, *Dogs of the British Islands*, 3; E. Laverack, *The Setter: With Notices of the Most Eminent Breeds now Extant; Instructions How to Breed, Rear, and Break, Dog Shows, Field Trials, General Management, etc.* (London: Longmans, Green, 1872), 10–14. For an obituary of Laverack, which discusses his work with setters, see "The Death of Mr. Laverack," *Bell's Life*, 14 Apr. 1877, 11.

59. Youatt, *The Dog*, 43–45; "The Spaniels," *Penny Magazine*, 13 Feb. 1841, 57–58; "Studbook of Sporting Dogs," *Field*, 1 Sept. 1860, 179; Richardson, *Dogs*, 88; J. Meyrick, *House Dogs and Sporting Dogs* (London: John Van Voorst, 1861), 45.

60. Meyrick, *House Dogs*, 42; T. Bewick, *A General History of Quadrupeds* (Newcastle upon Tyne: S. Hodgson, R. Beilby, and T. Bewick, 1800), 363; Richardson, *Dogs*, 89–90; Youatt, *The Dog*, 43–45.

61. "William Mansell," *Sporting Magazine*, Apr. 1807, 3; H. Dalziel, *British Dogs; their Varieties, History, Characteristics, Breeding, Management and Exhibition* (London: "The Bazaar" Office, 1879), 392; Meyrick, *House Dogs*, 44.

62. J. Mills, *Sportsman's Library* (Edinburgh, 1845), 323, 253; Forester, *The Dog*, 21.

63. "The Ring," *Bell's Life*, 21 Jan. 1838, 4; F. A. Butler, *A History of Boxing in Britain* (London: Barker, 1972); D. Brailsford, *Bareknuckles: A Social History of Prizefighting* (Cambridge: Lutterworth, 1988).

64. B. Harrison, "Animals and the State in Nineteenth-Century England," *English Historical Review* 88 (1973): 786–820; H. Kean, *Animal Rights: Political and Social Change in Britain since 1800* (London: Reaktion, 1998), 13–39; Pemberton, "The Rat-Catcher's Prank"; "Canine Fancy," *Bell's Life*, 21 Feb. 1847, 7; "Canine Fancy," *Bell's Life*, 28 Dec. 1858, 7.

65. There was a weekly column entitled "Canine Fancy." For example, see *Bell's Life*, 18 Apr. 1830, 3; 12 Apr. 1840, 4; 8 Feb. 1846, 7; 25 Apr. 1852, 6–7; 6 Jan. 1856, 7.

66. "Funeral of Jemmy Shaw," *Bell's Life,* 18 Jan. 1886, 3; J. Bondeson, *Amazing Dogs: A Cabinet of Canine Curiosities* (Stroud: Amberley, 2011), 18–22; "Canine Fancy," *Bell's Life,* 8 July 1849, 7.

67. W. Secord, *Dog Painting, 1840–1940: A Social History of the Dog in Art* (Woodbridge: Antique Collectors' Club, 1992), 160–62; E. W. Jaquet, *The Kennel Club: A History and Record of its Work* (London: The Kennel Gazette, 1905), 3; C. I. A. Ritchie, *The British Dog: Its History from Earliest Times* (London: Robert Hale, 1981), 187–88; "Canine Fancy," *Bell's Life,* 22 July 1855, 5.

68. "Canine Fancy," *Bell's Life,* 22 Apr. 1855, 6. Previously, Shaw ran his canine fancy events at the Blue Anchor, Bunhill Row, St. Luke's, London.

69. Ibid.; "Canine Fancy," *Bell's Life,* 16 Nov. 1845, 8. For an example of a report on dog racing, see *Bell's Life,* 28 Jan. 1855, 7.

70. "Canine Fancy," *Bell's Life,* 14 Aug. 1836, 3. See reports of fights in Wolverhampton and Birmingham, and a decade later in Derby, in *Bell's Life,* 31 May 1846, 6, and 14 Oct. 1857, 7.

71. "Canine Fancy," *Bell's Life,* 25 Jan. 1845, 7; "To the Canine Fancy and the Public in General," *Bell's Life,* 21 June 1857, 7.

72. Richardson, *Dogs,* 90; "The Spaniel Fancy," *Hampshire Advertiser and Salisbury Guardian,* 21 Mar. 1840, 2; "Canine Fancy," *Bell's Life,* 13 Aug. 1848, 3; "Canine Fancy," *Bell's Life,* 25 Mar. 1849, 7, and 9 Mar. 1856, 7.

73. "Obituary: Bill George," *Kennel Chronicle,* 30 June 1884, 61; "Mr. Punch's Visit to a Very Remarkable Place," *Punch* 11 (1846): 222. The accompanying article in *Punch* was based on an alleged meeting with "the famous Ben Thomas better known in the columns of *Bell's Life* as 'the father of the London Fancy.'" There is no mention of Ben Thomas in *Bell's Life.*

74. Advertisements, *Morning Post,* 24 Apr. 1856, 1; advertisements, *Royal Leamington Spa Courier,* 7 June 1856, 1.

75. H. Dalziel, "The Yorkshire Terrier," *Field,* 17 Nov. 1877, 575; V. Shaw, *Illustrated Book of the Dog* (London: Cassell, Petter, Galpin, 1881), 155–56; "Against Keeping Dogs," *Livesey's Moral Reformer,* 10 Mar. 1838, 74–75; "N," "Toy Dogs as House Dogs," *Field,* 10 Jan. 1857, 28; Howell, *Home and Astray,* 50–72.

76. K. Thomas, *Man and the Natural World, 1500–1800* (London: Allen Lane, 1983), 110–20. On lapdogs in fiction, see S. B. Palmer, "Slipping the Leash: Lady Bertram's Lapdog," *Persuasions* 25, no. 1 (2004); J. L. Wyett, "The Lap of Luxury: Lapdogs, Literature, and Social Meaning in the 'Long' Eighteenth Century," *LIT: Literature Interpretation Theory* 10 (1999): 275–301.

77. Thomas, *Man and the Natural World,* 117. This view is echoed by K. Kete, *The Beast in the Boudoir: Petkeeping in Nineteenth-Century Paris* (Berkeley: University of California Press, 1994). Also see L. Brown, *Homeless Dogs and Melancholy Apes: Humans and Other Animals in the Modern Literary Imagination* (London: Cornell University Press, 2010).

78. V. Woolf, *Flush: A Biography* (London: Hogarth, 1933); *Beeton's Home Book of Pets* (London: S. O. Beeton, 1861); J. Hamlett, *Material Relations: Domestic Interiors and Middle-Class Families in England, 1850–1910* (Manchester: Manchester University Press, 2010).

79. Richardson, *Dogs,* 90; Stonehenge, *The Dog,* 151; Youatt, *The Dog,* 44–45.

80. E. Longford, *Victoria R.I.* (London: Weidenfeld and Nicolson, 1998), 46; L. Strachey, *Queen Victoria* (London: Chatto and Windus, 1921), 73; Youatt, *The Dog,* 45.

81. H. Mayhew, *London Labour and the London Poor,* vol. 1 (London: Griffin, Bohn and Co., 1861), 358.

82. "The Street Dog," *Manchester Times,* 17 July 1858, 2; H. Smith, *Dogs,* 202–4; Youatt, *The Dog,* 67.

83. Youatt, *The Dog,* 68–70; Idstone [pseud.], *The Dog* (London: Cassell, Petter and Galpin, 1872), 31; Rogers, *A Complete Dictionary,* 25.

84. R. Jann, "Animal Analogies in the Construction of Class," *Nineteenth Century Studies* 8 (1994): 89–101; N. Pemberton and M. Worboys, *Rabies in Britain: Dogs, Disease, and Culture, 1830–2000* (London: Palgrave, 2012); Anon., "London Dogs," *Leisure Hour,* 23 Aug. 1860, 533–34; Rogers, *A Complete Dictionary,* 25.

85. Stonehenge, *The Dog,* 9.

86. The number of times certain terms are used in *The Dog:* "sort" (11), "sorts" (5), "strain" (47), "strains" (19), "type" (8), "types" (1), "race" (8), "races" (2), "variety" (32), "varieties" (25), "breed" (146), "breeds" (86).

87. Stonehenge, *The Dog,* 139.

88. Mastiffs had diminished in number in the first half of the nineteenth century, and men like Lukey saw themselves as "resuscitating" the dog. N. Dickin, *The Mastiff: Its History, Description and Standard* (London: Watmoughs, 1935), 51.

89. Dickin, *Mastiff,* 137–38.

### Chapter 2  •  *Adopting Breed, 1860–1867*

1. "It's 150 Years Since the First Ever Dog Show," *Our Dogs,* June 2009, www .dogworld.co.uk/product.php/52356/News/26-first-show. But see J. Sampson and M. M. Bins, "The Kennel Club and the Early History of Dog Shows and Breed Clubs," in *The Dog and its Genome,* ed. E. A. Ostrander, U. Giger, and K. Lindblad-Toh (Cold Spring Harbor, NY: Cold Spring Harbor Laboratory Press, 2006), 21–22.

2. Hugh Dalziel made this point in 1879. See H. Dalziel, *British Dogs: Their Varieties, History, Characteristics, Breeding, Management, and Exhibition* (London: "The Bazaar" Office, 1879), 172–73.

3. "The Exhibition of Dogs," *Birmingham Daily Post,* 3 Dec. 1860, 3. There had been an exhibition for sporting dogs only at the cattle and poultry show the previous year. "Exhibition of Sporting Dogs," *Birmingham Daily Post,* 30 Nov. 1859, 2. Also see A. Oliver, *From Little Acorns: The History of the Birmingham Dog Show Society, Estd. 1859* (Birmingham: National Dog Show Society, 1998), 7.

4. F. C. S. Pearce, *The Kennel Club Stud Book* (London: Field, 1874), 1.

5. Walsh had been a judge at the Newcastle upon Tyne show in 1859 and was in demand across the country throughout the 1860s. G. C. Boase, "Walsh, John Henry (1810–1888)," rev. J. Lock, *Oxford Dictionary of National Biography* (Oxford: Oxford University Press, 2004), www.oxforddnb.com.

6. M. Smith, "Scores on the Paws: How One Man Changed the Shape of Dogs to Come," Guardian, 7 Mar. 2013, www.theguardian.com.

7. W. Vamplew, *Industrialisation and Popular Sport in England in the Nineteenth Century* (Leicester: C.R.S.S./Leicester University Press, 1994).

8. R. Brailsford, "The Neglected State of the Pointers and Setters, and a Few Hints for Their Improvement," *Field,* 6 Feb. 1858, 121–22. On the Brailsford family, see Oliver, *From Little Acorns,* 7–9. The twelfth Earl of Derby had maintained a menagerie of 94 mammals and over 300 birds on the estate. He was a keen sportsman, favoring horse racing, hunting, cricket, and cockfighting, and founded the Derby and Oaks horse races. A. G. Crosby, "Stanley, Edward Smith, twelfth earl of Derby (1752–1834)," *Oxford Dictionary of National Biography.*

9. Brailsford, "The Neglected State," 122; *Field,* 20 Feb. 1858, 164; 6 Mar. 1858, 195; 13 Mar. 1858, 216–17; 20 Mar. 1858, 244; 27 Mar. 1858, 265–66; 10 Apr. 1858, 302; Charles Kent, "Berkeley, (George Charles) Grantley Fitzhardinge (1800–1881)," rev. J. Lock, *Oxford Dictionary of National Biography*; R. K. P., "The Proposed Exhibition," *Field,* 10 Apr. 1858, 302; J. J., "The Proposed Exhibition of Dogs," *Field,* 13 Mar. 1858, 216. The correspondence in the *Field* was an opportunity to speculate on the origins of types of dogs. Thus, R.K.P. stated that "the retriever is the mongrel of the Newfoundland and the spaniel or setter" and that "the staghound is the mongrel of two strains of hound grafted on to the bloodhound."

10. R. K. P., "The Proposed Exhibition of Sporting Dogs," *Field,* 10 Apr. 1858, 302; R. Brailsford, "The Proposed Exhibition of Dogs," *Field,* 27 Mar. 1858, 265.

11. D. Porter and S. Wagg, eds., *Amateurism in British Sport: It Matters Not Who Won Or Lost?* (Abingdon, UK: Routledge, 2008), 1–41; D. Loftus, "Entrepreneurialism or Gentlemanly Capitalism," in *The Victorian World,* ed. M. Hewitt (London: Routledge, 2012), 193–208.

12. R. Brailsford, "The Proposed Exhibition of Dogs," *Field,* 27 Mar. 1858, 265–66; R. Brailsford, "The National Show of Sporting Dogs," *Field,* 2 Apr. 1859, 260; "In Consequence of the Announcement of a Sporting Dog-Show," *Field,* 4 June 1859, 441.

13. G. C. Boase, "Walsh, John Henry (1810–1888)," rev. J. Lock, *Oxford Dictionary of National Biography*; Stonehenge, *The Greyhound: being a treatise on the art of breeding, rearing, and training greyhounds for public running; their diseases and treatment* (London: Longman, Brown, Green, and Longmans, 1853); Stonehenge, *Manual of British Rural Sports* (London: Routledge, 1856); J. H. Walsh, ed., *The English Cookery Book uniting a good style with economy, and adapted to all persons in every clime; containing many unpublished receipts in daily use by private families. Collected by a committee of ladies* (London: G. Routledge, 1858). *The Greyhound* was based on a series of articles Walsh had contributed to *Bell's Life of London and Sporting Chronicle*. See Stonehenge, "On the Breeding, Rearing, Training, and Diseases of the Greyhound.—No. 1," *Bell's Life,* 21 Nov. 1852, 5, and fifteen articles printed on 28 Nov. 1852; 2, 9, 16 and 23 Jan. 1853; 6, 13, 20 and 27 Feb. 1853; and 6 Mar. 1853.

14. R. N. Rose, *Field, 1853–1953* (London: Michael Joseph, 1953), 21–35, 53–62, 36–44. Entries to the championship had to be sent to Walsh's home, the Cedars, in Putney. "The Lawn Tennis Championship," *Field,* 9 June 1877, 689; *Sporting Gazette,* 23 June 1877, 591. Also see J. Barrett, *Wimbledon: The Official History of the Championships* (London: Collins Willow, 2001).

15. Editorial, "Sport & Sportsmen," *Field,* 6 Mar. 1858, 195.

16. P. Atkins, ed., *Animal Cities: Beastly Urban Histories* (Farnham: Ashgate, 2012); H. Vetlen, *Beastly London: A History of Animals in the City* (London: Reaktion, 2013).

17. "Sporting Dog and Poultry Show," *Newcastle Courant*, 1 July 1859, 3; "Exhibition of Poultry," *Newcastle Journal*, 10 Apr. 1847, 4; "Newcastle Sporting Dog Show," *Field*, 2 July 1859, 6; editorial, *Field*, 27 Aug. 1859, 173–74; "Sporting Dog and Poultry Show," 3; "Annual Meeting of the Cleveland Agricultural Society," *York Herald*, 10 Sept. 1859, 5; M. Clayton, *Peterborough Royal Foxhound Show: A History* (Shrewsbury: Quiller, 2006), 15–22; Clayton, *Peterborough*, 23–32.

18. "Important and National Show of Sporting Dogs," *Era*, 4 Dec. 1859, 4; editorial, *Field*, 14 Apr. 1860, 293; "Prize Dogs," *Berrow's Worcester Journal*, 3 Dec. 1859, 3. George Moore moved to create the society; Brailsford seconded. Moore was a local landowner, "Lord of the Manor at Appleby Parva," who also owned Newhouse Grange in Merevale, Leicestershire. Richard Brailsford's sons William and Frank were present at the April 1860 meeting, the former being a gamekeeper on the estate of Lord Lichfield at Shugborough in Shropshire. Oliver, *Little Acorns*, 10–11; editorial, *Field*, 14 Apr. 1860, 293.

19. "National Exhibition of Sporting Dogs," *Birmingham Daily Post*, 10 Apr. 1860, 2; editorial, *Field*, 14 Aug. 1860, 293.

20. W. Lort, "Mr Brailsford and the National Show of Dogs," *Field*, 21 Apr. 1860, 316. Lort became a respected and much in-demand judge at dog and agricultural shows across the country until his death in 1891. "Death of William Lort at Vaynol," *North Wales Chronicle*, 30 May 1891, 5; C. H. Lane, *Dog Shows and Doggy People* (London: Hutchinson, 1902), 269–70.

21. R. Brailsford, "Registry of Sporting Dogs," *Field*, 5 May 1860, 363.

22. See *Times*, 4 Dec. 1860, 9; *Manchester Times*, 8 Dec. 1860, 3; "Birmingham Cattle and Dog Show," *Leicestershire Mercury*, 8 Dec. 1860, 8; "The Dog Show at Birmingham," *Bell's Life*, 9 Dec. 1860, 8; "The Birmingham Dog Show," *Illustrated London News (ILN)*, 15 Dec. 1860, 1. The dogs shown were deerhounds, mastiffs, bloodhounds, Newfoundlands, pointers, bulldogs, toy terriers, and Maltese lion dogs.

23. "The Dog Show at Birmingham," *Bell's Life*, 9 Dec. 1860, 8. The Cochin China and Leghorn "buff" and "fluff" rose were varieties of chicken.

24. "Birmingham Cattle And Poultry Show And Exhibition Of Dogs," *Times*, 4 Dec. 1860, 9.

25. "The National Exhibition of Dogs," *Field*, 8 Dec. 1860, 477; "The Exhibition of Dogs," *Birmingham Journal*, 8 Dec. 1860, 6; "The National Exhibition of Dogs: Held at Birmingham," *Field*, 8 Dec. 1860, 477; "The Dog Show at Birmingham," *Bell's Life*, 9 Dec. 1860, 8. The London Fancy had formed a West End Canine Association and Club in the mid-1850s. "Canine Fancy," *Bell's Life*, 17 June 1855, 7, and 11 Nov. 1855, 7.

26. "Canine Fancy," *Bell's Life*, 30 Dec. 1860, 2. On the boxing match, see F. Keating, "Heenan v Sayers: The Fight that Changed Boxing Forever," *Guardian*, 17 Apr. 2010, www.guardian.co.uk.

27. "North of England Exhibition of Sporting Dogs," *Field*, 30 July 1861, 55.

28. Sportsman, "The Leeds Dog Show," *Field*, 10 Aug. 1861, 134; Stonehenge, *The Greyhound* (London: Longman, Green, Longman, Roberts, and Green, 1864), 342–50, 358. Also see An Essex Man, "The Leeds Dog Show," *Field*, 3 Aug. 1861, 115; "North of England," 55. Walsh was among the founders of the club. "Understandings" referred to legs and feet—what the animal stands on. The *OED* quotes Stonehenge's *Manual of British Rural Sports* as its source for the term: "'Discount' was . . . the perfection of a

strong, well-bred horse . . . if only his understandings had been sound." Stonehenge, *Manual,* 381–82.

29. S. Gunn, *The Public Culture of the Victorian Middle Class: Ritual and Authority in the English Industrial City, 1840–1914* (Manchester: Manchester University Press, 2000).

30. J. Secord, "Nature's Fancy: Charles Darwin and the Breeding of Pigeons," *Isis* 72 (1981): 162–86; "The Poultry Mania," *Leisure Hour,* 31 Mar. 1853, 221–23; J. Driver, "Poultry Mania," *History Today* 41, no. 8 (1991): 7–8; C. R. Stanley, *Windsor Home Park: the Aviary and Poultry Farm,* 1845, Royal Collection Trust, RL 19772, www .royalcollection.org.uk; "The Great Convocation of Poultry," *Household Words,* 19 Jan. 1852, 284; "The Manchester Poultry Show," *Manchester Times,* 28 Jan. 1855, 11; "Exhibition of Poultry in Manchester," *Manchester Courier,* 6 Jan. 1855, 10; "Crystal Palace Poultry Show," *Times,* 13 Feb. 1860, 5, and 25 Aug. 1860, 9; "Crystal Palace Poultry Show," *Field,* 1 Sept. 1860, 191.

31. "Proposed Exhibition of Dogs in the Metropolis," *Observer,* 11 Mar. 1861, 6.

32. R. Hale, *The Zoo: The Story of London Zoo* (London: Crowood, 2005); I. Charman, *The Zoo: The Wild and the Wonderful Tale of the Founding of London Zoo* (London: Viking 2016); "Proposed Exhibition of Dogs," 6.

33. "The Dog Show in Holborn," *Examiner,* 5 Oct. 1861, 12; "Exhibition of Sporting and Other Dogs," *Illustrated Times,* 12 Oct. 1861, 9; "The Great Exhibition of Sporting and Other Dogs," *Field,* 5 Oct. 1861, 306; "The So-called Great Dog Show in the Holborn Repository," *Bell's Life,* 6 Oct. 1861, 8; "Exhibition of Fancy and Other Dogs," *Field,* 31 May 1862, 494.

34. "North of England Show of Sporting and Other Dogs," *Field,* 28 June 1862, 578–79; Advertisement, "MONSTER DOG SHOW at the Agricultural Hall, Islington," *ILN,* 21 June 1862, 636; "The Agricultural Hall, Islington," *ILN,* 7 Dec. 1861, 584–86; "Dog Show at the New Agricultural Hall," *Morning Post,* 23 June 1862, 2.

35. "The International Dog Show," *Daily News,* 25 June 1862, 6; "The Dog Show," *London Review,* 28 June 1862, 592; "National Sports," *ILN,* 12 July 1862, 63; "International Dog Show," *York Herald,* 28 June 1862, 7; "Dog Show at the New Agricultural Hall," *Era,* 29 June 1862, 14. Brenda's 1862 value would be equivalent in today's money to around £75,000.

36. "The Dog Show," 591; "The Great Dog Show," *Saturday Review,* 28 June 1862, 739–40; "At the Dog Show," *Chamber's Journal of Popular Literature,* 9 Aug. 1862, 81–84; "Report," *Daily Telegraph,* 1 July 1862, 4.

37. N. Durbach, *Spectacle of Deformity: Freak Shows and Modern British Culture* (Berkeley: University of California Press, 2009); S. V. Everett, *George Wombwell: Celebrated Menagerist* (London: The George Wombwell Collection, 2016). One writer claimed that Appleby's show, which was termed a "Gathering of the Tykes" (*Tyke* is a popular term for a person from Yorkshire) had "a Barnum air about it from the taking." A. H. B., "A 'Fancy Article': The First Dog-Shows and its Results: Toy-Dog Breeders, Dog Dealers and Stealers," *New Sporting Magazine* 292 (Apr. 1865): 308.

38. "Bow! Wow! Wow!" *New York Sun,* 30 May 1862, 4; "The Great Dog Show at Barnum's," *New York Times,* 15 May 1862; "Barnum's Dog Show—4,000 Delegates to the Congress of Curs," *San Francisco Daily Evening Bulletin,* 16 June 1862, 1; "Barnum's American Museum," *Frank Leslie's Illustrated Newspaper,* 3 May 1862, 2. The

British press argued such events were inappropriate for a country in the midst of civil war.

39. "Sequel of the Late Dog Show," *Times,* 12 July 1862, 11; editorial, "Going to the Dogs," *Field,* 5 July 1862, 5; "The Suspension of Mr. T. D. Appleby," *Leeds Mercury,* 12 July 1862, 7; "The Islington Dog Show Speculation," *Manchester Guardian,* 12 July 1862, 6.

40. Editorial, "Going to the Dogs," 5; editorial, "Gone to the Dogs," *Field,* 19 July 1862, 53.

41. Old Towler, "Breeding up to Defects," *Field,* 19 July 1862, 56; Aberfeldy, "The Dog Show," *Field,* 26 July 1862, 88; "Looty" (illustrated feature), *Field,* 23 May 1863, 496–97. Also see S. Cheang, "Women, Pets, and Imperialism: The British Pekingese Dog and Nostalgia for Old China," *Journal of British Studies* 45 (2006): 359–87.

42. C. Dickens, "Two Dog-Shows," *All the Year Round,* 2 Aug. 1862, 495; Howell, *At Home,* 38. Often the essay is said to have been penned by Dickens himself, though Howell has attributed it to John Hollingshead. However, in his autobiography, Hollingshead only claims to have written on the Holloway Home for the *Morning Post* and *Good Words.* The website Dickens Journals Online (www. djo.org.uk) lists ninety-seven articles by Hollingshead in Dickens's publications, but not "Two Dog-Shows," which is unattributed. Indeed, he seems not to have written for Dickens's publications after 1861. Colam quoted in G. Jenkins, *A Home of Their Own: The Heart-Warming 150-Year History of Battersea Dogs* (London: Bantam, 2010), 66–67. Also see P. Howell, "June 1859/December 1860: The Dog Show and the Dogs' Home," BRANCH: *Britain, Representation and Nineteenth-Century History,* ed. D. F. Felluga, Extension of Romanticism and Victorianism on the Net, online.

43. Dickens, "Two Dog-Shows," 496, 497.

44. Ibid., 494.

45. Oliver, *From Little Acorns,* 10–17; "The Forthcoming National Dog Show," *Birmingham Daily Post,* 25 Oct. 1862, 2; "The Dog Show at Birmingham," *Field,* 6 Dec. 1862, 523–24; "The National Dog Show," *Birmingham Daily Post,* 1 Dec. 1862, 3. The second Manchester Poultry and Dog Show, held between Christmas and New Year in the great music hall of Belle Vue Zoological Gardens, was modeled on the Birmingham event. The local press deemed it successful, though it did not register nationally. "Manchester Poultry and Dog Show," *Manchester Guardian,* 30 Dec. 1862, 3.

46. L. Nead, *Victorian Babylon: People, Streets and Images in Nineteenth-Century London* (New Haven: Yale University Press, 2000), 109–39; "Grand National Dog Show," *Era,* 25 Jan. 1863, 5; "The Grand National Dog Show at Ashburnham Hall," *Era,* 22 Feb. 1863, 16; editorial, "Give a Dog a Bad Name, and ———," *Field,* 21 Feb. 1863, 165; editorial, "The 'Dawg' Show," *Field,* 28 Feb. 1863, 189–90; "Canine Fancy," *Bell's Life,* 22 Mar. 1863, 7.

47. Editorial, "The 'Dawg' Show," 190.

48. "The Cremorne Dog Show," *Field,* 21 Mar. 1863, 271; "The Cremorne Dog Show," *Field,* 28 Mar. 1863, 286.

49. "The Grand National Dog Show," *Era,* 29 Mar. 1863, 14; "The Cremorne Dog Show," *Morning Post,* 24 Mar. 1863, 6; "The Dog Show at Cremorne," *London Review* 6 (1863): 329; "The Dog Show at Chelsea," *Times,* 26 Mar. 1863, 14; "The Grand National

Dog Show," *Era,* 29 Mar. 1863, 14. Also see "The Dog Show," *Saturday Review* 15 (1863): 407–8; "The Dogs at Cremorne," *Spectator* 36 (1863): 1808–9; "Dogs in Public," *Once a Week* 8 (1863): 444–45. Other publications endorsing the event were the *Illustrated London News,* 28 Mar. 1863, 351; *Lady's Newspaper,* 28 Mar. 1863, 398; *Reynolds's Newspaper,* 29 Mar. 1863, 4; and the *Standard,* 31 Mar. 1863, 2.

50. "Dog Show at the Islington Agricultural Hall," *Penny Illustrated Paper,* 6 June 1863, 381; "The Islington Dog Show," *Era,* 31 May 1863, 13; "Islington, Paris and Poitou," *Field,* 30 May 1863, 515. At Islington, sporting dogs outnumbered nonsporting 1,000 to 700. The head of the show was Mr. Sidney.

51. "The Dog Show at the Islington Agricultural Hall," *Penny Illustrated Paper,* 6 June 1863, 13; "International Dog Show at Islington," *Sporting Gazette,* 4 June 1864, 4; "International Dog Show at Islington," *Times,* 3 June 1865, 9; "International Dog Show at the Agricultural Hall Islington," *Bell's Life,* 3 June 1865, 3; 17 June 1865, 7; 1 July 1865, 7; 8 July 1865, 8; "The Birmingham Dog Show of 1863," *Field,* 5 Dec. 1863, 557–58; Oliver, *From Little Acorns,* 19–23; "Poultry and Dog Show at Belle Vue Gardens," *Manchester Courier,* 26 Dec. 1863, 9; "The Christmas Dog and Poultry Show at Belle Vue Gardens," *Manchester Times,* 24 Dec. 1864, 5; "Manchester Poultry and Dog Show," *Manchester Courier,* 29 Dec. 1865, 3, and 30 Dec. 1865, 2.

52. A. H. B., "A "Fancy Article," 307–14, 376–81, 309 (italics in the original); J. Greenwood, *Unsentimental Journeys; or Byways of the Modern Babylon* (London: 1867), chap. 4, "A Dog Show," 25–31, 30; A. H. B., "A 'Fancy Article,'" 309.

53. "The Dog Show," *Pall Mall Gazette,* 5 June 1865, 11.

54. "Birmingham Dog Show," *Field,* 12 Dec. 1863, 582–83; "The Birmingham Dog Show of 1863," *Field,* 5 Dec. 1863, 557. Also see follow-up correspondence on 19 Dec. 1863, 607, and 26 Dec. 1863, 624.

55. W. W., "Rules for Judging a Dog Shows," *Field,* 11 June 1864, 407.

56. "Rules for Judging Dogs," *Field,* 25 June 1864, 427; "The Mule and Donkey Show," *Standard,* 10 Aug. 1864, 2; "A Dog and His Points," *Field,* 6 Aug. 1864, 87. Also see "The Donkey Movement," *London Review* 9 (1864): 145–46.

57. "Judging Dogs," *Field,* 9 July 1864, 25; 23 July 1864, 66; 30 July 1864, 77; D. Hume, "Judging at Dog Shows," *Field,* 6 Aug. 1864, 90; editorial, "Points of the St. Bernard," *Field,* 15 Oct. 1864, 272.

58. "A Dog and His Points," *Field,* 6 Aug. 1864, 87, italics in the original; Secord, "Nature's Fancy," 162–86; "The Judging at Shows of Domestic Animals," *Field,* 3 Dec. 1864, 383; editorial, "The Numerical Method of Judging Domestic Animals," *Field,* 18 Mar. 1865, 173.

59. "Birmingham National Dog Show," *Sporting Gazette,* 3 Dec. 1864, 947, "The Birmingham Dog Show," *Field,* 3 Dec. 1864, 396–97; 10 Dec. 1864, 410; 17 Dec. 1864, 428; 24 Dec. 1864, 441; "The Judging at Shows of Domestic Animals," *Field,* 31 Dec. 1864, 447.

60. "The Judging of Retrievers at Birmingham," *Field,* 26 Dec. 1863, 624.

61. Ibid.; Stonehenge, *The Dog,* 157–58; Huzzlebee, "Birmingham Dog Show," *Field,* 12 Dec. 1863, 583; Experientia, "Retrievers at Birmingham," *Field,* 2 Jan. 1864, 2.

62. Bow-wow, "Retrievers and Dog Shows," *Field,* 9 Jan. 1864, 35.

63. C. Lane, *Dog Shows,* 14. Lane wrote, "I should like to pay my humble tribute to the memory of my old friend Mr. John Douglas, one of the first to tread the thorny

path of a show manager, and whose colossal form and solid, self-contained demeanour took a considerable shock to disturb, and who enjoyed until his death the friendship and esteem of a large body of exhibitors, and, although a disciplinarian, was never faddish or unreasonable, and would go out of his way to oblige in anything within his power at least, this was my experience of him."

64. E. W. Jaquet, *The Kennel Club: A History and Record of its Work* (London: Kennel Gazette, 1905), 3 and 22; editorial, "Field Trial of Stud-Pointers and Setters," *Field,* 7 Apr. 1865, 143; T. W. B., "Trial of Dogs in the Field," *Field,* 8 July 1865, 50.

65. "Trial of Stud Pointers and Setters in the Field," *Field,* 22 Apr. 1865, 264.

66. G. S. Lowe, quoted in R. Leighton, *The New Book of the Dog,* vol. 1 (London: Cassell, 1907), 235.

67. "The International Dog Show at the Agricultural Hall, Islington," *Bell's Life,* 3 June 1865, 6.

68. Down Charge, "Trial of Stud Pointers and Setters in the Field," *Field,* 15 Apr. 1865, 256; "Trial of Dogs in the Field," *Field,* 19 Aug. 1865, 144, and 9 Sept. 1865, 200; editorial, "The Trial of Sporting Dogs in the Field," *Field,* 7 Oct. 1865, 251; One of the Judges, "The Trial of Dogs in the Field," *Field,* 15 July 1865, 67; The Other Judge, "The Trial of Dogs in the Field," *Field,* 15 July 1865, 68; Jaquet, *Kennel Club,* 6, 22–25, 33–34, 105–6, and 203.

69. "Judging Our Domestic Animals: The Numerical Method," *Field,* 18 Mar. 1865, 173, and 20 May 1865, 343; "The Model Pointer," *Field,* 9 Sept. 1865, 199. Walsh wrote the article and included Major seemingly on the strength of the drawing; he had not seen the dog in person.

70. "The Model Pointer," 199.

71. Ibid.; "The Weight of the Pointer," *Field,* 30 Sept. 1865, 238.

72. "The Black-tan Gordon Setter," *Field,* 7 Oct. 1865, 255; "Retrievers," *Field,* 28 Oct. 1865, 302; "Retrievers contd.—The Deer-hound," *Field,* 2 Dec. 1865, 394; "The Clumber Spaniel," *Field,* 16 Dec. 1865, 434–35; "The Fox-terrier," *Field,* 30 Dec. 1865, 482; Stonehenge, *The Dogs of the British Islands: Being a Series of Articles and Letters by Various Contributors, Reprinted from The Field Newspaper* (London: Horace Cox, 1867).

73. Stonehenge, *Dogs of the British Islands,* v–vii.

74. Stonehenge, *The Dog,* 86, 88, 89.

75. Stonehenge, *The Dog,* 129–35; *Dogs of the British Islands,* vi, 9–71.

76. Stonehenge, *The Greyhound* (1864); Stonehenge, *Dogs of the British Islands,* 90. The changes in greyhounds are discussed in detail in Stonehenge, *The Dog,* 24–37.

77. N. F. Blake, "Worde, Wynkyn de (d. 1534/5)," *Oxford Dictionary of National Biography.* There are longer versions of the rhyme. See Horme, "Coursing," *Sporting Review,* Feb. 1854, 116.

78. Stonehenge, *The Dog,* 42, 122–23, 125; Stonehenge, *Dogs of the British Islands,* 74.

79. Stonehenge, *Dogs of the British Islands,* iii and 60–63.

80. H. Ritvo, *The Animal Estate: The English and Other Creatures in the Victorian Age* (Cambridge, MA: Harvard University Press, 1987), 45–81.

81. A. H. B., "The 'Fancy Article,'" 307–14 and 376–81.

## Chapter 3  •  Showing Breed, 1867–1874

1. "B. Kerr, "The World of Thomas Pearce, Vicar of Morden, 1853–1882," *Proceedings of the Dorset Natural History and Archaeological Society* 94 (1972): 70–74; Idstone, *The Idstone Papers: A series of articles and desultory observation on spirit and things in general* (London: The Field, 1872); Idstone, *The Dog: with Simple Directions For His Treatment, and Notices of the Best Dogs of the Day* (London: The Field), 1872; Stonehenge, *The Dogs of the British Islands: Being a Series of Articles and Letters by Various Contributors* (London: Horace Cox, 1867), 19–30, 47–50, and 110–33; "The Dog Show," *Field,* 7 Dec. 1867, 464; 14 Dec. 1867, 488–89; 21 Dec. 1867, 512; 28 Dec. 1867, 532–33; 5 Dec. 1868, 452. In *Dogs of the British Islands,* there were twenty-three pages of letters on the fox terrier and others on setters and spaniels.

2. L. J., "The Birmingham Dog Show," *Field,* 19 Dec. 1868, 506.

3. M. Smith, "The St. Bernards at Birmingham," *Field,* 7 Dec. 1867, 464–65. According to the *Sporting Gazette,* the St. Bernards were "a wretched lot," and half the class were "parti-coloured butchers, curs and mongrel Mastiffs," and the judging was "admirably suited to its victims." Our Special Reporter, "Birmingham Dog Show," *Sporting Gazette,* 7 Dec. 1867, 3.

4. John Cumming Macdona was a founder member of the National Dog Club and the Kennel Club. He stood unsuccessfully for the Chesterfield seat for the Conservative Party in 1885, before being elected in 1892 for the Southwark constituency in 1892. He served for ten years, during which time he played an active role in the Kennel Club in London. As well as breeding dogs, he was an enthusiastic player of, and writer on, golf and a supporter of "automobilism," being active in the Automobile Club. *Manchester Courier,* 10 May 1907, 8. Macdona is best known today from his portrait, by Sir Leslie Ward, which was published in *Vanity Fair* on 8 February 1894.

5. "The St. Bernard's at Birmingham," *Field,* 14 Dec. 1867, 488; 21 Dec. 1867, 512; 28 Dec. 1867, 532; 4 Jan. 1868, 7; 11 Jan. 1868, 31; H. Dalziel, *British Dogs; Their Varieties, History and Characteristics* (London: "The Bazaar" Office, 1879), 248–51. The author of the chapter on the St. Bernard was Dalziel himself, writing under his pseudonym "Corsincon." Corsincon is a hill in the Scottish Borders, mentioned in a poem by Robert Burns. Also see H. Dalziel, *The St. Bernard: Its History, Points, Breeding, and Rearing* (London: L. Upcott Gill, 1888).

6. On display at the Naturhistorisches Museum, Bern, see www.nmbe.ch. Also see L. E. Thorsen, "A Dog of Myth and Matter: Barry the Saint Bernard in Bern," in *Animals on Display: The Creaturely in Museums, Zoos, and Natural History,* ed. L. E. Thorsen et al. (University Park: Penn State University Press, 2013), 128–52.

7. Stonehenge, *Dogs of the British Islands,* 76–78. Macdona claimed to be the expert, having introduced the St. Bernard into Britain. On Tell's pedigree and performances, see F. C. S. Pearce, ed., *The Kennel Club Stud Book: A Record of Dog Shows and Field Trials* (London: The Field, 1874), 503–4.

8. "The Birmingham Dog Show," *Sporting Gazette,* 5 Dec. 1868, 4. In today's money, the equivalent of Tell's value would be more than £800,000.

9. A breeder, "The Mastiffs at Birmingham," *Field,* 7 Dec. 1867, 465.

10. "Dog Shows and their Judges," *Field,* 18 Jan. 1868, 49; T. Pearce, "The Birmingham Dog Show," *Field,* 7 Nov. 1868, 375; Vox, "Judging at Dog Shows," *Field,* 7 Nov. 1868, 375.

11. T. Wootton, "The Fox Terriers at Birmingham," *Field,* 4 Jan. 1868, 7; Jus, "Dogs and Dog Shows," *Field,* 14 Dec. 1867, 488–89; "Canine Fancy," *Bell's Life,* 14 Dec. 1867, 7.

12. Letters on the results at Birmingham in the *Field,* 7 Dec. 1867, 465; 21 Dec. 1867, 512; 4 Jan. 1868, 7 and 512; E. Nichols, "Awards at the Late Birmingham Dog Show," *Field,* 28 Dec. 1867, 532; C. H. Lane, *All About Dogs: A Book For Doggy People* (London: John Lane, 1901), 27–31; "Comment: The Bloodhounds at Birmingham," *Field,* 11 Jan. 1868, 7.

13. M. Hutchings, "In the beginning . . . was the National Dog Club," *Our Dogs,* 2001, www.ourdogs.co.uk; "The Shooting Dog Club," *Field,* 14 Nov. 1868, 391; C. H. Lane, *Dog Shows and Doggy People* (London: Hutchinson, 1902), 200–204. Richard John Lloyd Price was the son of Richard John Price and Charlotte Lloyd. He "inherited the Rhiwlas Estate at the age of seventeen years on the death of his grandfather, Richard Watkin Price, in 1860. He was a lifelong sportsman and lover of animals and something of an entrepreneur with a highly developed sense of marketing. He wrote several books and amongst his business ventures were the establishment of the Welsh Whisky Distillery at Fron-goch and the Rhiwlas Brush Works." Gwynedd Archives, https://archives.library.wales.

14. Editorial, "The Shooting Dog Club," *Field,* 12 Dec. 1868, 475; *Field,* 12 Dec. 1868, 475. The members were listed: R. T. L. Price (acting chairman), F. R. Bevan (Weston Grove, Southhampton), Rev. F. W. Adey (Markyate, Bucks.), Richard Llewellin Purcell (Exeter College Oxford), J. S. Philips (Kibworth, Leics.), Geo. Potts (Jedburgh, Roxburghshire, Scotland), Wm. Lort (King's Norton, Worcs.), Rev. J. Cumming Macdona (West Kirby, Lancs.), Mr. Francis Brailsford (Knowsley, Lancs.), J. H. Whitehouse (Ipsley Court, Warks.), E. Laverack (Manchester), Sam. Long (Clifton, Bristol), Rev. J. W. Mellor (Ryde, Isle of Wight), G. T. Bucknell (Witney, Oxon.).

15. Editorial, "The National Dog Club Show and its Shows," *Field,* 27 Feb. 1869, 160; 29 May 1869, 443–44; editorial, "The London Show of the National Dog Club," *Field,* 15 May 1869, 399.

16. "The Islington Dog Show," *Daily News,* 2 June 1869, 5; "The National Dog Club," *Bell's Life* 5 June 1869, 8; "The Islington Dog Show," *ILN,* 12 June 1869, 593–94; "The National Dog Club's First Exhibition," *Field,* 5 June 1869, 466–67; "Prize Dogs in the National Dog Show at Islington," *ILN,* 12 June 1869, 20; "The National Dog Club's First Exhibition," *Field,* 5 June 1869, 466–67; Stonehenge, *Dogs of the British Islands,* 74; "The Newfoundland Dog," *Field,* 12 June 1869, 490; "Newfoundlands at London Show," *Field,* 19 June 1869, 514; 3 July 1869, 13; Interloper, "Newfoundlands," *Field,* 10 July 1869, 34. In the fourth edition of *The Dogs of the British Islands published in 1882,* Walsh noted that judges still "refuse to recognise" their merits and "relegate" them to a separate class (p. 179). On Landseer's dog paintings, see W. Secord, *Dog Painting: A Social History of the Dog in Art* (Woodbridge, Suffolk: Antique Collector's Club, 1992), 17–18, 37–43, 196–97, 204–7, and 233–41; W. C. Monkhouse, *Landseer's Animal Illustrations: Including a Concise Art History of Sir Edwin Landseer,* ed. J. Batty (Alton, UK: Nimrod, 1900).

17. Stonehenge, *Dogs of the British Islands* (1867), 57–60; S. Tenison Mosse and W. J. Mellor, "Dandie Dinmont Terriers," *Field,* 2 Jan. 1869, 16–17; C. Collins, "Dandie

Dinmont Terriers," *Field,* 9 Jan. 1869, 48; A Border Sportsman, "Historical Account of the Origin, Progress and Decline of the Pepper-and-Mustard Terrier," *Field,* 23 Jan. 1869, 90; E. B. S. and S. Tenison Mosse, "What is a Dandie Dinmont?" *Field,* 6 Mar. 1869, 198.

18. Editorial, "The Islington Horse and Dog Shows," *Field,* 12 June 1869, 489; editorial, "The London Show of the National Dog Club," *Field,* 15 May 1869, 399.

19. "The National Dog Club," *Field,* 27 Nov. 1869, 454; Bevan, "The National Dog Club," *Field,* 12 Feb. 1870, 149; S. Lang, "The National Dog Club," *Field,* 18 Dec. 1869, 522; "National Dog Club," *Field,* 15 Jan. 1870, 66; 22 Jan. 1870, 74; 12 Feb. 1870, 149; 19 Feb. 1870, 174; 5 Mar. 1870, 207; "The National Dog Club's Field Trials for Pointers and Setters," *Hampshire Advertiser,* 4 May 1870, 4.

20. "Crystal Palace," *Standard,* 22 June 1870, 3; *Debrett's Illustrated Peerage* (London: Dean and Son, 1876), 194–95. The 10th Earl Ferrers was Sewallis Edward Shirley (1847–1912) and a contemporary of Sewallis Evelyn Shirley.

21. "The Dog Show," *Birmingham Daily Post,* 2 Dec. 1867, 6; "South Durham and North Yorkshire Horse and Foal and Sporting Dog Show," *York Herald,* 25 Aug. 1860, 5; "South Durham and North Yorkshire Horse and Dog Show," *York Herald,* 7 Aug. 1869, 5; "The Birmingham Dog Show," *Bell's Life,* 1 Dec. 1869, 3. The Darlington Horse and Foal Show was founded in 1783 and became the South Durham and North Yorkshire Horse and Foal and Sporting Dog Show in August 1860 when dogs were first included.

22. Legacies of British Slave-ownership, www.ucl.ac.uk; C. Hamlin, "Murchison, Charles (1830–1879)," *Oxford Dictionary of National Biography* (Oxford: Oxford University Press, 2004), www.oxforddnb.com; Law Report, "Murchison v. Southgate," *Times,* 10 May 1872, 13. Alexander Murchison graduated from Marischal College Aberdeen in 1813 and went to work in Jamaica. He became owner of the Springfield and Guininault estates in the Vere district, where he also was a trustee of the free school. He moved back to Elgin in 1830, where he died in 1845. J. H. Murchison was secretary to companies in South and North Wales, and in the 1880s promoted the prospects of investment in South African gold fields. His publications included *Review of British Mining, with a brief outline of the position and prospects of some of the principal dividend and progressive Mines* (London: 1858) and *The Conservatives and "Liberals," their principles and policy* (London: Saunders, Otley, 1866). Concerning his showing and breeding of dogs, see "South Durham and North York Horse and Dog Show," *Newcastle Courant,* 6 Aug. 1869, 2; *Western Times,* 20 Aug. 1869, 5; *Bradford Observer,* 6 Sept. 1869, 4; *Bell's Life,* 1 Dec. 1869, 3; Lane, *Dog Shows,* 7, 153, 291, 296, and 392.

23. "National Dog Show," *Western Mail,* 23 June 1870, 2; "Crystal Palace Exhibition of Dogs," *Field,* 4 June 1870, 486; 11 June 1870, 495; 25 June 1870, 533–34; Worstead, "Crystal Palace Dog Show," *Bell's Life,* 3 Apr. 1870, 8; "The Crystal Palace Dog Show," *Bell's Life,* 21 May 1870, 6, and 28 May 1870, 7 and 8; MacGregor, "The Crystal Palace Dog Show," *Bell's Life,* 1 June 1870, 3. There were only two clergy on the committee, F. W. Adye and John Cumming Macdona, and one judge, Rev. Thomas Pearce (Idstone). In the event, the standards were not published.

24. K. Nichols and S. V. Turner, eds., *After 1851: The Material and Visual Cultures of the Crystal Palace at Sydenham* (Manchester: Manchester University Press, 2017); "National Dog Show," *Western Mail* (23 June 1870), 2.

25. Advertisements, "Crystal Palace," *Times,* 21 June 1870, 1; "Crystal Palace," *Standard,* 22 June 1870, 3; "National Dog Show," *Western Mail,* 23 June 1870, 2; The Rambler, "Dog Days," *Times,* 26 June 1870, 2; J. Pigott, *Palace of the People: The Crystal Palace at Sydenham, 1854–1936* (London: C. Hurst, 2004).

26. "National Dog Show," *Western Mail,* 23 June 1870, 2; "Crystal Palace Dog Show," *Times,* 23 June 1870, 12; "National Dog Show at Crystal Palace," *Sporting Gazette,* 25 June 1870, 3; "Crystal Palace," *Standard,* 22 June 1970, 3; "The Crystal Palace Dog Show," *ILN,* 2 July 1870, 20; "The Crystal Palace Dog Show," *Graphic,* 2 July 1870, 7–8; "Dog Show at the Crystal Palace," *Birmingham Daily Post,* 23 June 1870, 5; "Dog Shows," *Bell's Life,* 25 June 1870, 5.

27. "The Crystal Palace Dog Show," *Bell's Life,* 25 June 1870, 6. One correspondent compared it to the infamous London show in 1862, which was referred to as "the notorious Appleby *fiasco.*"

28. WC, "Exhibition of Sporting Dog, and Gentlemen Breeders, &c.," *Bell's Life,* 4 June 1870, 6; Thomas, "Exhibition of Sporting Dog, and Gentlemen Breeders, &c.," *Bell's Life,* 11 June 1870, 6; A Looker On, "The Crystal Palace Dog Show," *Bell's Life,* 6 July 1870, 3; Dash, "The Crystal Palace Dog Show," *Bell's Life,* 9 July 1870, 6. Also see letters from Exhibitor, Spear, J.W.G., and Not an Unsuccessful Exhibitor, "The Crystal Palace Dog Show," *Bell's Life,* 9 July 1870, 8.

29. W. Shaw, "To the Editor," *Bell's Life,* 6 July 1870, 3; Thomas, "The Crystal Palace Dog Show," 5.

30. One of the Committee, "The Late Crystal Palace Dog Show," *Bell's Life,* 13 July 1870, 2; "The Recent Crystal Palace Dog Show," *Sporting Gazette,* 23 July 1870, 552; "Sporting Dog Shows," *Bell's Life,* 30 July 1870, 6.

31. M. A., "The Birmingham Show," *Field,* 5 Nov. 1870, 406; T. Wooton, "Mode of Judging at Birmingham Dog Show," *Field,* 26 Nov. 1870, 467; "The Birmingham Show," *Field,* 5 Nov. 1870, 406; Carrier, "Judging at Dog Shows," *Field,* 2 Nov. 1870, 417; J. H. Murchison, "Mode of Judging at Birmingham Dog Show," *Field,* 12 Nov. 1870, 417–18, 26 Nov. 1870, 467; and 15 Oct. 1870, 330.

32. "Birmingham Cattle and Poultry Show and the Exhibition of Dogs," *Times,* 28 Nov. 1870, 12; "The Dog Show," *Birmingham Daily Post,* 28 Nov. 1870, 5, and 30 Nov. 1870, 6; "Among the Dogs," *Observer,* 4 June 1871, 7; Mastiff, "On Breeding Mastiffs," *Field,* 21 Jan. 1871, 52; 28 Jan. 1871, 60; 6 Feb. 1871, 80.

33. "The Dog Show," *Times,* 3 June 1871, 5; "Crystal Palace—Grand Exhibition of Sporting and Other Dogs," *Bell's Life,* 3 June 1871, 3; "Dog Show at the Crystal Palace," *Sporting Gazette,* 3 June 1871, 390; "Crystal Palace Dog Show," *Field,* 10 June 1871, 482–83, and 24 June 1871, 509; "Dog Show at the Crystal Palace," *Observer,* 4 June 1871, 7; "Amongst the Dogs at the Crystal Palace," *Gentleman's Magazine* 1 (1871): 478–84; "Among the Dogs," *Manchester Guardian,* 7; "Crystal Palace Dog Show," *Bell's Life,* 10 June 1871, 5, and 17 June 1871, 10.

34. Editorial, "Judging at Dog Shows," *Field,* 17 June 1871, 485.

35. R. Nicholls, *The Belle Vue Story* (Manchester: Neil Richardson, 1992); "The Cat Show at the Crystal Palace," *Morning Post,* 13 July 1871, 6; "The Cats at the Crystal Palace," *Pall Mall Gazette,* 14 July 1871, 7. A second cat show was held in December 1871. *Times,* 4 Dec. 1871, 12. The leading figure in the development of cat shows

was Harrison Weir; see J. Batty, *Harrison Weir: Artist, Author and Poultryman* (London: Beech, 2003).

36. "The Dog Show," *Times*, 6 June 1872, 10; "The Crystal Palace Dog Show," *Standard*, 5 June 1872, 5; "Crystal Palace Dog Show," *Bell's Life*, 8 June 1872, 11. The show had over one thousand entries in ninety-five classes and ran at only a small financial loss.

37. "Grand National Dog Show at Nottingham," *Daily News*, 2 Oct. 1872, 3; "Biography of John Chaworth Musters," University of Nottingham Special Collections, www.nottingham.ac.uk; "Nottingham v. Birmingham and Crystal Palace Dog Shows," *Field*, 5 Aug. 1872, 183; 12 Aug. 1872, 212; 19 Aug. 1872, 231; *Field*, 24 Aug. 1872, 183; "Nottingham Goose Fair and National Dog Show," *Sheffield Telegraph*, 3 Oct. 1872, 3; editorial, "Nottingham Dog Show," *Field*, 31 Aug. 1872, 205–6; "Nottingham Dog Show Preface," quoted in *Field*, 24 Aug. 1872, 183.

38. "The Grand National Dog Show at Nottingham," *Daily News*, 2 Oct. 1872; "Nottingham Dog Show," *Bell's Life*, 5 Oct. 1872, 12; "The Nottingham Dog Show," *Derby Mercury*, 9 Oct. 1872, 8; "The Nottingham Dog Show," *Field*, 26 Oct. 1872, 409; "Nottingham v. Birmingham and Crystal Palace Dog Shows," *Field*, 7 Sept. 1872, 231; "Nottingham Dog Show," 19 Oct. 1872, 391, and 9 Nov. 1871, 472; "The Grand National Dog Show: Nottingham, 1872," *Bell's Life*, 12 Oct. 1872, 11; "Nottingham Dog Show," *Field*, 12 Oct. 1872, 231; "The Nottingham Dog Show," *Derby Mercury*, 9 Oct. 1872, 8. There were also some supportive remarks in the sporting press. Draco and Dingle belonged to Major J. A. Cowen of Blaydon-on-Tyne.

39. "Devon and Cornwall Field Trials," *Field*, 26 Apr. 1873, 386; "The Dog Show," *Times*, 20 June 1873, 4. The trials were held over three days, 21–23 April, and the results published in Pearce, *Kennel Club Stud Book*, 172–75. The twelve attending the first meeting were Sewallis Shirley, Sam Lang, H. T. Mendel, Major Platt, T. W. Hazelhurst, Mr. Whitehouse, William Lort, George Brewis, John Cumming Macdona, J. H. Dawes, C. W. Hodge, and Frank Adcock. Kennel Club Library, London, Members' Book, 1873–1879.

40. E. W. Jaquet, *The Kennel Club: A History and Record of its Work* (London: The Kennel Gazette, 1905), 5.

41. "The Crystal Palace Dog Show," *Field*, 21 June 1873, 596–97; "The Dog Show at the Crystal Palace," *Bell's Life*, 21 June 1873, 11. Only one press report mentioned that the Kennel Club organized the show: "The Dog Show at the Crystal Palace," *Standard*, 18 June 1873, 4. It was not mentioned in the *Field* or *Bell's Life*.

42. Fixed Idea, "The Late Manchester Dog Show," *Field*, 31 May 1873, 531, and "Crystal Palace Dog Show," *Field*, 5 July 1873, 17; O. R. K., "The Anatomy of Dog Shows," *Field*, 5 July 1873, 17.

43. A Silly Goose, "The Crystal Palace Dog Show," *Field*, 19 July 1873, 59; editorial, "Crystal Palace Dog Show," *Field*, 5 July 1873; Frank, "Dog Shows and Show Dogs," *Field*, 19 July 1873, 59; editorial, "Dog Shows and Dog Trials," 26 July 1873, 85–86. Also see correspondence on 21 June 1873, 596–97; 28 June 1875, 647; and 19 July 1873, 59.

44. "Country Notes," *Bell's Life*, 12 July 1873, 6; J. G. V. Wakerley, "The National Canine Society," *Bell's Life*, 19 July 1873, 11; "Country Notes," *Bell's Life*, 26 July 1873,

6; J. G. V. Wakerley, "National Canine Society," *Bell's Life,* 2 Aug. 1873, 10; "Country Notes," *Bell's Life,* 9 Aug. 1873, 6; J. G. V. Wakerley, "The Grand National Canine Society," *Bell's Life,* 16 Aug. 1873, 3, and 23 Aug. 1873, 10.

45. "The Great Nottingham Dog Show," *Bell's Life,* 4 Oct. 1873, 7; "Country Notes," *Bell's Life,* 11 Oct. 1873, 4; "The Nottingham Dog Show," *Times,* 3 Oct. 1874, 11.

46. "The Birmingham Dog Show," *Field,* 6 Dec. 1873, 596; M. B. Wynn, "A Mastiff Club," *Field,* 28 June 1873, 648; "The Proposed Mastiff Club," *Field,* 5 July 1873, 17; Phylocyon, "Judging at Dog Shows," *Field,* 20 Dec. 1873, 627; Critic, "Judging Mastiffs," *Field,* 6 Dec. 1873, 3, 597; Caractacus, "Management of Dogs Shows," *Field,* 20 Dec. 1873, 626; Jack-a-Lantern, "The Canine World," *Field,* 21 Feb. 1874, 189. Wynn wrote what is still regarded as an important book on the breed, *The History of the Mastiff* (Melton Mowbray: William Loxley, 1886).

47. Caractacus, "Letter," *Field,* 7 Feb. 1874, 142; Dr. Stables, "On Dog Shows," *Field,* 21 Feb. 1874, 188–89; Caractacus, "Letter," *Field,* 7 Mar. 1874, 222, and 21 Mar. 1874, 285; A. S. De Fivas, "On Dog Shows," *Field,* 28 Feb. 1874, 202; "The Kennel Club," *Field,* 21 Mar. 1874, 288. See the listing for the 1875 Kennel Club June show in *Field,* 17 Apr. 1875, 391.

48. "The Crystal Palace Dog Show," *Field,* 13 June 1874, 574; A. S. De Fivas, "The Management of Dog Shows," *Field,* 27 June 1874, 631.

49. J. M. D. Brown, "The Management of Dog Shows," *Field,* 18 July 1874, 61; "Mr Douglas's Testimonial," *Field,* 12 June 1875, 595; "Death of Mr John Douglas," *Field,* 24 Aug. 1895, 347, and "The Death of John Douglas" *Stock-Keeper & Fanciers' Chronicle,* 23 Aug. 1895, 145; Caractacus, "The Management of Dog Shows," *Field,* 27 June 1874, 630–631; "The Crystal Palace Dog Show," *Sporting Gazette,* 13 June 1874, 535; "The Late Crystal Palace Show," *Bell's Life,* 20 June 1874, 3.

### Chapter 4   •   Governing Breed

1. The National American Kennel Club was founded in 1876, and the American Kennel Club in 1884. The French Société Centrale Canine began in 1881, while in Germany specialist breed clubs flourished before they were brought together in an umbrella organization in 1906, Das Kartell der stammbuchführenden Spezialklubs für Jagd- und Nutzhunde (Cartel of breed clubs holding a stud book for hunting and utility dogs).

2. There were two other authors who had competing books, George R. Jesse and Henry Webb. However, neither had the connections in sporting circles, and their books did not have the same status. G. R. Jesse, *Researches into the History of British Dogs* (London: Robert Hardwicke, 1866); H. Webb, *Dogs: Their Points, Whims, Instincts and Peculiarities* (London: Dean and Son, 1876). Lewis Clements, who published under the pseudonyms "Wildfowler" and "Snapshot," was the leading authority on shooting and gundogs. Wildfowler/Snapshot, *Shooting Adventures, Canine Lore and Sea-Fishing Trips,* vols. 1 and 2 (London: Chapman Hall, 1879).

3. F. C. S. Pearce, *The Kennel Club Stud Book* (London: The Field, 1874), v; A. Milne-Smith, *London Clubland: A Cultural History of Class and Gender in Late-Victorian Britain* (New York: Palgrave Macmillan, 2011).

4. Pearce, *The Kennel Club Stud Book,* i–x, quote on v; B. Lightman, ed., *Global Spencerism: The Communication and Appropriation of a British Evolutionist* (Leiden: Brill, 2016).

5. Ibid., 503–4. Tell was dog number 2458: "Pedigree: By *Hero* (deceased from the celebrated *Barry* at the Hospice) out of *Diane.*"

6. "The Kennel Club Stud Book," *Field,* 10 Feb. 1877, 153.

7. Kvoy, "The Field Trial Derby," *Field,* 16 Jan. 1875, 42; J. G. V. Wakeley, "The National Canine Society," *Bell's Life,* 19 July 1873, 11; Our Correspondent, "Country Notes," *Bell's Life,* 23 Aug. 1871, 6; F. Adcock, "The Bulldog Toro," *Field,* 17 Jan. 1874, 70–71; 24 Jan. 1874, 70–71; 31 Jan. 1874, 118; *Sporting Gazette,* 14 July 1877, 662; Stonehenge, "The Bull Dog," *Field,* 17 Mar. 1877, 295–96, and correspondence on 331, 386, 413–14, 447. Adcock's attempt to enlarge the bulldog dated from 1874.

8. Few bulldog aficionados agreed with Adcock's views or his methods, and he stepped beyond the pale when he purchased Toro, who weighed ninety pounds and with whom he hoped to further enlarge the British bulldog. His hope turned out to be in vain. "The Kennel Club Dog Show at Islington," *Field,* 7 July 1877, 7, and 21 July 1877, 67; Stonehenge, *The Dogs of the British Islands,* 4th ed. (London: Horace Cox, 1882), 164.

9. H. Dalziel, *British Dogs: Their Varieties, History, Characteristics, Breeding, Management, and Exhibition* (London: "The Bazaar" Office, 1879), 222; "The Bulldog Club's First Annual Show," *Sporting Gazette,* 17 June 1876, 10; "The Bulldog Club," *Sporting Gazette,* 17 Apr. 1875, 13; "The Bulldog Club," *Sporting Gazette,* 26 Feb. 1876, 1; "The New Bulldog Club," *Field,* 10 July 1875, 32; Stonehenge, "The Bulldog," 296. There had been a number of clubs within the London Fancy since the 1860s, for example, the Small Toy Bulldog Club and New Bulldog Club.

10. A. Oliver, *From Little Acorns: The History of the Birmingham Dog Show Society, Est. 1859* (Birmingham: Birmingham Dog Show Society, 1998), 30; "A Canine Court of Appeal," *Field,* 12 Dec. 1874, 639; "Canine Fancy," *Bell's Life,* 15 Dec. 1877, 12, and 22 Dec. 1877, 12.

11. "The Alexandra Palace Dog Show," *Bell's Life,* 22 Dec. 1877, 11.

12. "Editorial," *Live Stock Journal and Fancier's Gazette,* 25 July 1879, 74.

13. "The Bulldog Bumble," *Field,* 1 July 1876, 20; *Field,* 20 Dec. 1879, 846; 24 Jan. 1880, 96; 31 Jan. 1880, 124; 7 Feb. 1880, 140; 14 Feb. 1880, 191; "Obituary: William Lort," *Stock-Keeper and Fancier's Chronicle* (*SK&FC*), 29 May 1891, 411; H. Dalziel, "The Kennel Club and the Mode of Appointing Judges," *Field,* 27 Dec. 1879, 869.

14. Kuon, "The Kennel Club and the Mode of Appointing Judges," *Field,* 3 Jan. 1880, 16; S. E. Shirley, "The Kennel Club and the Mode of Appointing Judges," *Field,* 3 Jan. 1880, 17; "The Kennel Club and the Mode of Appointing Judges," *Field,* 10 Jan. 1880, 28; 24 Jan. 1880, 96; 31 Jan. 1880, 124; 7 Feb. 1880, 140; W. Allison, "The Kennel Club," *Field,* 31 Jan. 1880, 124.

15. Editorial, "The New Kennel Club Rules for Dog Shows," *Field,* 7 Apr. 1880, 492, and 8 May 1880, 568; S. E. Shirley, "The New Kennel Club Rules for Dog Shows," *Field,* 15 May 1880, 600; 31 July 1880, 180; 7 Aug. 1880, 209; 14 Aug. 1880, 257.

16. See the summary of meetings in E. W. Jaquet, *The Kennel Club: A History and Record of its Work* (London: The Kennel Gazette, 1905), 27–59. The dispute over the

judging of fox terriers at the club's show in June 1876 is discussed in "The Kennel Club," *Field,* 10 June 1876, 647; 17 June 1876, 677; 1 July 1876, 20–21; 8 July 1876, 44.

17. Editorial, "Kennel Club Rule," *Field,* 18 Apr. 1882, 453; S. E. Shirley, "Darlington Show and the Kennel Club," *Field,* 13 Aug. 1881, 231; "The Kennel Club and the Darlington Show," *SK&FC,* 9 Sept. 1881, 8–9; J. Marlow, *Captain Boycott and the Irish* (London: André Deutsch, 1973).

18. "The Kennel Club on Its High Horse," *SK&FC,* 28 July 1882, 52; J. H. Murchison, "The Kennel Club," *Field,* 1 July 1882, 3. Also see correspondence, *Field,* 8 July 1882, 44; 15 July 1882, 82; 22 July 1882, 122; 5 Aug. 1882, 104; 12 Aug. 1882, 248–49. An editorial in the *Kennel Chronicle* in 1882 accused the Kennel Club of "tyranny." "The Kennel Club and Dog Shows," *Kennel Chronicle* 3 (1882): 89.

19. *Northern Echo,* 24 June 1882, 3; The Kennel Club New Rules," *SK&FC,* 14 July 1882, 19, and 21 July 1882, 35; "The Kennel Club New Rules," *SK&FC,* 14 July 1882, 19, and 21 July 1882, 35.

20. "Birmingham Dog Show Boycotted," *Field,* 21 Oct. 1882, 567, and 28 Oct. 1882, 618; "Birmingham and the Kennel Club 'Stud Book,'" *Field,* 4 Nov. 1882, 637, and 11 Nov. 1882, 674; J. H. Murchison, "The Kennel Club and Its Opponents," *Field,* 25 Nov. 1882, 781; "The British Barnum," in F. Jackson, *Crufts* (London: Pelham, 1990), 11–32. Also see "The Alexandra Palace Dog Show," *Bell's Life,* 17 June 1882, 3; "The Crystal Palace Dog Show," *Bell's Life,* 20 Jan. 1883, 5.

21. "The New Canine Confederation," *SK&FC,* 15 Dec. 1882, 390; "The Kennel Club and Its Opponents," *Field,* 18 Nov. 1882, 732; editorial, "Supervision of Dog Shows by the K.C.," *Field,* 25 Nov. 1882, 745; editorial, "A Final Appeal to the Kennel Club," *Field,* 30 Dec. 1882, 911; "Proposed Formation of a National Canine Society," *Birmingham Daily Post,* 28 Nov. 1882, 4; "Proposed National Canine Club and Register," *Field,* 2 Dec. 1882, 816; J. C. Macdona, "Dog Shows and the Kennel Club," *Field,* 1 July 1882, 3; "The Birmingham Dog Show," *Bell's Life,* 2 Dec. 1882, 9. In the 1 July 1882 issue of the *Field,* Walsh had crossed swords with J. H. Murchison over the Kennel Club's new policies.

22. "The National Dog Club," *Times,* 18 Jan. 1883, 7; Oliver, *From Little Acorns,* 40; editorial, "The National Dog Club and the K.C.," *Field,* 17 Feb. 1883, 201; V. Shaw, *The National Dog Club Stud Book, containing prize lists, . . . accounts of field trials, and pedigrees of winning dogs, in 1882* (London: Horace Cox, 1883); V. Shaw, "National Dog Club," *Field,* 31 Mar. 1883, 43.

23. "The Use of Bloodhounds in Tracking Criminals," *North-Eastern Daily Gazette,* 13 Feb. 1889, 4. Also see N. Pemberton, "'Bloodhounds as Detectives': Dogs, Slum Stench and Late-Victorian Murder Investigation," *Cultural and Social History* 10 (2013): 69–91.

24. "Dog Show at Aston Lower Grounds," *Birmingham Daily Post,* 7 May 1883, 5; "Dog Show at Sheffield," *Sheffield & Rotherham Independent,* 21 Oct. 1885, 2. The Sheffield and Hallamshire Kennel Club had held a show in the same venue in April 1883. "Sheffield Dog Show," *Sheffield & Rotherham Independent,* 16 Apr. 1883, 3.

25. J. H. Salter, "The Kennel Club and Its Opponents," *Field,* 6 Jan. 1883, 25; editorial, "The Proposed National Club and the Kennel Club," *Field,* 13 Jan. 1882, 35–36. Salter hoped that Birmingham and the Kennel Club could make friendly terms, but if not he wrote that he was "quite prepared to see the issue to its bitter end,

and to stand or fall with my ship." J. H. Salter, "The Kennel Club's Last Circular," *Field,* 18 Jan. 1883, 60.

26. John Henry Salter was a general practitioner in the Essex village of Tolleshunt D'Arcy. He had fought as a bareknuckle fighter under the name of Jack O'Reilly, was active in many other sports, had shot big game, and, allegedly, was out at daybreak every morning to shoot wild duck on the Essex marshes. He often shot abroad, especially in Russia. (He spoke Russian.) There were stories, unverified, that the tsar on his visits to Britain shot with Salter. R. Innes-Smith, "Salter, John Henry (1841–1932)," *Oxford Dictionary of National Biography* (Oxford: Oxford University Press, 2004), www.oxforddnb.com; J. O. Thompson, ed., *Dr Salter of Tolleshunt D'Arcy in the county of Essex, medical man, freemason, sportsman, sporting-dog breeder and horticulturalist: his diary and reminiscences from the year 1849 to the year 1932* (London: John Lane, 1933).

27. "Darlington Dog Show," *Field,* 4 Aug. 1883, 156; "Birmingham Dog Show," *Field,* 3 Dec. 1883, 792–93; "Notes on a Dog Show," *Graphic,* 8 Dec. 1883, 5, 7; "The National Dog Show," *Times,* 4 Dec. 1883, 7; "The Birmingham Dog Show," *Bell's Life,* 8 Dec. 1883, 9.

28. "National Dog Show," *Times,* 1 Dec. 1885, 17; "Darlington Dog Show," *Field,* 14 June 1884, 812; "Birmingham Dog Show," *Field,* 6 Dec. 1884, 796–98; "The Birmingham Dog Show," *Horse and Hound,* 5 Dec. 1885, 669; "Birmingham Dog Show," *Kennel Gazette,* 4 Dec. 1885, 261; "The National Dog Show" *Birmingham Daily Post,* 30 Nov. 1885, 5. For a detailed account of the rapprochement between Birmingham and the Kennel Club, from the perspective of the Birmingham Society, see Oliver, *From Little Acorns,* 39–46. The Kennel Club view is set out in Jaquet, *Kennel Club,* 83–85.

29. Jaquet, *Kennel Club,* 26. Figures compiled from Jaquet, *Kennel Club,* and the dog shows listed in the 1885 indexes of the *Field* (vols. 65 and 66).

30. Dalziel, *British Dogs*; Vero Shaw, *The Illustrated Book of the Dog* (London: Cassell, Petter, Galpin, 1881).

31. It is unclear whether Dalziel had any veterinary or medical training, though his occupation as a commercial traveler in drugs suggests that he might have at least started a medical qualification. H. Dalziel, *The Diseases of Dogs, their pathology, diagnosis, and treatment: to which is added a complete dictionary of canine "materia medica"* (London: "The Bazaar" Office, 1874); H. Dalziel, *The Diseases of Horses, their pathology, diagnosis and treatment; to which is added a complete dictionary of equine "materia medica"* (London: "The Bazaar" Office, 1879).

32. Shaw, *Illustrated Book,* 4; Dalziel, *British Dogs,* 396. Authors of chapters on individual breeds in Dalziel's book include Capt. G. A. Graham on the Irish wolfhound, Frank Adcock on the bulldog, and Lewis Clements (aka "Snapshot" and "Wildfowler") on the basset hound.

33. J. A. Venn, *Alumni Cantabrigienses,* part 2, vol. 5 (Cambridge: Cambridge University Press, 1953), 482; Oliver, *From Little Acorns,* 40; Dalziel, *British Dogs,* 14.

34. G. S. Woods, "Stables, William Gordon (1837x40–1910)," rev. G. Arnold, *Oxford Dictionary of National Biography.*

35. C. H. Lane, *Dog Shows and Doggy People* (London: Hutchinson, 1902), 184–86; H. Furniss, *The Works of Charles Burton Barber* (London: Cassell, 1896); "Arthur Wardle (1864–1949)," www.rehs.com/arthur_wardle.htm; William Secord Gallery,

Arthur Wardle, www.dogpainting.com. Barber's most famous work was *Queen Victoria with John Brown* (1894). On the development of color lithography, see R. Verrhoogt, *Art in Reproduction: Nineteenth-century Prints After Lawrence Alma-tadema, Jozef Israels, and Ary Scheffer* (Amsterdam: Amsterdam University Press, 2007), 81–118.

36. Harry Wyndham Carter, who styled himself with the invented title of "Earl of Winchcombe," kept basset hounds. He was convicted in 1886 and sentenced to five years penal servitude, during which he was transferred to a lunatic asylum. He escaped from a private asylum in December 1893 and was soon in trouble again. Convicted in 1894 for threatening the life of Queen Victoria, he was declared insane at trial and sent to Broadmoor, where he spent the next forty-three years. He died in December 1937. *Leamington Spa Courier and Warwickshire Standard,* 24 Feb. 1894, 3: "Sent Threatening Letter to Queen Victoria: Man's Death in Asylum," *Evening Telegraph and Post,* 4 Dec. 1937, 7.

37. "Common Pleas Division," *Daily News,* 5 June 1880, 6; "Common Pleas Division," *Times,* 9 June 1880, 6; "An Important Dog Case," *Northern Echo,* 10 June 1880, 4; "Alleged Libel on a Dog," *Reynolds News,* 13 June 1880, 8; "Alleged Libel of a Dog," *Illustrated Police News,* 12 June 1880, 3; "High Court of Justice, June 4," *Field,* 12 June 1880, 746.

38. Editorial, "Dog Shows and the Sporting Press," *Kennel Gazette,* July 1880, 63–64; editorial, "Dog Shows and the Sporting Press," *Kennel Gazette,* Aug. 1880, 84–85.

39. "Recent Improvements in Our Breeds," *Field,* 24 Feb. 1877, 214; J. H. Walsh, *The Dogs of the British Islands,* 3rd ed. (London: "The Field" Office, 1878); Stonehenge, "The Modern Pointer," *Field,* 28 Apr. 1877, 507–8.

40. "The Bulldog Club," *Sporting Gazette,* 31 Aug. 1878, 18; Stonehenge, *The Dog in Health and Disease,* 4th ed. (London: Longman, Green, 1887), v.

41. "The Dublin Dog Show," *Bell's Life,* 13 Apr. 1878, 12; "The Scottish Kennel Club," *Edinburgh Evening News,* 19 Jan. 1882, 2; Shaw, *Illustrated Book,* 118–19; "Canine Notes: On Dog Shows," *Illustrated Sporting and Dramatic News,* 20 Nov. 1875, 15; "The Dandie Dinmont Club," *Field,* 19 Feb. 1876, 205; "The Dandie Dinmont Club Standard," *Field,* 18 Mar. 1876, 354. There was animated correspondence on the breed in subsequent years: "The Dandie Dinmont Points," *Field,* 5 May 1877, 539, 737, 768; "The Dandie Dinmont," *Field,* 16 Nov. 1878, 627, 660, 712, 741, 757, 785.

42. "South Durham and North Yorkshire Horse and Dog Show," *York Herald,* 27 July 1867, 5; "Agricultural Society," *Sunderland Daily Echo,* 18 Aug. 1876, 3–4; "The Bedlington Terrier," *Field,* 8 Jan. 1870, 47, and 12 Feb. 1870, 149; "What Is a Bedlington?," *Field,* 26 Feb. 1870, 192; Carrier, "Manufacturing a Bedlington," *Field,* 15 Jan. 1870, 46 and 75. Bedlington is a town situated fourteen miles north of Newcastle upon Tyne. In the nineteenth century it had iron works and coal mines, and its own longstanding poultry and pigeon show. The Bedlington terrier was a new, albeit briefly described, addition to the second edition of Stonehenge, *The Dogs of the British Islands,* 2nd ed. (London: Horace Cox, 1872), 123.

43. Shaw, *Illustrated Book,* 126; Dalziel, *British Dogs,* 367; Walsh, *Dogs of the British Islands* (1882), 232; Doctor, "What Constitutes a Skye Terrier?" *Field,* 14

Mar. 1874, 266; Canine, "What Constitutes a Skye Terrier?" *Field,* 28 Mar. 1874, 317; Lee, *Modern Dogs,* 244–68.

44. "Specialist Clubs," *SK&FC,* 16 Apr. 1880, 311; "New Dog Clubs," *Kennel Chronicle and Pedigree Register,* Nov. 1880, 1; Hertfordshire Farmer, "Functions of Specialist Clubs," *Kennel Review* 1 (Oct. 1882): 2; H. Wyndham Carter, "Specialist Clubs," *Kennel Review* 1 (Oct. 1882): 1–2. The *Kennel Review* became a vehicle for opposition to the Kennel Club until it ceased publication in 1884.

45. "Notes of a Dog Show," 5; "St Bernard's Club Show—The First," *Times,* 3 Nov. 1882, 3; "St Bernard Club Show," *Bell's Life,* 4 Nov. 1882, 11.

46. "The Evolution of Dog Shows," *SK&FC,* 8 May 1881, 358. See correspondence on the National Spaniel Club in the *Field,* beginning with T. Jacobs, "The Proposed Spaniel Club," *Field,* 11 Nov. 1882, 674, and correspondence in the following weeks on pp. 732, 781, 816, 896, 937; it continued in the early months of 1883. See A. W. Langdale, "The National Spaniel Club," *Field,* 7 Jan. 1883, 61, and correspondence in the following months on pp. 434, 448, 530, and 542.

47. "The Evolution," 358.

48. "Fido's Lessons," *Illustrated London News,* 5 Sept. 1885, 13–14; "The Yelping French Poodle!," *Funny Folks,* 5 Sept. 1885, 281; "Prodigy and Poodle," *Punch,* 26 Sept. 1885, 154.

49. "Hydrophobia at Bradford," *Leeds Mercury,* 26 Mar. 1886, 4; "Cruel Torture of a Dog," *Sheffield & Rotherham Independent,* 27 May 1887, 7; Petty Sessions, *Essex Newsman,* 7 Mar. 1885, 2.

50. "Savage Dogs in Bradford," *Huddersfield Chronicle,* 1 May 1886, 1.

51. Ibid.

52. "Dog Shows and the Kennel Club," *Field,* 14 Jan. 1888, 35.

### Chapter 5 • *Improving Breed I: Experience*

1. B. B., "The Past and Future of Dog Shows," *Kennel Review* 1 (1882–83): 185–87.

2. E. W. Jaquet, *The Kennel Club: A History and Record of Its Work* (London: Kennel Gazette, 1905), 19–20.

3. H. St. James Stephens, ed., *Kennel Club Calendar and Stud Book* (London: The Kennel Club, 1900).

4. Jaquet, *Kennel Club,* 20. Also see "Description of the Different Varieties of Spaniel," *Field,* 22 Jan. 1887, 94, and 29 Jan. 1887, 151.

5. H. Dalziel, *British Dogs; Their Varieties, History and Characteristics* (London: "The Bazaar" Office, 1879), 431.

6. V. Shaw, *Illustrated Book of the Dog* (London: Cassell, Petter, Galpin, 1881), 155.

7. R. B. Lee, *History and Description of the Modern Dogs of Great Britain and Ireland: The Terriers* (London: Horace Cox, 1894), 339–42.

8. Idstone, *The Dog, with Simple Directions for His Treatment* (London: Cassell, Petter, and Galpin, 1872), 208–9; Stonehenge, *The Dogs of the British Islands,* 2nd ed. (London: Horace Cox, 1872), 108.

9. Dalziel, "The Yorkshire Terrier," in Stonehenge, *The Dogs of the British Islands,* 3rd ed. (London: "The Field" Office, 1878), 211; G. Stables, *Ladies' Dogs as Companions* (London: Dean and Sons, 1879); Shaw, *Illustrated Book,* 155. The leading breeder in the

1860s was Mrs. M. A. Foster, whose Huddersfield Ben won many prizes and was a successful stud dog, until he was run over by a cart and killed in 1871.

10. Dalziel, *British Dogs*, 433. In the third edition of *The Dog* (1879), Walsh also doubted the Yorkshire terrier could have been changed so quickly by selection, and he speculated that the lengthened coat had come from crossing with Maltese terriers, and the color from King Charles spaniels. Stonehenge, *The Dog in Health and Disease*, 3rd ed. (London: Longman, Green, 1879), 125.

11. Stonehenge, *The Dog* (1879), 126; Dalziel, *British Dogs*, 434, 435.

12. F. G. W. Crafer, "The Bulldog," in Dalziel, *British Dogs*, 222.

13. "Exhibition of Bulldogs," *Morning Post*, 22 May 1884, 6; "Bulldog Show," *Times*, 30 May 1888, 3; Dalziel, *British Dogs*, 222; Shaw, *Illustrated Book*, 86, 83.

14. Shaw, *Illustrated Book*, 90.

15. P. L. G. R. S. T., "Bulldogs," *Kennel Gazette*, Jan. 1892, 6; F. W. Crowther, "Bulldogs," *Kennel Gazette*, Jan. 1893, 11.

16. H. St. John Cooper, *Bulldogs and Bulldog Men* (London: Jarrold and Sons, 1908). In Manchester in 1892, a British Bulldog Club was founded to counter the influence of the established, London-based club. Its primary goal was "to promote and encourage the breeding of pure Bulldogs in all of the world, but more especially in the provinces of England, to Ireland, and in Scotland."

17. E. Farman, "Bulldogs," *Kennel Gazette*, Jan. 1894, 5; E. Farman, *The Bulldog: A Monograph* (London: Stock Keeper Company, 1899); F. W. Taylor, "Bulldogs," *Kennel Gazette*, Jan. 1891, 12. Also see W. F. Jeffries, "Bulldogs," *Kennel Gazette*, Jan. 1900, 11–12.

18. P. Bew, *Ireland: The Politics of Enmity, 1789–2006* (Oxford: Oxford University Press, 2007); T. W. Moody and F. X. Martin, eds., *The Course of Irish History* (Cork: Mercier, 1967); W. Youatt, *The Dog* (London: Longman, Brown, Green, and Longmans, 1854), 40; Stonehenge, *The Dog in Health and Disease* (London: Longman, Brown, Green, and Longman, 1859), 38; "Captain George Augustus Graham," Dursley Glos Web, www.dursleyglos.org.uk; G. A. Graham, "The Irish Wolfhound," in Dalziel, *British Dogs*, 32. Garnier had eccentric views on mastiffs and St. Bernards. He maintained that old English mastiffs and St. Bernards were the same dog but that the former had become extinct. R. B. Lee, *A History and Description of the Modern Dogs of Great Britain and Ireland (Non-Sporting Division)*, 3rd ed. (London: Horace Cox, 1899), 18–19.

19. Graham, "The Irish Wolfhound," 34. Oliver Goldsmith is best remembered for his novel *The Vicar of Wakefield* (1766), the play *She Stoops to Conquer* (1769), and the poem, "An Elegy on the Death of a Mad Dog" (1766). J. Dussinger, "Oliver Goldsmith (1728?–1774)," *Oxford Dictionary of National Biography* (Oxford: Oxford University Press, 2004), www.oxforddnb.com.

20. G. A. Graham, "The Irish Wolfhound," *Livestock Journal and Fancier's Gazette* (*LSJFG*), 31 Jan. 1879, 88. Also see R. Leighton, *The New Book of the Dog* (London: Cassell, 1907), 160.

21. G. A. Graham, *The Irish Wolfhound* (Dursley: privately printed, 1885), 2; G. W. Hickman, "The Irish Wolfhound," *LSJFG*, 30 Apr. 1880, 356; M. B. Wynn, "The Irish Wolfhound," *LSJFG*, 14 Mar. 1879, 207; Capt. G. A. Graham, "The Irish Wolfhound"; Dalziel, *British Dogs*, 40, 41.

22. "The Dog Show," *Freeman's Journal,* 2 Apr. 1879, 3.

23. "The 'Manchester Martyrs,'" *Freeman's Journal,* 24 Nov. 1879, 3.

24. Dalziel, *British Dogs,* 32–42; G. A. Graham, "The Irish Wolfhound," *LSJFG,* 4 Apr 1879, 269; H. D. Richardson, *The Dog: Its Origin, Natural History and Varieties* (Dublin: James McGlashan, 1847), 335–41; G. A. Graham, "The Irish Wolfhound," *LSJFG,* 16 May 1879, 387. Graham also relied upon the views of Mr. A. McNeill of Colonsay. Later works traced the history of the Irish wolfhound from "the dawn of our history." R. M. Scott, "The History, Character and Description of The Irish Wolfhound," in *The Irish Wolfhound Annual* (London: Irish Wolfhound Association, 1925), relied heavily on Father Edmund Hogan, S.J., *The History of the Irish Wolf Dog* (Dublin: Sealy, 1897).

25. Capt. G. A. Graham, "The Irish Wolfhound," in Dalziel, *British Dogs,* 32.

26. Shaw, *Illustrated Book,* 217.

27. M. B. Wynn, "The Irish Wolfhound," *LSJFG,* 16 May 1879, 387; G. W. Hickman, "The Irish Wolfhound," *LSJFG,* 16 May 1879, 388; G. Hickman, "The Irish Wolfhound," *LSJFG,* 28 May 1880; Stonehenge, *The Dog* (1879), 37; "The Irish Wolfhound," *Field,* 27 June 1885, 840; G. A. Graham, "The Irish Wolfhound," *Field,* 4 July 1885, 3. On the alleged outcrosses with the Scottish deerhound, mastiff, and Great Dane, see Stonehenge, *The Dog* (1879), 37.

28. G. A. Graham, "Irish Wolfhounds," *Kennel Gazette,* Jan. 1891, 3–4; G. A. Graham, "Irish Wolfhounds," *Kennel Gazette,* Jan. 1892, 7; J. B. Waldy, "Irish Wolfhounds," *Kennel Gazette,* Jan. 1899, 11.

29. "People Involved in the Breed's Resuscitation," www.irishwolfhounds.org /others.htm.

30. "Captain Graham and the Resuscitation of the Breed," Irish Wolfhound History, www.irishwolfhounds.org/graham.htm.

31. Stonehenge, *The Dog* (1859), 171–98, and Stonehenge, *The Dog* (1879), 251–74.

32. Stonehenge, *The Dog* (1859), 175.

33. Ibid., 172.

34. M. E. Derry, *Masterminding Nature: The Breeding of Animals, 1750–2010* (Toronto: University of Toronto Press, 2015), 18.

35. Ibid., 173; M. Sommer, "How Cultural Is Heritage? Humanity's Black Sheep from Charles Darwin to Jack London," in *A Cultural History of Heredity III: 19th and Early 20th Centuries,* ed. S. Müller-Wille and H.-J. Rheinberger (Berlin, 2005), 233–53, www.mpiwgberlin.mpg.de/Preprints/P294.PDF; N. Stepan, *The Idea of Race in Science: Great Britain, 1800–1960* (London: Macmillan, 1982), 83–100; R. A. Nye, "Heredity or Milieu: The Foundations of Modern European Criminological Theory," *Isis* 67 (1976): 334–55.

36. Stonehenge, *The Dog* (1879), 265, 258, 257–64.

37. Dalziel, *British Dogs,* 454.

38. Stonehenge, *The Dog* (1859), 172–73; "Breeding and Dogs at Stud," *SK&FC,* 27 Jan. 1882, 53, and 3 Feb. 1882, 68; Stonehenge, *The Dog* (1859), 175, and Stonehenge, *The Dog* (1887), esp. 236.

39. Shaw, *Illustrated Book,* 215. The sixth point was a restatement of prepotency.

40. Ibid., 524–55; C. Mazzoni, *Maternal Impressions: Pregnancy and Childbirth in Literature and Theory* (Ithaca: Cornell University Press, 2002).

41. W. B. Tegetmeier and W. W. Boulton, *Breeding for Colour and the Physiology of Breeding* (1874). We could not locate a copy of this publication, though it is widely cited in Victorian dog books.

42. W. D. Drury, *British Dogs: Their Points, Selection, and Show Preparation* (London: L. Upcott Gill, 1903), 315.

43. C. H. Lane, *All About Dogs; A Book for Doggy People* (London: John Lane, 1900), 81. Lane describes Jacob as "the most successful breeder and exhibitor of [black spaniels], during the last twenty years."

44. Shaw, *Illustrated Book*, 445.

45. Ibid.

46. G. Stables, *The Common Sense of Dog Doctoring* (London: Spratt's Patent, 1884), 85; Shaw, *Illustrated Book*, 520.

47. C. Darwin, *The Variation of Animals and Plants under Domestication*, vol. 1 (London: John Murray, 1868); E. Laverack, *The Setter: With notices of the most eminent breeds now extant; instructions how to breed, rear, and break; dog shows, field trials, general management, etc.* (London: Longman, Green, 1872), 25–31; Caractacus, "Observations on Breeding," *SK&FC*, 2 Sept. 1881, 7.

48. J. Woodroffe Hill, "In-and-In Breeding," *SK&FC*, 9 Sept. 1881, 9; "In-and-in Breeding," *SK&FC*, 14 Oct. 1881, 9; 21 Oct. 1881, 10; 28 Oct. 1881, 9–10; and 4 Nov. 1881, 11.

49. R. J. Lee, "Maternal Impression," *British Medical Journal* i (1875): 167–69; J. Clapperton, "Maternal Impressions," *British Medical Journal* i (1875): 169–70.

50. Dalziel, *British Dogs*, 462–63.

51. R. W. Burkhardt, "Closing the Door on Lord Morton's Mare: The Rise and Fall of Telegony," *Studies in the History of Biology* 3 (1979): 1–21; Dalziel, *British Dogs*, 461; O. Morton, "A Communication of a Singular Fact in Natural History," *Philosophical Transactions of the Royal Society of London* 111 (1821): 20–22; J. Endersby, *A Guinea Pig's History of Biology* (London: Random House, 2007), 17–23; C. Sengoopta, *Otto Weininger: Sex, Science, and Self in Imperial Vienna* (London: Chicago University Press, 2000), 131–37; M. Carlson, "Ibsen, Strindberg, and Telegony," *PMLA* 100 (1985): 774–82; R. Lewinsohn, *A History of Sexual Customs* (London: Longman Green, 1958), 204–6; S. Kern, *A Cultural History of Causality: Science, Murder Novels, and Systems of Thought* (Princeton: Princeton University Press, 2004), 47; Shaw, *Illustrated Book*, 461.

52. Este, "The Influence of the First or Previous Sire," *Kennel Gazette*, Nov. 1890, 265–67; H. Ritvo, "The Animal Connection," in *The Boundaries of Humanity: Humans, Animals, Machines*, ed. J. J. Sheehan and M. Sosna (Berkeley: University of California Press, 1991), 76–80; A. Weismann, *Germ-plasm: A Theory of Heredity* (London: Walter Scott, 1893), 383–86; R. J. Wood, "'The Sheep Breeders' View of Heredity (1723–1843)," in *A Cultural History of Heredity*, vol. 2, *18th and 19th Centuries*, ed. H.-J. Rheinberger and S. Müller-Wille (Chicago: University of Chicago Press, 2007), 21–46; J. C. Ewart, *The Penycuik Experiments* (London: A. and C. Black, 1899); F. H. A. "James Cossar Ewart, 1851–1933," *Obituary Notices of Fellows of the Royal Society* 1 (1934): 189–95; H. Ritvo, *The Platypus and the Mermaid: And Other Figments of the Classifying Imagination* (Cambridge, MA: Harvard University Press, 1997), 110–11. In 1895, W. B. Tegetmeier wrote a book that engaged with Ewart's work: *Horses, Asses,*

*Zebras, Mules and Mule Breeding* (London: Horace Cox, 1895). Also see D. Barnaby, ed., *Letters to Mr Tegetmeier: From J. Cossar Ewart and Others to the Editor of the Field at the Turn of the Nineteenth Century* (Timperley: ZSGM, 2004).

53. G. L. Norgate, "Lee, Rawdon Briggs (1845–1908)," rev. J. Lock, *Oxford Dictionary of National Biography*; R. B. Lee, *History and Description of the Fox-Terrier* (London: Horace Cox, 1890); *History and Description of the Collie or Sheep Dog in his British Varieties* (London: Horace Cox, 1890); *History and Description of the Modern Dogs of Great Britain and Ireland: Sporting Division* (London: Horace Cox, 1893); *History and Description of the Modern Dogs of Great Britain and Ireland: Non-Sporting Division* (London: Horace Cox, 1894); and *Modern Dogs: The Terriers* (1894).

54. Lee, *History and Description of the Collie,* 14–16.

### Chapter 6 • *Improving Breed II: Science*

1. Editorial, "Scientific Breeding," *Kennel Gazette,* Apr. 1880, 1–2; editorial, "Veterinarians and Dogs," *Kennel Gazette,* June 1880, 1–2.

2. C. Darwin, *On the Origin of Species by Means of Natural Selection, or the Preservation of Favoured Races in the Struggle for Life* (London: John Murray, 1859); J. van Wyhe, "Mind the Gap: Did Darwin Avoid Publishing His Theory for Many Years?," *Notes and Records of the Royal Society* 61 (2007): 177–205; C. Darwin, *The Variation of Animals and Plants under Domestication,* vol. 1 (London: John Murray, 1868); P. J. Vorzimmer, "Darwin's Questions about the Breeding of Animals (1839)," *Journal of the History of Biology* 2 (1969): 269–81; E. Mayr, *One Long Argument: Charles Darwin and the Genesis of Modern Evolutionary Thought* (Cambridge: Harvard University Press, 1991); M. Ruse, "Charles Darwin and Artificial Selection," *Journal of the History of Ideas* 36 (1975): 339–50; B. Theunissen, "Darwin and His Pigeons: The Analogy Between Artificial and Natural Selection Revisited," *Journal of the History of Biology* 45 (2012): 179–212. On recent debates about the relationship between artificial and natural selection, see A. Sullivan-Clarke, "On the Causal Efficacy of Natural Selection: A Response to Richards' Critique of the Standard Interpretation," *Studies in History and Philosophy of Biological and Biomedical Sciences* 44 (2013): 745–55.

3. Darwin, *On the Origin of Species,* 34–35.

4. Darwin, *The Variation of Animals and Plants* (1868), 1:33; Darwin, *The Variation of Animals and Plants under Domestication,* vol. 1 (London: John Murray, 1875), 41, 38, 29, 34–37, 44, 45. In the first edition Darwin wrote of "types of dogs" (p. 33) rather than "races of dogs."

5. J. Endersby, "Darwin on Generation, Pangenesis and Sexual Selection," in *The Cambridge Companion to Darwin,* ed. J. Hodge and G. Radick (Cambridge: Cambridge University Press, 2003), 69–91; G. L. Geison, "Darwin and Heredity: The Evolution of His Hypothesis of Pangenesis," *Journal of the History of Medicine and Allied Sciences* 24 (1969): 375–411; P. J. Bowler, *The Mendelian Revolution: The Emergence of Hereditarian Concepts in Modern Science and Society* (London: Athlone, 1989), 37–44, 65–66, and 88–89.

6. P. White, *Thomas Huxley: Making the "Man of Science"* (Cambridge: Cambridge University Press, 2003), 58–62; A. Desmond, *Huxley: The Devil's Disciple* (London: Michael Joseph, 1994); A. Desmond, "Huxley, Thomas Henry (1825–1895)," *Oxford*

*Dictionary of National Biography* (Oxford: Oxford University Press, 2004), www
.oxforddnb.com. Huxley's lectures at the Royal Institution were titled "Dogs and
the Problems Associated with Them." See report in the *Country Gentleman,* 10
Apr. 1880, 371.

7. Imperial College Archives (ICA), Huxley Letters, 10 May 1880. Huxley had
been working specifically on South American dogs and published on this work. See
T. H. Huxley, "On the Cranial and Dental Characters of the Canidæ," *Proceedings of
the Zoological Society of London* 48 (1880): 238–88; "On the Epipubis in the Dog and
Fox," *Proceedings of the Royal Society of London* 30 (1879–80): 162–63.

8. ICA, Huxley Papers, *Biology Lectures,* vol. 39, *Working Men's Lectures: The
Carnivora, Cats, dogs, etc.,* p. 116; "Royal Institution Lectures: Dogs," *Illustrated
London News,* 17 Apr. 1880, 378.

9. L. Huxley, *Life and Letters of Thomas Henry Huxley* (London: Macmillan, 1903),
275; ICA, Huxley Papers, *Biology Lectures,* 117; G. R. Jesse, *Researches into the History
of British Dogs* (London: Robert Hardwicke, 1866), 84 and 271.

10. E. Millais, *The Theory and Practice of Rational Breeding* (London: The Fancier's
Gazette, 1889); M. Worboys, "Millais, Sir Everett, second baronet (1856–1897),"
*Oxford Dictionary of National Biography*; S. Fagence Cooper, *The Model Wife: The
Passionate Lives of Effie Gray, Ruskin, and Millais* (London: Gerald Duckworth, 2010),
176–78; W. J. O'Connor, *British Physiologists, 1885–1914: A Biographical Dictionary*
(Manchester: Manchester University Press, 1991), 534.

11. E. Millais, "An 'Artificial Impregnation' by a Dog Breeder," *Veterinary Journal*
18 (1884): 256; H. Dalziel, *British Dogs: Their Varieties, History and Characteristics*
(London: "The Bazaar" Office, 1879), 462.

12. Millais, "Artificial Impregnation," 256–58; W. Heape, "The Artificial Insemi-
nation of Mammals and Subsequent Possible Fertilisation or Impregnation of Their
Ova," *Proceedings of the Royal Society of London* 61 (1897): 52; J. R. Campbell et al.,
*Animal Sciences: The Biology, Care, and Production of Domestic Animals* (Long Grove,
IL: Wayland, 2003), 242; J. D. Biggars, "Walter Heape, FRS: A Pioneer of Reproductive
Biology," *Journal of Reproduction and Fertility* 93 (1991): 173–86. On Heape's work,
see F. H. A. Marshall, "Walter Heape: 1855–1929 (Obituary)," *Proceedings of the Royal
Society,* B, 106 (1930): xv–xvii. On the history of artificial insemination in livestock at
this time, see S. Wilmot, "From 'Public Service' to Artificial Insemination: Animal
Breeding Science and Reproductive Research in Early Twentieth-century Britain,"
*Studies in History and Philosophy of Biological and Biomedical Sciences* 3 (2007):
411–41.

13. Millais, *Theory and Practice,* viii, 97, vii.

14. Ibid., 41–42, 55; W. D. Drury, *British Dogs: Their Points, Selection, and Show
Preparation* (London: L. Upcott Gill, 1903), 211; E. Millais, "Basset Bloodhounds," *Dog
Owners' Annual 1897* (Manchester: "Our Dogs," 1897), 17–30; Millais, *Theory and
Practice,* 55.

15. In fact, one bitch that he paired with Model was Model's great-granddaughter.
Millais later described this pairing as "in and in breeding with a vengeance," though
others would have termed it linebreeding as the dogs were not very close relations.
Quoted in Balogh, "The Early History of the Basset Hound in England," www
.huntingbassets.com/articles/earlyhistory.php.

16. F. Galton, "The Average Contribution of Each Several Ancestor to the Total Heritage of the Offspring," *Proceedings of the Royal Society of London* 61 (1897): 401–13; M. Bulmer, "Galton's Law of Ancestral Heredity," *Heredity* 81 (1998): 579–85; R. G. Swinburne, "Galton's Law: Formulation and Development," *Annals of Science* 21 (1965): 15–31.

17. F. Galton, *Inquiries into Human Faculty and Its Development* (London: Macmillan, 1883), 17; "Hereditary Talent and Character," *Macmillan's Magazine* 12 (1865): 157–66 and 318–27; "A Theory of Heredity," *Journal of the Anthropological Institute* 5 (1875): 329–48; "Regression towards Mediocrity in Hereditary Stature," *Journal of the Anthropological Institute* 15 (1885): 246–63; and *Natural Inheritance* (London: Macmillan, 1889). For the context of Galton's early writings, see J. C. Waller, "Ideas of Heredity, Reproduction and Eugenics in Britain, 1800–1875," *Studies in History and Philosophy of Biological and Biomedical Sciences* 32 (2001): 457–89.

18. Galton, "Hereditary Talent," 158.

19. G. J. Romanes, "The Darwinian Theory of Instinct," *Nineteenth Century* 16 (1884): 442–44; Roger Smith, "Romanes, George John (1848–1894)," *Oxford Dictionary of National Biography*; P. Bowler, *The Non-Darwinian Revolution* (Baltimore: Johns Hopkins University Press, 1988); C. Lloyd Morgan, *Animal Behaviour* (London: Edward Arnold, 1900), 134–37.

20. Millais, *Theory and Practice*, 32–40.

21. C. H. Lane, *Dog Shows and Doggy People* (London: Hutchinson, 1902), 178–80. Marples had previously contributed to the *Kennel Review*, the *Stock-Keeper and Fanciers' Chronicle*, and *Canine World*, and set up the *British Fancier*.

22. E. Millais, "The Pathogenic Microbe of Distemper in Dogs, and Its Use for Protective Inoculation," *British Medical Journal* i (1890): 856–60; M. Worboys, "Louis Pasteur," in *The Scientists: An Epic of Discovery*, ed. A. Robinson (London: Thames and Hudson, 2012), 252–57.

23. E. Millais, *Two Problems of Reproduction* (Manchester: "Our Dogs," 1895); Heape, "Artificial Insemination," 57. Millais was understood to be skeptical of telegony; however, he subsequently wrote to the *British Fancier* to correct that impression. He stated, "I see no reason to disbelieve in the phenomenon, for in my lecture I particularly allude to the fact where cases of recurrent telegony exist." He went on to emphasize that the "influence" of the first sire soon dissipates. E. Millais, "Two Problems of Reproduction," *British Fancier*, 15 Mar. 1895, 237.

24. Millais, "Two Problems," 9.

25. Millais quoted in Various Authors, *The Basset Hound: A Complete Anthology, 1860–1940* (Alcester, Warks., UK: Vintage Dog Books, 2010), 34–35.

26. E. Millais, "Basset Bloodhounds: Their Origins, Raison D'Etre and Value," *Dog Owner's Annual 1895* (Manchester: "Our Dogs," 1895), 17–30; Millais, *Two Problems*, 23. On other breeds, see, M. Rosenberg, "Golden Retrievers Are White, Pit Bulls are Black, and Chihuahuas Are Hispanic: Representations of Breeds of Dog and Issues of Race in Popular Culture," in *Making Animal Meaning (The Animal Turn)*, ed. L. Kalof and G. M. Montgomery (East Lansing: Michigan State University Press, 2011), 113–26; A. Burton, *The Trouble with Empire: Challenges to Modern British Imperialism* (Oxford: Oxford University Press, 2015), 2–4 and 176.

27. "Progress in England," Meet the Basset Hounds, www.meet-the-basset-breeds .co.uk/basset-hounds; "The Basset Hound: A Short History," Basset Hound Owners Club, www.bassethoundowners.org.uk/breedhistory3.html. See "Save the Basset," http://savethebasset.altervista.org/home-english.html. The origins of "save the Basset" are explained thus: "The transmission of a television program conducted by the BBC in the United Kingdom, showing presumed health problems in certain dog breeds (basset hound excluded), has let animalists and the BBC to join forces and put pressure on the English Kennel Club. Due to this action taken by the animalists and the BBC, the English Kennel Club has modified the standard of certain breeds, also those dog breeds which the KC has the faculty of valid modification also for the FCI (Fédération Cynologique Internationale)."

28. Dalziel, *British Dogs,* 453–87; V. Shaw, *The Illustrated Book of the Dog* (London: Cassell, Petter, Galpin, 1881), 532–606; A. Woods and S. Matthews, "'Little, If at All, Removed from the Illiterate Farrier or Cow-leech': The English Veterinary Surgeon, c. 1860–85, and the Campaign for Veterinary Reform," *Medical History* 54 (2010): 54, 29–54.

29. G. Stables, *The Commonsense of Dog Doctoring* (London: Tucker, Johnson, 1884), 124; "Spratt's Sanatorium for Dogs," *Stock-Keeper and Fanciers' Chronicle* (*SK&FC*), 17 Feb. 1893, 145–48.

30. P. Howell, *At Home and Astray: The Domestic Dog in Victorian Britain* (Charlottesville: University of Virginia Press, 2015); N. Pemberton and M. Worboys, *Rabies in Britain: Dogs, Disease, and Culture, 1830–2000* (London: Palgrave, 2012), 140–56.

31. W. Hunting, "The 'Distemper' of Dogs," *Bailey's Monthly Magazine of Sports and Pastimes,* 1 July 1876, 20; J. Woodroffe Hill, *The Management and Diseases of the Dog* (London: Sonnenschein, 1888), 311; Stonehenge, *The Dogs of the British Islands,* 4th ed. (London: Horace Cox, 1882), 17–19.

32. H. Dalziel, *The Greyhound* (London: Upcott Hill, 1887), end pages; "Gillard's Compound for Distemper," *County Gentleman: Sporting Gazette and Agricultural Journal,* 24 Feb. 1883, 198. Heald's sold the following: hepatic aperient balls, tonic condition balls, cough balls, worm powders, mange specific, ear canker lotion, tick lotion, dog soap, and medicine chests, containing all these remedies. Also see *Dog Owners' Annual 1895* (Manchester: "Our Dogs," 1895), 8.

33. Idstone, *The Dog, with Simple Directions for his Treatment* (London: Cassell, Petter and Galpin, 1872), 4; W. Youatt, *The Dog,* 2nd ed. (London: Longman, Brown, Green, and Longman 1854), 231–33; J. H. Steel, *A Treatise on Diseases of the Dogs* (London: Longman, Green, 1888), 7.

34. "Kennel Club Rules: Revised January 1884," *Kennel Club Calendar and Stud Book* (London: Club, 1884), xxvii.

35. "The Manchester Dog Show," *Manchester Courier,* 24 May 1888, 6; advertisement, *SK&FC,* 27 Jun 1890, xvii.

36. Millais, "The Pathogenic Microbe," 856–60; "Resignation of Mr. Everett Millais from the Kennel Club," *British Fancier,* 6 Jan. 1893, 609; "Canine Hygiene," *British Medical Journal* i (1893): 144.

37. E. Millais, "A Distemper Bill," *SK&FC,* 20 Mar 1891, 233; "Distemper at Dog Shows," *Canine World,* 20 Mar. 1890, 243; *Dog Owners' Annual 1891* (Manchester:

"Our Dogs," 1891), 86–87; E. Millais, "To Muzzle or Not To Muzzle? II," *Dog Owner's Annual 1890* (Manchester: "Our Dogs," 1890), 27–36; E. Millais, "The Kennel Club," *Dog Owners' Annual 1893* (Manchester: "Our Dogs," 1893), 50–85; "Death Of Everett Millais," *Field,* 11 Sept. 1897, 445.

38. Stonehenge, *The Dog in Health and Disease,* 4th ed. (London: Longman, Green, 1887), 271–81; J. Herriot, *If Only They Could Talk* (London: Pan, 1970). See E. Millais, "My Kennels," *Our Dogs Annual 1895* (Manchester: "Our Dogs," 1895), 67–87.

39. "Feeding," *SK&FC,* 13 Jan. 1882, 20; Stonehenge, *The Dog* (1887), 289; Shaw, *Illustrated Book,* 244.

40. Lane, *Dog Shows,* 224–25; F. Jackson, *Crufts* (London: Pelham, 1990); K. C. Grier, *Pets in America: A History* (Chapel Hill: University of North Carolina Press, 2006), 281–86; "Randle's Dog Biscuits," 2s. per stone, *Norfolk Chronicle and Norwich Gazette,* 29 Jan. 1859, 1; "W. H. Quick's Stores," *North Devon Journal,* 17 Feb. 1859, 1; *Morning Post,* 18 Mar. 1859, 1; "Dog Biscuits of Superior Quality," *Standard,* 21 Oct. 1859, 1. James Spratt was born in Exeter, Devon, into a naval family in 1809, moved with his wife Elizabeth to the United States, and settled in Cincinnati. For an advertisement for his lightning rods, see the *Chelmsford Chronicle,* 14 June 1861, 2.

41. "Keep Your Dogs," *Standard,* 5 Aug 1864, 1. 1d. was one penny.

42. "Hamlin's Invaluable 'POULTRY POWDER,' " *Western Gazette,* 31 Oct. 1913, 9; F. Eisner, *Yearbook of Pharmacy* (London: J. and A. Churchill, 1877), 700; Lane, *Dog Shows,* 84–85; John Martin, "Cruft, Charles Alfred (1852–1938)," *Oxford Dictionary of National Biography.*

43. Stonehenge, *The Dogs of the British Islands* (London: Horace Cox, 1872), 287 and end pages; Idstone, *The Dog,* end pages. Also see H. P. Webb, *Dogs: Their Points, Whims, Instincts and Peculiarities* (London: Dean and Son, 1876); advertisement, *Field,* 21 June 1871, 623. "Spratt's Patent," *Kennel Club Stud Book* (1874), end pages; "Spratt's Patent," *SK&FC,* 25 Nov. 1892, xii, and 9 June 1893, xx.

44. J. Hartley, ed., *The Selected Letters of Charles Dickens* (Oxford: Oxford University Press, 2012), 396. This quote is also cited in H. MacGregor, "A Dog's Dinner," *Field,* 4 Dec. 1948, 192.

45. Stables, *The Commonsense of Dog Doctoring,* 121.

46. "Spratt's Fibrine Cakes," *Field,* 3 June 1871, 458.

47. "Purveyors to the Royal Kennels at Sandringham," *SK&FC,* 2 Dec. 1893, 608.

48. The veterinarian was William Woodroffe Hill; see "Champion Dogs," *Ludgate Illustrated Magazine,* July 1894, 251–52. "Wylam v. Lloyd," *Times,* 15 Jan. 1875, 10; "Spratt's Patent—Injunctions," *Standard,* 30 June 1875, 1. See advertisements in the *Country Gentleman,* 17 Jan. 1880, 1; Lee, *Modern Dogs: Sporting,* end pages. For the history of branding, see T. da Silva Lopes and M. Casson, "Entrepreneurship and the Development of Global Brands," *Business History Review* 81, no. 4 (2007): 651–80.

49. "The Rise and Progress of Dog Shows," *SK&FC,* 27 July 1895, 57–58.

50. R. B. Lee, *A History and Description of the Modern Dogs of Great Britain and Ireland: Sporting Division* (London: Horace Cox, 1894); *A History and Description of the Modern Dogs of Great Britain and Ireland: Non-Sporting Division* (London: Horace Cox, 1894); *A History and Description of the Modern Dogs of Great Britain and Ireland: The Terriers* (London: Horace Cox, 1994); and *A History and Description of the Modern Dogs of Great Britain and Ireland: Sporting Division,* 2nd ed. (London: Horace Cox,

1897), vi (quote); V. Shaw, "Dogdom's Deterioration," *National Observer* 11 (1894): 286–87.

## Chapter 7  •  Whither Breed

1. George du Maurier was on the staff of *Punch* and wrote the best-selling novel *Trilby* (1894). He was a close friend of Henry James, and their relationship was the subject of David Lodge's novel *Author, Author* (London: Secker and Warburg, 2004). L. Ormond, "Du Maurier, George Louis Palmella Busson," *Oxford Dictionary of National Biography* (Oxford: Oxford University Press, 2004), www.oxforddnb.com.

2. H. Ritvo, *The Animal Estate: The English and Other Creatures in the Victorian Age* (Cambridge, MA: Harvard University Press, 1987), 113–14.

3. "Dog Shows and the Kennel Club," *Field,* 14 Jan. 1888, 55. Walsh died on 12 February 1888. "In Memoriam," *Field,* 18 Feb. 1888, 205.

4. "Dog Shows and the Kennel Club," *Field,* 11 Jan 1888, 87.

5. This was the fate of the working Jack Russell Terrier, which was only accepted as a breed in 2015.

6. H. St. James Stephen, *The Kennel Club Calendar and Stud Book* (London: Kennel Club, 1884), viii–ix.

7. Stonehenge, *The Dogs of the British Islands,* 4th ed. (London: Horace Cox, 1882), 93–94, 95.

8. E. W. Jaquet, *The Kennel Club: A History and Record of Its Work* (London: The Kennel Gazette, 1905), 112.

9. H. Dalziel, *British Dogs: Describing the History, Characteristics, Breeding, Management and Exhibition of the Various Breeds of Dogs Established in Great Britain,* vol. 1, 2nd ed. (London: L. Upcott Gill, 1888), 390.

10. One who does not believe in Clubs, "Show v. Working Spaniels," *Stock-Keeper and Fanciers' Chronicle (SK&FC),* 13 Apr. 1888, 224; Midnight, "Show v. Working Spaniels," *SK&FC,* 20 Apr. 1888, 241.

11. A Member of the Spaniel Club, "Show v. Working Spaniels," *SK&FC,* 20 Apr. 1888, 241; T. Jacobs, "Show v. Working Spaniels," *SK&FC,* 27 Apr. 1888, 254. In the 1900s, a rapprochement between different parties saw eight breeds recognized by the Kennel Club: Irish Water Spaniels, Water Spaniels other than Irish, Clumber Spaniels, Sussex Spaniels, Field Spaniels, English Springers, Welsh Springers, and Cocker Spaniels. R. Leighton, *New Book of the Dog* (London: Cassell, 1907), 269.

12. Anon. [Vero Shaw], "Dogdom's Deterioration," *National Observer,* 11 May 1894, 656–57.

13. Vero Shaw, "English Dogs," *National Observer,* 5 Oct. 1895, 579, 581. On the wider cultural climate, see E. Chamberlin and S. L. Gilman, eds., *Degeneration: The Dark Side of Progress* (New York: Columbia University Press, 1985); D. Pick, *Faces of Degeneration: A European Disorder, c.1848–c.1918* (Cambridge: Cambridge University Press, 1989), and D. Glover, *Literature, Immigration, and Diaspora in Fin-de-Siècle England: A Cultural History of the 1905 Aliens Act* (Cambridge: Cambridge University Press, 2012).

14. Editorial, "To the Kennel Club," *British Fancier,* 31 July 1896, 40.

15. C. L. Lane, *Doggy Shows and Doggy People* (London: Hutchinson, 1902), 80; editorial, "What Dog Shows Have Done," *Kennel Gazette,* July 1890, 173.

16. "Past and Future of Dog Shows," *Kennel Review,* May 1883, 185; "Pug Dog Show," *Daily News,* 17 June 1885, 6; "Toy Dog Show at the Aquarium," *Bell's Life,* 8 Oct. 1885, 3; "Terrier Show at the Westminster Aquarium," *Field,* 13 Mar. 1886, 318. Also see report in *Bell's Life,* 11 Mar. 1886, 3. There were 538 entries, judged in 57 classes. The breeds were: Smooth Fox Terriers; Wire-haired Fox Terriers; Irish Terriers; Welsh, or Black and Tan Wire-haired terriers; Bedlington Terriers; Smooth-haired Terriers; Skye Terriers; Hard-haired Scottish Terriers; Dandie Dinmont Terriers; Airedale Terriers; Bull Terriers; Fox Terriers Rough; and Yorkshire Terriers. F. Jackson, *Crufts* (London: Pelham, 1990), 53–16; F. Pelham, *Crufts: The Official History* (London: Pelham, 1990). The possessive apostrophe was dropped in 1990.

17. "Pugs and Toy Dogs," *Daily News,* 20 Oct. 1886, 6; "Toy Dog Show," *North Eastern Daily Gazette,* 20 Oct. 1886, 2; "The Toy Craze," *British Fancier,* 18 Sept. 1896, 205.

18. "Pug Dog Show," *Daily News,* 17 June 1885, 6; "A Pug Dog Show," *Morning Post,* 17 June 1885, 3. G. T. Clark, "The Weight of Toy Spaniels," *SK&FC,* 3 Feb. 1888, 70; "Show of Toy Dogs," *Field,* 10 Dec. 1887, 900; P. Morris, *Walter Potter's Curious World of Taxidermy* (London: Constable and Robinson, 2013); *County Gentleman,* 24 Dec. 1887, 1727–28; editorial, "Sporting and 'Ladies' Dogs," *British Fancier,* 31 July 1896, 47. The report in the *County Gentleman* "heartily congratulated" Mr. Cruft on the event.

19. "Crystal Place Dog Show," *Morning Post,* 14 Jan. 1885, 4; H. Dalziel, "Dogs and Their Management," *Harper's New Monthly Magazine,* Mar. 1886, 583–94. To make matters worse, they had been joined by "another class": men who "ignore the eternal law of duty to their neighbour, and through selfishness, ignorance or idleness, leave their dogs untaught and undisciplined to exercise their worst instincts" (584).

20. T. Collins, *Sport in Capitalist Society: A Short History* (Abingdon, UK: Routledge, 2013), 28–31.

21. Editorial, "What Dog Shows Have Done," 173.

22. "The Utility of Dog Shows," *Field,* 17 Nov. 1894, 746–47; Ædipus, "Canine Progression," *Kennel Gazette,* Sept. 1894, 221; F. Redmond, "Fox-terriers," *SK&FC,* 7 Nov. 1890, 390.

23. H. Cox, "A Danger to Dogs," *SK&FC,* 16 Apr. 1897, 327. On Harding Cox, see Jaquet, *Kennel Club,* 30–31.

24. W. F. Jefferies, "Bulldogs," *Kennel Gazette,* Jan. 1900, 11; C. Smith, "Bulldogs," *SK&FC,* 29 Jan. 1897, 96; G. Johnson, "Bulldogs Past and Present," *British Fancier,* 16 Oct. 1896, 303; J. Sidney Turner, Mastiffs, *Kennel Gazette,* Jan. 1899, 12.

25. Stonehenge, *The Dogs of the British Islands* (1882), 266. In the fourth edition, the description written by Frank Adcock appears in the appendix with other foreign dogs, with the added note: "In spite of Mr. Adcock's urgent pleading for this breed, I cannot consider it as one of 'The Dogs of the British Islands.' 'STONEHENGE.'"

26. R. Hood-Wright, "Great Danes," *Kennel Gazette,* Feb. 1900, 62.

27. F. A. Manning, "The History and Growth of the Kennel Club," *Dog Owners' Annual* (1890), 147–48.

28. Anon., "Dogdom's Deterioration," 657; C. Huff, "Victorian Exhibitionism and Eugenics: The Case of Francis Galton and the 1899 Crystal Palace Dog Show,"

*Victorian Review* 28 (2002): 1–20. A. Bashford and P. Levine, eds., *The Oxford Handbook of the History of Eugenics* (Oxford: Oxford University Press, 2012), does not include in its index entries for animals, dogs, livestock, poultry, or sheep, nor for Robert Bakewell. G. R. Searle, "Eugenics: The Early Years," in R. A. Peel, ed., *Essays in the History of Eugenics* (London: Galton Institute, 1998), 20–35; G. R. Searle, *Eugenics and Politics in Britain, 1900–1914* (Leyden: Nordhoff, 1976).

29. The history of the struggles of the Irish Terrier Club to have their regulation implemented was given at the Kennel Club general meeting on 10 February 1893. *British Fancier,* 17 Mar. 1893, 162. Also see *Kennel Gazette,* Mar. 1893, 68–69. Jaquet, *Kennel Club,* 52–54 and 121; Este, "The Question of Cropping Dogs' Ears," *Kennel Gazette,* Jan. 1889, 6–7; Humanity, "Mutilated Dogs," *Times,* 19 Dec. 1888, 4. Also see E. Landseer, "Letter," *Times,* 11 July 1862, 9. Este, *Kennel Gazette,* Dec. 1887, 40–41.

30. "One Who Has Gone to the Dogs," *Daily Post,* 12 July 1862, 5; B. Harrison, *Peaceable Kingdom: Stability and Change in Modern Britain* (Oxford: Oxford University Press, 1982). See A. W. Moss, *Valiant Crusade: The History of the R.S.P.C.A.* (London: Cassell, 1961). "National Dog Show at Islington," *ILN,* 5 July 1862, 13. Landseer's status in dog affairs was indicated when he was called, alongside the leading veterinarian George Fleming, to give evidence on dogs' suffering in a prosecution in 1869. "Canine Fancy," *Bell's Life,* 11 Feb. 1865, 7. On later views, see "Cutting Dogs' Ears," *Birmingham Daily Post,* 27 May 1869, 6; "Cruelty by Cropping a Dog's Ears," *Liverpool Mercury,* 28 Sept. 1874, 3.

31. Fusilier, "The Cropping Question," *SK&FC,* 4 Jan. 1889, 4.

32. Stonehenge, *The Dogs of the British Islands, being a series of articles and letters by various contributors, reprinted from The Field newspaper* (London: Horace Cox, 1867), 143; H. Dalziel, *British Dogs: Their Varieties, History, Characteristics, Breeding, Management and Exhibition* (London: "The Bazaar" Office, 1879), 272 and 331.

33. "The Cropping Question," *SK&FC,* 7 Sept. 1888, 180.

34. See letters from Anxious to Learn; Lead, Kindly Lightly; and Rat-Trap, "The Cropping Question," *SK&FC,* 4 Jan. 1889, 4–5, and follow-up correspondence on 11 Jan 1889, 21–22; 18 Jan. 1889, 35–36; 25 Jan. 1889, 48; 1 Feb. 1889, 63–64.

35. Humanity, "Mutilated Dogs," *Times,* 19 Dec. 1888, 4; "The Cropping Question," *SK&FC,* 23 Nov. 1888; 30 Nov. 1888, 387; 21 Dec. 1888, 432; 28 Dec. 1888, 450–51; "The Cropping Question," *SK&FC,* 28 Sept. 1888, 233; "Bull-terriers and the Cropping Question," 30 Nov. 1888, 387.

36. "Kennel Club Committee Meeting," *Kennel Gazette,* Feb. 1889, 38; Kennel Club Committee Meeting, "The Cropping Question," *Kennel Gazette,* Mar. 1889, 59; "The Cropping of Dogs," *SK&FC,* 1 Feb. 1889, 63–64.

37. "Kennel Club Committee Meeting," *Kennel Gazette,* Mar. 1889, 59–60; "The Cropping of Irish Terriers," *Huddersfield Daily Chronicle,* 18 Mar. 1889, 3; "The Cropping Question," *SK&FC,* 22 Mar. 1889, 162; Este, "The Cropping Question," *Kennel Gazette,* Aug. 1889, 197; "Kennel Club Committee Meeting," 60.

38. The figures were as follows. For v. Against: Irish Terriers, 17 v. 23; Great Danes, 1 v. 6; Bull Terriers, 41 v. 7; Black-and-tan Terriers, 23 v. 21; White English Terriers, 21 v. 16; Toy Terriers, 6 v. 4. Jaquet, *Kennel Club,* 123.

39. "Kennel Club Committee Meeting," *Kennel Gazette,* July 1889, 161–64; also see Jaquet, *Kennel Club,* 121–22.

40. A. F. Russell, "Shall Trimming be Recognised?," *SK&FC*, 25 May 1888, 324; V, "Shall Trimming Be Recognised?" *SK&FC*, 6 Apr. 1888, 208; One Who Knows, "Shall Trimming Be Recognised?," *SK&FC*, 30 Mar. 1888, 196; "'Trimming': Where Does It Begin and 'Faking' Start," *British Fancier*, 1 Sept. 1893, 175.

41. H. Dalziel, *British Dogs: Describing the History, Characteristics, Points, Club Standards, and General Management of the Various Breeds of Dogs established in Great Britain*, 2nd ed. (London: L. Upcott Gill, 1889), 231–33. Also called the Old White English Terrier.

42. Stonehenge, *Dogs of the British Islands* (1882), 57–60; R. B. Lee, *Modern Dogs: The Terriers* (London: Horace Cox, 1894), 62–63, 53, 60–61, and 67. Andrew Lees Bell was a Terrier breeder and "expert" from Dunfermline. See A. Lees Bell, "Smooth Irish Terriers," *SK&FC*, 31 Jan. 1890, 84.

43. "Dog Fanciers Prosecuted," *Morning Post*, 8 Dec. 1894, 7; "The Cropping of Dogs' Ears: Horrible Barbarity," *Daily News*, 29 Dec. 1894, 6; "At Worship-street," *Times*, 29 Dec. 1894, 3.

44. This complaint was longstanding. See An Old Exhibitor, "The Kennel Club," *Canine World*, 7 Nov. 1890, 505.

45. W. Moores, "Founder of Our Dogs: The Life of Theo Marples" and "The Start of Our Dogs," *Our Dogs*, www.ourdogs.co.uk.

46. Editorial, "The Cropping of Dogs' Ears," *Our Dogs*, 2 Feb. 1895, 87, and 16 Feb. 1895, 127–28; "The Cropping of Dogs' Ears," *Our Dogs*, 26 Jan. 1895, 72; editorial, "Cropping Dogs' Ears," *Field*, 2 Mar. 1895, 254.

47. "Cropping the Ears of Dogs," *Animal World*, Feb. 1895, 18–19, and Mar. 1895, 36; "The Dog-Cropping Case," *Standard*, 19 Jan. 1895, 2; "Police," *Times*, 26 Jan. 1895, 4; Merthyr Guest, "Cropping Dogs' Ears," *Field*, 2 Feb. 1895, 122; "At Worship-street," *Times*, 2 Feb. 1895, 14. The Bull-terrier Club had met informally after the Liverpool Dog Show in January 1895 and announced its "conversion." "The Bull-terrier Club's Conversion on the Cropping Question," *Our Dogs*, 2 Feb. 1895, 87.

48. Jaquet, *Kennel Club*, 177; John Rylands Special Collections, University of Manchester, Papers of George R. Sims, GRS 10/2/1 Scrapbook, c. 1895–1898; "Cropping Dogs' Ears," *Times*, 25 Jan. 1895, 5; "Cropping," *Our Dogs*, 2 Feb. 1895, 87; "Cropping Dogs' Ears," *Times*, 1 Feb. 1895, 7; "The Cropping of Dogs' Ears: Magistrate's Decision," *SK&FC*, 1 Feb. 1895, 88–89; *Our Dogs*, 2 Feb. 1895, 87–88; "Kennel Club: General Meeting Held 26 Feb 1895," *Kennel Gazette*, Mar. 1895, 59; *Our Dogs*, 16 Feb. 1895, 128.

49. "Cropping Dogs' Ears," *Times*, 8 Feb. 1895, 5; *Morning Post*, 8 Feb. 1895, 5; *Pall Mall Gazette*, 8 Feb. 1895, 2; *Evening Telegraph, and Star*, 8 Feb. 1895, 2; *Edinburgh Evening News*, 8 Feb. 1895, 2; *Country Gentleman*, 9 Feb. 1895, 182; "Toby to H.R.H.," *Punch*, 16 Feb. 1895, 81; Jaquet, *Kennel Club*, 176; "Whispers of the Fancy," *SK&FC*, 8 Feb. 1895, 115; "Cropping Ears," *SK&FC*, 15 Feb. 1889, 135; "Consummation of our Ear-Cropping Efforts," *Animal World*, Apr. 1895, 52. Also see G. W. Murdoch, "Cropping and Baffled," *Animal World*, May 1895, 66–67.

50. Jaquet, *Kennel Club*, 177; "Cruft's Dog Show," *Daily News*, 14 Feb. 1895, 3; "The Great Dane Club," *British Fancier*, 10 Jan. 1896, 728; "The Great Dane," *SK&FC*, 9 Jan. 1897, 26.

51. "The Humanity of Docking," *British Fancier,* 4 Sept. 1896, 161; "Cropping," *Our Dogs,* 2 Feb. 1895, 87.

52. "General Meeting of the Kennel Club Held 27 January 1895," *Kennel Gazette,* Mar. 1895, 60–2; "Kennel Club General Meeting Held 4 Mar 1896," *Kennel Gazette,* Mar. 1896, 67; "Report," *British Fancier,* 13 Mar. 1896, 215.

53. The Schipperkes Club disputed that docking was cruel and that they "gouged out" the stump, but the accusation stuck. "Docking Schipperkes," *SK&FC,* 27 Oct. 1896, 349.

54. "Kennel Club Sub-Committee Meeting—Docking Committee," *Kennel Gazette,* Oct. 1896, 276–7; E. B. Joachim, "Docking Schipperkes," *SK&FC,* 23 Oct. 1896, 349. The Kennel Club finally banned the docking of show dogs in Britain in 2007, though dogs docked previously could continue to be entered into events.

55. H. Warnes, "The Trimming Question," *SK&FC,* 2 Apr. 1897, 291. Warnes was a Bedlington Terrier aficionado.

56. Ouida, "Commonsense as to Dogs," *Morning Post,* 12 Jan. 1886, 3; H. Killoran, "Ramée, Marie Louise de la [Ouida] (1839–1908)," *Oxford Dictionary of National Biography.* Killoran describes her devotion to her pets: "From 1904 until her death de la Ramée lived in squalor with many adopted stray dogs in her tenement at Viareggio. Her dedication to these pets was perhaps the best index of her increasing eccentricity, as she indulged them with luxuries while starving herself; the local people called her 'Crazy Lady with the Dogs.'" Ouida, "Dogs," *Times,* 9 Oct. 1889, 3.

57. *Hearth and Home,* 25 Oct. 1894, 853; "Cropping Dogs' Ears: A Barbarous Practice of the Past," *Ladies Kennel Journal,* Feb. 1895, 203; Jaquet, *Kennel Club,* 172–73.

58. Hon. Sec LKA (Mrs. Stennard-Robinson), "The Ladies Kennel Association," *Ladies Kennel Journal* 1, no. 1 (Dec. 1894): 1; "In the Witness Box: The Ladies' Kennel Association," *Sunday Times,* 9 June 1895, 2; "What the Papers Say about the L.K.A.," *Ladies Kennel Journal* 1, no. 1 (Dec. 1894): 3–7; D. M. Griffiths, "Cornwell, Alice Ann (1852–1932)," *Oxford Dictionary of National Biography.* On Alice Cornwell's standing in the world of business in the late 1880s, see "'Princess Midas' in London: An Interview with Miss Alice Cornwell," *Adelaide Advertiser,* 6 June 1889, 5.

59. "The Australian 'Princess Midas,'" *Te Ahora News,* 21 Apr. 1888, 5. One newspaper article referred to her as "the Australian heroine of the London Stock Exchange." "The Swish Family Robinson," *SK&FC,* 7 Aug. 1896, 99; "The Kennel," *Sydney Morning Herald,* 16 June 1900, 12; Lane, *Dog Shows,* 210–11. Lane indicated that "Mrs. Stennard Robinson has shown an amount of ability, perseverance, and resource which are almost incredible, and quite unexampled in the history of this country." He diplomatically added, "Of course, matters have not always run smoothly—it is not to be expected they could; but the manner in which this lady has overcome obstacles and opposition, and brought many and important undertakings to a successful issue, has compelled admiration even from those who might differ from her on some matters of detail" (210).

60. "Kennel Gossip," *Nottinghamshire Guardian,* 14 July 1894, 7; *Hearth and Home,* 1 Nov. 1894, 893; "Rules and Regulations," *Ladies Kennel Journal* 1 (Dec. 1894): 7. The canine activities of Kathleen Florence May (née Candy), Duchess of Newcastle

(1872–1955) are discussed in Lane, *Dog Shows,* 192. The duchess is quoted in "'K. Newcastle' and the L.K.A.," *Ladies Kennel Journal* 2, no. 1 (June 1895): 23–24.

61. Grayson, *The Ladies Kennel Association,* 4; "Whispers of the Fancy," *SK&FC,* 14 June 1895, 453–54; "In the Witness Box," 2; K. Newcastle, "The Ladies' Show at Ranelagh," *SK&FC,* 21 June 1895, 483; A. Stennard-Robinson, "The Ladies' Show at Ranelagh," *SK&FC,* 28 June 1895, 501; "The Ladies' Show at Ranelagh," *SK&FC,* 12 July 1895, 23; F. B. Lord, "The Ladies' Show at Ranelagh," *SK&FC,* 28 June 1895, 502. See follow-up correspondence in *SK&FC,* 5 July 1895, 5, and 12 July 1895, 23.

62. "Women in the Kennel World," *British Fancier,* 12 Dec. 1896, 590–91; Anon., "On Show: Dogs," *Strand Magazine,* 13 May 1897, 582; Sad Dog, "Kennel Rule X," *SK&FC,* 31 July 1896, 81; "From Husband to Wife," *British Fancier,* 7 Aug. 1896, 71.

63. "Women in the Kennel World," *British Fancier,* 12 June 1896, 590–91; editorial, "The Affairs of the L.K.A.," *British Fancier,* 17 July 1896, 752; Letters on "Affairs of the L.K.A.," *British Fancier,* 17 July 1896, 761–62, and 24 July 1896, 16; *Ladies Kennel Journal* report quoted in "Miss Slaughter and the L.K.A. Committee," *SK&FC,* 21 Aug. 1896, 153.

64. "The New Reign of Terror—Or 'A Reign of Courtesy,'" *SK&FC,* 17 July 1896, 41; "The New Reign of Terror: The Robinson Joke," *SK&FC,* 21 Aug. 1896, 156; "The Ladies Kennel Association," *Field,* 21 May 1898, 776; "Ourselves and the Kennel Club," *Ladies Kennel Journal* 4 (July 1896): 42; "Ourselves and the Kennel Club," *Ladies Kennel Journal* 4 (Sept.–Oct. 1896): 188; "The Scurrilous Attack on the Hon. Sec. and the LKA Published by *Our Dogs,*" *Ladies Kennel Journal* 3 (June 1896): 345–46; Cerberus, "The LKA v. The Kennel Club," *Our Dogs,* 13 June 1896, 564, and 20 June 1896, 599.

65. Jaquet, *Kennel Club,* 207, 229, 236, and 239; "Kennel Club Ladies Branch," *Field,* 17 July 1897, 124, and 24 July 1897, 153; "Professionals' Pets," *Era,* 18 Dec. 1897, 20; "The Dogs of Celebrities," *Strand Magazine,* July 1894, 396–404; "Fashionable Dogs: Ladies' Kennel Association Show at Earl's Court," *Daily Mail,* 16 Dec. 1897, 3.

66. P. Grayson, *The Ladies Kennel Association: The First Hundred Years* (Ashford, Kent, UK: Dog World, 2004), 5–6.

67. Ouida, "British Dog Shows," *Standard,* 29 July 1899, 4.

68. Leiothrix, "British Dog Shows," *Standard,* 3 Aug. 1899, 6.

69. "This Morning's News," *Daily News,* 1 July 1891, 5.

70. Frances Power Cobbe, quoted in *Ladies Kennel Journal* 1, no. 1 (Jan. 1895), 123; Lane, *Dog Shows,* 200–204.

71. "The Borzoi, commonly called Russian Wolfhound," *SK&FC,* 7 Nov. 1890, 388–89; "Borzoi Points," *SK&FC,* 1 Jan. 1892, 10; "Russian Wolfhounds in England," *SK&FC,* 9 Apr. 1897, 308–9; W. D. Drury, *British Dogs: Their Points, Selection, and Show Preparation* 3rd ed. (London: Upcott and Gill, 1903), 130.

72. At the Kennel Club show in June 1892, the three highest entries were Fox Terriers (224), St. Bernards (123), and Sheepdogs (107). See *SK&FC,* 3 June 1892, 536.

73. G. Stables, *Sable and White: The Autobiography of a Show Collie* (London: Jarrold and Sons, 1893). The novel is discussed in T. Coslett, *Talking Animals: British Children's Fiction, 1786–1914* (Aldershot: Ashgate, 2006), 67–69 and 86–87.

74. "Collie News," *British Fancier,* 21 July 1896, 133; J. Munro, "My Collie: Natural History Anecdotes," *Leisure Hour,* Jan. 1888, 66; E. Healey, "Coutts, Angela Georgina Burdett-, suo jure Baroness Burdett-Coutts (1814–1906)," *Oxford Dictionary of National Biography;* "Lady Burdett-Coutt's Collie," *London Journal,* 3 Nov. 1894, 359.

75. In her book *Bred for Perfection: Shorthorn Cattle, Collies, and Arabian Horses since 1800* (Baltimore: Johns Hopkins University Press, 2003), historian Margaret Derry chose the Collie as an exemplary illustration of the issues in the breeding of modern dogs, in terms of the value of pedigree, fancy points versus working ability, and the very meaning of breed. Her discussion is transatlantic and mainly on the twentieth century; ours is for Britain at the end of the Victorian period.

76. H. Dalziel, *The Collie: Its History, Points, and Breeding, Etc.,* Monographs on British Dogs (London: Upcott Hill, 1888); R. B. Lee, *The Collie or Sheepdog* (London: Horace Cox 1890); "1893—A Retrospect," *SK&FC,* 29 Dec. 1893, 602–3; Derry, *Bred for Perfection,* 73–76; "J. P. Morgan and His Collies," Collies of the Meadow, https://colliesofthemeadow.wordpress.com/2011/11/03/jp-morgan-and-collies/; "Some Valuable Dogs," *Ludgate,* Sept. 1897, 454. Today the equivalent of the price paid for Ormskirk Herald would be around £166,600. Lee noted that the Collie "occupies a position second to none of the canine race in pecuniary value" (v).

77. S. Moorehouse, *The British Sheepdog* (London: 1950); Lee, *The Collie,* iv; Lee, *Modern Dogs: Non-Sporting* (1899), 116.

78. W. Youatt, *The Dog* (London: Society for the Diffusion of Useful Knowledge, 1845), 59–60; Idstone, *The Dog, with Simple Directions for his Treatment* (London: Cassell, Petter, and Galpin, 1872), 221–24, 217; Lee, *The Collie,* 81.

79. Stonehenge, *The Dogs of the British Islands,* 3rd ed. (London: Horace Cox, 1878), 186, 189.

80. "The Crystal Palace Dog Show," *Field,* 21 June 1873, 597; "The Crystal Palace Dog Show," *Field,* 13 June 1874, 574–75; "The Crystal Palace Dog Show," *Field,* 10 June 1876, 647. Trefoil's son, Charlemagne, was mated with Maud, who carried the true Highland blood of the "Cockie," who dominated shows in the early 1870s; see Lee, *Collie,* 46–47.

81. A. M. Urdank, "The Rationalisation of Rural Sport: British Sheepdog Trials, 1873–1946," *Rural History* 17 (2006): 65–82; E. W. H. Jones, *Sheep-dog Trials and the Sheep-dog* (Brecon: E. Poole, 1892).

82. Dalziel, *British Dogs* (1879), 195, 196, 199, 202.

83. G. L. Norgate, "Lee, Rawdon Briggs (1845–1908)," rev. J. Lock, *Oxford Dictionary of National Biography;* "The Bull-Dog Show," *Daily News,* 28 May 1885, 6; "The Collie Club Show," *Daily News,* 3 June 1885, 3; Foy, "Collies and Their Clubs," *SK&FC,* 2 May 1890, 315, and 9 May 1890, 332; R. G. Jordan, "Head of the Collies," *SK&FC,* 31 Jan 1890, 83; "Collies: What Is the Correct Shape of Head?," *SK&FC,* 13 Mar. 1891, 217.

84. W. Arkwright, "The Fancier versus the Collie," *Kennel Gazette,* Aug. 1888, 173; Jaquet, *Kennel Club,* 27. Arkwright was a trustee and leading figure in the Kennel Club. His book *The Pointer and His Predecessors* (London: Arthur L. Humphreys, 1902) is still regarded as authoritative. His obituary in the *Times* (25 Feb. 1925) describes him as "an author, an expert on Oriental porcelains, a traveller, a sportsman, and a very great horticulturalist." Arkwright, who was injured in a riding accident in the 1870s,

has been identified as the inspiration for Clifford Chatterley in *Lady Chatterley's Lover*. D. H. Lawrence, *Lady Chatterley's Lover and A Propos of "Lady Chatterley's Lover,"* ed. M. Squires (Cambridge: Cambridge University Press, 1993), 339.

85. Arkwright, "The Fancier," 173.

86. Ibid.

87. Excelsior, "The Collie Proper v. the Collie of Fancy," *SK&FC*, 24 Aug. 1888, 143; A Breeder, "Collies," *SK&FC*, 17 Jan. 1890, 42; J. Lawson, "The Size of the Collie," *SK&FC*, 7 June 1889, 362; R. W. Jordan, "Head of the Collie," *SK&FC*, 31 Jan. 1890, 83; J. Lawson, "The Size of the Collie," *SK&FC*, 7 June 1890, 362; W. Rust, "Show Collies as Workers," *SK&FC*, 20 Apr. 1888, 240; "Dogdom in 1892: A Retrospect," *SK&FC*, 13 Jan. 1893, 634; "Report," *British Fancier*, 24 Sept. 1893, 539.

88. G. Hill, "The Diseases of the Fancy," *SK&FC*, 18 Mar. 1898, 207.

89. "The Great Collie Ear Trial," *SK&FC*, 15 Dec. 1898, 596, and 20 Jan. 1899, 41; "The Training of Dogs' Ears," *Edinburgh Evening News*, 15 Dec. 1898, 3; M. Reed, "Gordon, Henry Panmure (1837–1902)," *Oxford Dictionary of National Biography*.

90. "The Training of Dogs' Ears," *Edinburgh Evening News*, 16 Dec. 1898, 3.

91. Jaquet, *Kennel Club*, 249; "The Crystal Palace Dog Show," *Country Life Illustrated*, 28 Oct. 1899, 540; "Cruft's Dog Show," *Country Gentleman*, 11 Feb. 1899, 185; "Cruft's International Dog Show," *Hearth and Home*, 1 Mar. 1900, 690; "Cruft's Dog Show," *Country Life Illustrated*, 18 Feb. 1899, 220; Jaquet, *Kennel Club*, 21.

92. "Bi-Annual General Meeting of the Kennel Club, 4 October 1899," *Kennel Gazette*, Oct. 1899, 413 and 423–25. Some thought that the liberty to crossbreed would "spell progress in some breeds of dogs which sadly require a little strengthening in constitution" (413).

93. Jaquet, *Kennel Club*, 233–34.

94. Stephen, *The Kennel Club Calendar*, xi.

### Conclusion

1. M. E. Derry, *Bred for Perfection: Shorthorn Cattle, Collies, and Arabian Horses since 1800* (Baltimore: Johns Hopkins University Press, 2003).

2. H. G. Parker et al., "Genomic Analyses Reveal the Influence of Geographic Origin, Migration, and Hybridization on Modern Dog Breed Development," *Cell Reports* 19 (2017): 705; G. Larson et al., "Rethinking Dog Domestication by Integrating Genetics, Archeology, and Biogeography," *PNAS* 109 (2012): 8878–83, quotes on 8879 and 8880, www.pnas.org/cgi/doi/10.1073/pnas.1203005109.

3. An editorial in *Our Dogs* in April 1974 posed the question "Breeder or Dealer?," *Our Dogs*, 18 Apr. 1974, 2.

4. "Pedigree Dogs Exposed: The Blog," http://pedigreedogsexposed.blogspot.co .uk; M. Brandow, *A Matter of Breeding: A Biting History of Pedigree Dogs and How the Quest for Status Has Harmed Man's Best Friend* (Boston: Beacon, 2015).

5. P. Bateson, F. W. Nicholas, D. R. Sagan, and C. M. Wadr, "Guest Editorial: Analysis of the Canine Genome and Canine Health: Bridging a Gap," *Veterinary Journal* 196 (2013): 1–3.

6. P. Bateson, "Independent Enquiry into Dog Breeding" (published privately, 2010), 9, 21–22. Only available at www.researchgate.net/publication/266277147 _Independent_Inquiry_into_Dog_Breeding.

7. N. Durbach, *Spectacle of Deformity: Freak Shows and Modern British Culture* (Berkeley: University of California Press, 2009).

8. See crufts.org.uk.

9. C. Dickens, *Little Dorrit* (1855–57); A. Trollope, *The Way We Live Now* (1875); see also J. Taylor, *Boardroom Scandal: The Criminalisation of Company Fraud in Nineteenth-Century Britain* (Oxford: Oxford University Press, 2013).

10. H. Ritvo, *The Animal Estate: The English and Other Creatures in the Victorian Age* (Cambridge, MA: Harvard University Press, 1987).

11. M. Daunton, "Gentlemanly Capitalism and British Industry, 1820–1914," *Past and Present* 122 (1989): 119–58.

Adcock, Frank, 179; Abbess, 118; improving bulldogs, 121; Toro, 121

agricultural shows, 12, 54–56, 59–61, 64, 72, 80, 88

Albert, Prince, 47

Alexandra, Princess (Princess of Wales), 74–75, 128, 207, 209, 212. *See also* Alexandra, Queen

Alexandra, Queen, 207–8

Alexandra Palace dog shows, 78, 122, 132

Allison, William, 123

Animal Cruelty Act (1835), 45, 196

Appleby, T. Dawkins, 67–69

Arkwright, William, 215–17

artificial insemination, 160, 164–66, 170

Bakewell, Robert, 10–12, 36, 156

Barber, Charles Burton, 131

Barnum, P. T., 69, 224

Barrett Browning, Elizabeth, Flush, 49

basset hounds, 141, 164; Basset Hound Club, 135, 166; breeding, 164–72; Fino IV, 167; Fino de Paris, 167

battue shooting, 41, 44, 56

Bedlington terriers, 97, 116

Bekoff, Marc, 16

Bell, Andrew Lee, 199

*Bell's Life of London and Sporting Chronicle*, 17, 41, 46–47

Berkeley, Grantley F., 56

Birmingham dog shows, 54, 59–65, 75, 79, 88, 91–92, 95–96, 99–110; committee, 122, 126–27, 137; Curzon Hall, 75; First Exhibition of Sporting and Other Dogs

(1860), 54; open versus closed judging, 93–94, 103–4

Birmingham Dog Show Society. *See* National Dog Show Society

black-and-tan terriers, 46, 95, 102, 122, 156; Black-and-Tan Terrier Club, 188, 198; cropping of, 197, 202

Black field spaniels, 188

Blaine, Delabere, 27–28, 157

Blenheim spaniels, 29, 44, 48–49, 187

blood: bloodlines, 40, 64, 118, 133, 146, 151–56, 158, 171, 193; as hereditary matter, 11–12, 51, 61, 80, 83, 88, 139, 143, 152–55, 157, 160, 163, 189, 213; pure (good), 44, 49, 64, 80, 85, 88, 103, 121, 139–40, 145–46, 166, 177, 194–95, 219–20, 223–24. *See also* heredity

bloodhounds, 53, 95, 125, 153, 170–72; Dickens on, 71–72; Draco and Dingle, 107; Druid, 72; judging, 95; origin, 33; Rufus, 119–20

Borzois, 13, 191, 212

bouledogues français, 219

Brailsford, Francis, 95

Brailsford, Frederick, 73

Brailsford, Richard, 55–56, 60–61

Brandow, Michael, 16, 223

breaking, 3, 42–43

breed, 8, 10; other terms for, 19, 23, 27, 36, 52, 57, 167

breed clubs, 18, 115–16, 125, 129, 133, 196, 223; breed club shows, 136–37; dates of foundation, 135; emergence, 133–34, 193; relations with Kennel Club, 133–34, 138, 140–41, 188–89, 198. *See also individual breeds* breeding: artificial insemination, 160, 164–66, 170; art versus science, 20, 140, 155–57, 164, 222; livestock and poultry, 62. *See also* blood: bloodlines; crossbreeds; inbreeding; prepotency

breeds. *See* deterioration of breeds; improvement of breeds; proliferation of breeds; *specific breeds*

breed standards, 1–2, 4, 7–9, 13, 16–20, 76–77; exaggerated points, 20, 85–86, 136, 140, 159, 183, 185, 190, 193; numerical values, 77–79, 81, 83, 96; points, 7–8, 19, 38, 47, 60, 64–65, 77–88, 93, 95–96, 102–4, 133, 136, 140, 150, 159, 183, 185, 187, 189–91, 193–94, 203, 219

*British Fancier*, 190, 200, 209, 213

British Kennel Association, 126

Brough, Edwin, 118–19

Buffon, Comte de, 2, 25–26, 29; *Histoire Naturelle*, 2, 25

bulldogs, 4, 27, 30, 32, 40, 46–48, 57, 62–63, 70–71, 95, 105, 107, 181, 194–95; Bull Dog Club, 133–36, 145, 190; Bulldog Show, 215; character, 51, 53; conformation, 86, 121, 130, 140, 152, 159, 162; cropping, 202; crossbreeding with, 152–53, 188; deterioration, 70, 118, 219; recovery, 143–45; size, 121; stud book, 118

bull terriers, 46, 53, 99, 130; Bull-terrier Club, 133, 198; cropping, 197–98, 200–202

Butler, Charles, 38–39

Butler, Francis, 42

Butler, Samuel, 177

Byron, Lord, 3

Caius, John, 25–27

Carling, Mary, 200–201

Carling, Robert, 200–201

Carter, Henry Wyndham, 129, 135

clumber spaniels, 44, 84, 186–87, 207

Cobbe, Frances Power, 212

cocker spaniels, 44, 154, 186–88; Cocker Spaniel Club, 135

Colam, John, 71, 201, 244n42

collies, 16, 136, 141, 185, 213–18; books on, 159, 213; Collie Club, 135, 215, 217; conformation changes, 217; Great Ear Trial, 218–19; grooming, 199; Old English sheepdog, 214; prick versus drop ears, 218–19; *Sable and ·*

*White*, 212–13; Scotch, 214; Shepherd's dog, 214; Trefoil, 215; working or show breed, 213–14

Committee of Gentlemen, 98–105; Crystal Palace Dog Show (1870), 100–103; forerunner of the Kennel Club, 110; Second Great National Exhibition (1871), 104–5; Third Great National Exhibition (1872), 107

Cooper, Harry, 200–201

Cooper, Jilly, 1–2

coursing, 37–41

Cox, Harding, 194

Cremorne Dog Show, 74

cropping of ears, 13, 20, 69, 176, 185, 226; controversy in 1889, 196–99; controversy in 1894–95, 200–204

crossbreeds, 1, 8, 80, 219–20; designer, 1, 171. *See also* breeding; mongrels

Cruft, Charles: as British Barnum, 125; career, 178; Cruft's Dog Show, 219; First Great Terrier Show, 190; manager of Spratt's Patent, 125

Crufts Dog Show, in twenty-first century, 224

Crystal Palace: dog shows, 100–105, 107–8, 110; exhibitions and entertainments, 65, 101

Cuban mastiff, 62, 71

curs, 11, 24, 35, 50–52, 60, 71, 103, 174–75, 189

Curzon, Viscount, 60

Cuvier, Georges, 32

Cuvier, Georges-Frédéric, 32

dachshunds, 130, 135, 140, 164, 188–89

Dalmatians, 197

Dalziel, Hugh, 48, 115, 121–23, 129–32, 143, 173, 187; antecedent impressions, 157; biography, 129, 173; *British Dogs*, 129, 143–44, 147–48, 153, 187; on bulldogs, 143–44; on collies, 213–15, 215; critic of the Kennel Club, 123, 187; dog lovers and dog dealers, 153, 193; supporting ban on cropping, 196, 200; telegony, 158; on Yorkshire terriers, 141–42

Dandie Dinmont terriers, 88, 96–97; Dandie Dinmont Club, 134; deterioration, 188; in *Guy Mannering*, 96

Darlington Show Committee, 124–26, 128; Darlington Dog Show, 75, 99–100, 124–26, 249n21

Darwin, Charles, 20, 43, 129, 160, 164; on
    artificial selection, 160–62, 182; on breeding
    domesticated animals, 159–61; Darwinism,
    117, 152, 163, 166; on origin of dogs, 161–62;
    *On the Origin of Species*, 62, 85, 161–62;
    pangenesis, 163, 182; *Variation of Animals
    and Plants*, 156, 162
Dawes, Joshua Horton, 117
dealers, 8, 66, 73, 95, 181, 190, 197, 200, 215,
    225; and profit, 19, 65, 73, 95, 153, 184, 193;
    set against gentlemen, 54, 102, 118, 122, 139;
    and social class, 42, 50, 66, 76, 88, 111
deerhounds, 84
degeneration. *See* deterioration of breeds
Derry, Margaret, 16, 272n75
deterioration of breeds, 80, 91, 167, 184–85,
    188–89, 195; bulldogs, 145; collies, 213;
    countering, 167–70, 219; due to inbreeding,
    51, 139, 145, 152, 175, 186, 191, 199, 219
Dickens, Charles, 225; and character of Bill
    Sikes, 48; dog's dinner recipe, 180; The Two
    Dog-Shows, 65, 70–72, 244n42
diet, 176–82; omnivores or carnivores, 161, 177,
    180
*Dink's Sportsman's Vade Mecum*, 33, 45
distemper, 21, 68, 148, 161, 169, 173–75
docking tails, 13, 20, 185, 197, 202–4
dog fancies. *See* Fancy, the; new fancy
dog fighting, 13
dog food brands: Chamberlin and Smith,
    181–82; Petherbridge's Meat and Oatmeal
    Biscuits, 181; Slater's Meat Biscuit for Dogs,
    181; W. G. Clarke's Buffalo Meat Biscuits,
    181–82. *See also* Spratt's Patent Company
doggy people, 2, 12, 17, 73, 108, 226; dog lovers,
    122, 189, 193, 204; exhibitors at shows, 2, 13,
    19, 59, 64–67, 76–79, 89, 91–92, 98, 101–5,
    108, 123–25, 132–33, 137–39, 160, 176–77, 190,
    193, 198, 200. *See also* dealers; new fancy
*Dog Owners' Annual*, 169
dog shows: attendance, 74, 97, 104, 127, 209;
    commercialism, 56–57, 64–65, 69, 76, 90,
    103, 111, 122; commercial viability, 55, 66, 72,
    90, 98, 107, 210; complaints by exhibitors,
    55, 67, 69, 73, 76–77, 91, 94, 101–2, 105, 108,
    118, 132, 175, 190, 215, 220; faking at, 197–99;
    judges, 18, 38, 47, 55, 62–65, 71, 73, 76–81, 85,
    89, 91–96, 100–103, 108–10, 117–18, 122–24,

132–36, 139, 203, 207, 212, 215, 217, 220, 223,
    236; judging at, 47, 55–59, 60, 62, 64–65, 67,
    76–79, 87, 91–99, 100–105, 110, 117, 122–24,
    127–29, 132–36, 197, 200, 222; respectability
    of, 90, 101–103, 111, 118, 122, 137; sanitary
    measures at, 175–76; size and number, 75, 91,
    104–5, 108, 128, 145, 187, 191, 219. *See also*
    breed standards
Dogue de Bordeaux, 189
Douglas, John, 81, 107–8, 110, 179
Durham, 127

Edwards, Sydenham, 27, 121, Plate 2, 234n10
elk-hounds, 219
English setters, 43
Esquilant, Frederick C., 78
eugenics, 14, 195–96

Fancy, the, 24, 45–48, 54, 116, 223; beauty
    shows, 24, 45–48, 72, 76, 86; in Birming-
    ham, 46, in London, 46, 62–63, 66, 72–73,
    76, 86, 102, 111, 118, 136, 144; in Manchester,
    46–47; ratting, 24, 45–48, 63, 72, 76; toy
    dogs, 46–48
Farman, Edgar, 145, 202
Fédération Cynologique Internationale, 2
*Field*, 17, 55; Country House, 59; sports
    reporting, 58–59. *See also under* Walsh,
    John Henry
field trials, 79–83
foxhounds, 11, 23, 35–37, 42, 59–60, 64, 67, 133,
    180
fox hunting, 35–37
fox terriers, 84, 94–95, 99–100, 122, 194, 207
French bulldogs. *See* bouledogues français

Galton, Francis, 20, 164–65, 182; ancestral law,
    160, 167; eugenics, 167–68; intelligence in
    dogs, 167–68, 212
genome, dog, 221–22
George, Bill, 48, 118
Gesner, Conrad, 25
Gillard, Frank, 174
Gordon setters, 43, 214
Graham, George Augustus, 146–50, 154
Great Collie Ear Trial, 218–19
Great Danes, 14, 26, 135, 141, 149–50, 191, 195,
    198–99, 202

Greenwood, James, 76

greyhounds, 24–26, 30, 37–41, 53, 100, 162, 180–81; Bedlamite, 64, 237n49; breeding, 64; conformation, 86–87; varieties c. 1845, 32

gun dogs, 24, 41–45, 78, 81–82, 95, 98, 133, 137, 184. *See also under* pointers; retrievers; *under* setters

Hamilton Smith, Charles, 33, 50

Handley, Samuel, 108, 110

Haraway, Donna, 16

harriers, 26

Heape, Walter, 165–66, 170

heredity, 4, 11–12, 28, 157, 160–64. *See also* blood: bloodlines

Hill, Gregory, 217

Hollingshead, John, 71

Homans, John, 16

Home for Lost and Starving Dogs, Holloway, 70–71

Howell, Philip, 14, 16

Hunting, William, 174

Huntly, Marquis of, 117

Hutchinson, W. N., 33, 43

Huxley, Thomas Henry, 160, 163

hybrids, 26, 56; dog-fox hybrid, 98; hybrid vigor, 40, 51. *See also* crossbreeds

Idstone. See Pearce, Thomas

improvement of breeds: by crossbreeding, 70, 139–40, 151, 152, 155; by field trials, 98; by inbreeding, 139–40, 145, 148–49, 151; by selection, 83, 139–40, 151; by show competitions, 70, 103, 193. *See also* breed clubs

inbreeding, 2, 4, 10, 14, 85, 151–56, 186, 193, 199, 223–24, 226; deterioration of dogs, 51, 139, 145, 152, 175, 186, 191, 199, 219; fixing features, 4, 139, 162; in wild animals, 40, 161–62. *See also* blood: bloodlines; prepotency

Irish greyhounds, 26–27

Irish Kennel Club, 134

Irish setters, 43

Irish terriers, 134, 196–97

Irish water spaniels, 44, 186

Irish wolfhound, 143, 146–54; as nationalist symbol, 147–48; recovery (resuscitation), 140, 146–51; as wolf-dog, 31, 146–47

Irving, Henry, 210, 213

Jefferies, Walter F., 194

Jesse, Edward, 34–35

Jesse, George, 129, 164

*Kennel Chronicle*, 134–35

Kennel Club: banning cropping in 1895, 202; as canine court of appeal, 111; establishment of, 107–11, 116–17; improving the nation's dogs, 111, 115, 117, 123, 133, 160, 183, 186; as Jockey Club of the canine world, 64, 89, 110, 115, 122–29, 189–90; Ladies Branch (KCLB), 210; rapprochement with National Dog Show Society in 1885, 125; refusing to set breed standards, 18, 133, 138, 220; relations with Ladies Kennel Association, 205–9; rules for shows and trials, 109, 115–17, 122–28, 175, 197, 201–2, 219; and social class, 116–37, 223–26; supporting cropping in 1889, 196–99

Kennel Club Stud Book, 179, 219; pedigrees, 117–20; and first editor Frank Pearce, 117

*Kennel Gazette*, 124, 128, 131, 145, 150–51, 160, 176, 190, 193–94

King Charles spaniels, 44, 47–50, 187, 223; Toy Spaniel Club, 187

Krehl, George, 172, 196, 203

Labradors, 1, 16, 230n9

Ladies Kennel Association (LKA), 13, 205–11; celebrities at shows, 210; charity dog shows, 210; Duchess of Newcastle's opposition to, 208–9; establishment of, 205; misogyny toward, 207–10; relations with Kennel Club, 205–9; shows, 208–10

Lamarck, Jean-Baptiste, 43

Landseer, Edwin, 3–4, 33, 67, 69, 96, 197

Landseer Newfoundlands, 4–5, 96

land spaniels, 44

lapdogs, 24, 49–50. *See also* toy dogs

Lapland dogs, 26

Larson, Greger, 221–22

Laverack, Edward, 43, 95, 110, 156

Lee, Rawdon B., 4, 159, 183, 199, 201; on collies, 213–14; *Modern Dogs*, 4–6, 183

Leeds, 59, 64

Lhasa terriers, 219

linebreeding, 156, 219, 262n15. *See also* inbreeding

lion dog, 44
livestock breeds and breeding, 8, 10–12, 15–19, 23–25, 54, 61, 88, 139, 151, 154, 223
*Livestock Journal and Fancier's Gazette*, 130–32, 149
London dog shows: Alexandra Palace, 78, 122, 132; Cremorne Gardens (1862), 72; Islington Agricultural Hall, 66–68, 70–71, 74–77, 81–82, 97, 101, 167; Westminster Aquarium, 215. *See also under* Crystal Palace
London Fancy. *See* Fancy, The
Lort, William, 44, 61, 95, 108, 110, 123
Loudon, Jane, 34
Low, David, 33
Low, George, 124
Lyell, Charles, 43

Macdona, John Cumming, 92–95, 108, 116, 118, 125, 136
Major, as first modern dog, 55, 83–84
Maltese terriers, 191
Manchester: Belle Vue Gardens, 75, 105; dog shows, 75, 95, 105, 108, 124, 175; Fancy, 46–47; Poultry Show, 65, 88
Manchester terriers, 122, 197, 199. *See also* black-and-tan terriers
Manning, F. A., 195
Marples, Theo, 169, 200; as founder of *Dog Owners' Annual* and *Our Dogs*, 200
Marshall, R., *An Early Canine Meeting*, Plate 3
masters of foxhounds, 24, 35–37, 133
mastiffs, 2, 24, 28, 29, 52–53, 93
Maurier, George Du, 184, 266n1
Mayhew, Edward, 32
Mayhew, Henry, 50
Meynell, Hugo, 35–37
Meyrick, John, 44
Millais, Everett, 20, 160–70; on artificial insemination, 164–66, 170; bacteriology of distemper, 169; biography, 164; critic of Kennel Club, 176; Experiment 12, 170–71; on inbreeding and crossbreeding, 166–72; introducing basset hound to Britain, 164–66; on telegony, 164–66, 170; *Theory and Practice of Rational Breeding*, 166–68; *Two Problems of Reproduction*, 169–70
Mills, John, 37, 42, 45

mongrels, 1–2, 27, 50–52, 174–75, 188. *See also* crossbreeds
Montgomery, James, 218
Moore, R. H., 131
Morgan, Conwy Lloyd, 168
Morgan, J. P., 213
Murchison, John Henry (J. H.), 98–100, 106, 108, 116–17, 125, 179
Musters, John Chaworth, 105

Nansen, Fridtjof, 179
National Canine Association, 65–66
National Canine Society (NCS), 91, 105–7, 109
National Coursing Club, 64
National Dog Club (NDC), 91, 95–98
National Dog Show Society, 60–61, 75, 79, 91, 100, 128; judging at shows, 93–94; rivalry with Kennel Club, 109, 128. *See also* Birmingham dog shows
Newcastle, Duchess of, 207–9, 212
Newcastle upon Tyne Sporting Dog Show (1859), 54, 59
new fancy, 90, 137–38, 139–40, 143, 153, 225; amateurs versus professionals in, 7, 16, 57, 65, 76, 90, 103, 122, 173, 176, 193, 197; dog lovers versus exhibitors, 122, 193, 204; respectability in, 102–3, 111, 122, 202–5, 225. *See also* breed clubs; dealers; dog shows
Newfoundlands, 17, 29, 45, 53, 62, 67, 73, 80, 92, 106, 140, 153, 175, Plate 1; Boatswain, 3–4; change in conformation, 2–6, 87, 96, 162; crossing with, 53, 80, 153; origin, 24, 29–30; at shows, 62, 95–96. *See also* Landseer Newfoundlands
Nimrod, 31
Norfolk spaniels, 44, 84, 187

Oakley, Annie, 179
Old English terriers, 197
otterhounds, 97
Ouida (Maria Louise de la Ramée), 204, 211

Panmure Gordon, Harry, 218–19
Pearce, Frank, 117
Pearce, Thomas (Idstone), 41, 91, 94, 117, 119–20; *The Dog*, 179, 214
pedigree dogs, 7, 14, 16, 20, 222–24, 227
Pedigree Dogs Exposed (blog), 223

pedigrees, 2, 7, 17; value in breeding, 80,
    85–86, 143, 154, 165, 193–95, 222–23.
    *See also* blood: bloodlines; breeding;
    stud books
Pekinese dogs, 70, 177
People for the Ethical Treatment of Animals
    (PETA), 223–24
pigeons: breeding, 65, 78, 88, 131, 151, 216;
    shows, 19, 34, 65, 88, 134
Platt, John, 107, 117
pointers, 29, 30, 34, 55–57, 73, 98, 102, 133,
    136, 153, 161, 180, 191; changes in confor-
    mation, 86–87; classification, 24, 32;
    crossbreeding with, 153; field trials,
    79–83; first modern breed, 55, 83–85; as
    gun dogs, 41–45, 133; as show dogs, 59–61,
    184–85, 216, 219
Pomeranians, 14, 26, 137
poodles, 1, 137, 141, 153, 203
poultry mania, 65
poultry shows, 54, 56, 65, 75, 111
prepotency, 10–11, 85, 121, 140, 152, 169–71, 177.
    *See also* breeding; inbreeding
press, dog, 13–15, 115, 131–32, 226. *See also*
    *individual titles*
Price, Richard Lloyd, 95, 110, 116, 123, 212;
    Rhiwlas Estate, 95, 116, 248n13
Prince of Wales, 74, 77, 96, 116, 202, 206
proliferation of breeds, 2, 7–8, 19, 139–43,
    184–87; spaniels, 141, 186–88; terriers, 97,
    141–43
pugs, 24, 27, 95, 131, 135, 149, 157, 175, 190–91,
    197, 205, 210, 223
*Punch*, Dog Fashions for 1889, 184

Quorn Hunt, 35–36

rabies, 52, 173, 176, 195, 204
Radcliffe, Frederick Delmé, 36
retrievers, 23–24, 45, 59, 73, 77, 79–81, 87, 104,
    106, 128, 133, 136, 141, 185, 191, 216
Richardson, H. D., 33–34
Ritvo, Harriet, 10, 15, 35, 88, 184, 225
Robinson, Philip, 205
Rogers, James, 33
Rogerson, G. Russell, 95
Romanes, George, 164, 168
Rotherham, Charles, 179

Royal Society for the Prevention of Cruelty to
    Animals (RSPCA), 28, 32, 71, 131, 176,
    196–203, 210
Russell, Edmund, 16
Russell, Nicholas, 9

Salter, John Henry, 126–27, 207, 255n26
Samoyeds, 219
Sanitas Company, 175, 182
Schipperkes, 141, 189, 203
Scotch terriers, 29, 31–32, 62, 116, 134, 142;
    Scotch Terrier Club, 134
Scottish deerhound, 30, 146–49
Scottish Kennel Club, 128, 134
Segrave, Lord, 36
Séguin, Édouard, 40
setters, 32, 102, 184, 186, 191; crossbreeding
    with, 80, 153; and field trials, 80–81, 82; as
    gun dogs, 41–45, 133, 177; as show dogs,
    184–85, 219
Sewell, Alfred J., 173
Shaw, Jemmy, 46, 102, 118
Shaw, Vero, 115, 118, 123, 132, 141, 159, 199;
    biography, 126, 129–30; on breeding, 153–54;
    on cropping, 200–201; "Dogdom's
    Deterioration," 188–89, 195; *The Illustrated
    Book of the Dog*, 9, 129–31, 187
sheep dogs, 24, 50, 213–14: trials, 213
shepherd's dog, as original dog, 26
Sherrington, Charles Scott, 164
Shirley, Sewallis: biography, 98–99; on
    collies, 214–15; on cropping, 197–98; as
    Kennel Club president, 107–10, 123–24; and
    politics, 98, 116; as show judge, 108, 110,
    126, 128
shock dog, 44
Shooting Dog Club (SDC), 95
Shorthose, John, 59, 106, 125
Sikes, Bill, 48, 225
Sims, George, 201
Skye terriers, 24, 48, 88, 97, 134, 141–42, 199
Smith, E. T., 73–74
Smith, Matthias, 92
spaniels, 24, 26, 32–33, 60, 141, 186–88,
    203, 226; as gun dogs, 41–45; Spaniel Club,
    136, 187, 188. *See also individual spaniel
    breeds*
Spanish pointers, 42–43, 85

Spencer, Herbert, 167

sporting dogs, 8, 38, 51, 129, 145, 155, 216; at all-breed shows, 60–66 69–70, 72–73, 95–96, 141, 190; breeding, 51, 79, 140, 145, 152; deterioration, 186–88; field trials, 80–83; National Show (1859), 55–58. *See also* pointers; retrievers; setters; spaniels

Spratt, James, 177

Spratt's Patent Company, 173–82; impact on dog feeding, 161, 180; marketing, 178–81; Patent-X biscuits, 125, 177–80; sanatorium, 173; support of Kennel Club, 177, veterinary medicines, 173

springer spaniels, 44

Stables, Gordon, 129, 199; advice on feeding, 180, 196; biography, 130; novel *Sable and White*, 212–13; veterinary advice, 172–73

St. Bernards, 24, 71, 78, 100, 122, 135–36, 140, 149, 179, 186, 191; Barry, 92; history, 92–93; Tell, 92, 94, 118

stealing (dognapping), 49

Stennard-Robinson, Alice (née Cornwell); biography, 205–7, misogyny toward, 210; role in Ladies Kennel Association, 205–10

St John's Newfoundland, 80

*Stock-Keeper and Fancier's Chronicle*, 130–32, 156, 187, 190, 196, 198, 200–202, 207, 217

Stonehenge. *See* Walsh, John Henry

stud books, 11, 61, 91, 96, 117–18, 121, 125–26, 130, 133, 219–20; Jockey Club, 40, 64, 117; private, 61. *See also* pedigrees

Sussex spaniels, 44

Talbot hound, 33, 35

Taunton, Walter K., 126

terriers. *See individual breeds*

*Thacker's Courser's Annual Remembrancer and Stud Book*, 40

Thomas, Keith, 5, 7, 49

toy dogs, 44, 62, 75, 86, 102, 131, 181, 209, 214; in the Fancy, 46–48; and ladies, 25, 49–50, 70, 129–30, 177; rise in numbers, 190–91; terriers, 62, 197; Toy Dog Club, 190; toy dog shows, 191–92, 205, 210

trade in show-dogs, 66–67, 79, 86, 94, 121–22, 129, 132, 136, 154, 218; attitudes toward trade, 7–8, 101–3, 121, 153, 193, 224–26; prices paid for dogs, 67, 88, 92, 213, 243n35, 247n8, 272n76; stud fees, 7, 53, 88, 94, 132, 139, 152. *See also* dealers; doggy people

Trollope, Anthony, 225

truffle dogs, 84

Turner, Sidney (Este), 176; on distemper, 176; on mastiffs, 195; opposing docking, 203; proposing ban on cropping in 1889, 196–98

turnspits, 27, 234n11

Victoria, Queen, 47, 49–50, 70, 117, 131; Dash, 49–50; Pekinese Looty, 70

Vyner, Robert, 31

Wakerley, J. G. V., 106–9

Wallen, Martin, 11

Walsh, John Henry, 4, 19, 39, 54–55; biography, 57–59, 241n13; on breed clubs, 133; on breeding dogs, 151–53; on distemper, 174–75; *The Dog in Health and Disease*, 52, 58, 80, 85–87, 131–33, 149, 151–53, 172, 177, 214; on dog shows, 69–70, 72–73, 78, 97, 104–5, 186–87, 190; *The Dogs of the British Islands*, 55, 84–88, 91–92, 94, 96, 104, 131, 133–34, 179, 181, 187, 195, 199, 214; as editor of the *Field*, 58–59, 62, 129, 133, 159, 186; on field trials, 81–82; *The Greyhound*, 39–41, 86–87; judging with numerical points, 79, 83, 93; on the Kennel Club, 110, 122–26, 138, 186, 190; *A Manual of British Rural Sports*, 57; proposing conformation standards, 78–79, 83–87; supporting National Dog Club, 95–96, 104, 126

Walton, John R., 10

Wardle, Arthur, 5, 131

Warnes, Harold, 204

water spaniels, 26, 186

Watson, Andrew, 218

Welsh spaniels, 187

whippets, 38, 207

Whitbread, Samuel, 81

white English terrier, 199, 201–2

Whitehouse, J. H., 116–17

Wight, Alf, (James Herriot), 177

wolf-dog crosses, 26

Wombwell, George, 69
working dogs, 55, 215
Wynn, Malcolm B., 110, 149

Yarborough, Lord, 36
Yarrell's Law, 163, 170
Yorkshire terriers, 48, 116, 130–31, 140–43

Youatt, William, 3–6, 30–32, 115; on collies, 214; on cross breeds as mongrels, 50–51; *The Dog*, 3–4, 28, 31, 33, 52; on origins of domesticated dogs, 29–30; on pointers, 42

Zoological Society of London, 66